SPECIFICATION OF DRUG SUBSTANCES AND PRODUCTS

SPECIFICATION OF DRUG SUBSTANCES AND PRODUCTS

DEVELOPMENT AND VALIDATION OF ANALYTICAL METHODS

Edited by

Christopher M. Riley,
Riley and Rabel Consulting Services,
Maryville, MO, USA

Thomas W. Rosanske,
T.W. Rosanske Consulting
Overland Park, KS, USA

Shelley R. Rabel Riley,
Department of Natural Sciences, Northwest Missouri State University, Maryville, MO, USA;
Riley and Rabel Consulting Services,
Maryville, MO, USA

ELSEVIER

AMSTERDAM • BOSTON • HEIDELBERG • LONDON • NEW YORK • OXFORD
PARIS • SAN DIEGO • SAN FRANCISCO • SYDNEY • TOKYO

Elsevier
Radarweg 29, PO Box 211, 1000 AE Amsterdam, The Netherlands
The Boulevard, Langford Lane, Kidlington, Oxford, OX5 1GB, UK
225 Wyman Street, Waltham, MA 02451, USA

British Library Cataloguing in Publication Data
A catalogue record for this book is available from the British Library

Library of Congress Cataloging-in-Publication Data
A catalog record for this book is available from the Library of Congress

ISBN: 978-0-08-098350-9

For information on all Elsevier publications visit our web site at store.elsevier.com

14 15 16 17 18 10 9 8 7 6 5 4 3 2 1

Contents

PART 4 SPECIFIC TESTS: DRUG PRODUCT

PART 5 PHARMACOPEIAL METHODS

PART 6 MICROBIAL METHODS

PART 7 BIOLOGICAL FLUIDS

List of Contributors

Daniel W. Armstrong
Department of Chemistry and Biochemistry, University of Texas at Arlington, Arlington, TX, USA

James Bergum
Statistical Consultant, BergumSTATS, LLC, Howell, NJ, USA

Beth Ann Brescia
BioMonitoring NA, EMD Millipore Corporation, Billerica, MA, USA

Peter Bryan
Mendham, NJ, USA

Todd L. Cecil
United States Pharmacopeial Convention, Inc., Rockville, Maryland, USA

Ping Chen
Analytical Chemistry, SSCI, A Division of Aptuit, West Lafayette, IN, USA

Eliza N. Fung
Bioanalytical Sciences, Bristol-Myers Squibb, Princeton, NJ, USA

Vivian A. Gray
V. A. Gray Consulting, Inc., Hockessin, DE, USA

Brian He
Analytical and Bioanalytical Development, Bristol-Myers Squibb, New Brunswick, NJ, USA

Laureen E. Little
Quality Services, Palm Desert, CA, USA

David K. Lloyd
Analytical and Bioanalytical Development, Bristol-Myers Squibb, New Brunswick, NJ, USA

James V. McArdle
McArdle & Associates, LLC, Carlsbad, CA, USA

Kurt L. Moyer
NSF Pharmalytica, Bristol, CT, USA

Bradford J. Mueller
Incyte Corporation Experimental Station, Wilmington, DE, USA

Nilusha L.T. Padivitage
Department of Chemistry and Biochemistry, University of Texas at Arlington, Arlington, TX, USA

Ernest Parente
Mallinckrodt Pharmaceuticals, St. Louis, MO, USA

Robyn L. Phelps
PharmAdvance Consulting, Inc., Sequim, WA, USA

Shelley R. Rabel Riley
Department of Natural Sciences, Northwest Missouri State University, Maryville, MO, USA; Riley and Rabel Consulting Services, Maryville, MO, USA

Christopher M. Riley
Riley and Rabel Consulting Services, Maryville, MO, USA

Thomas W. Rosanske
T.W. Rosanske Consulting, Overland Park, KS, USA

James Scull
NSF Pharmalytica, Bristol, CT, USA

Krzysztof Selinger
Forest Research Institute, Farmingdale, NY, USA,

Eric B. Sheinin
Sheinin & Associates LLC North Potomac, MD, USA

Pamela A. Smith
Analytical Chemistry, SSCI, A Division of Aptuit, West Lafayette, IN, USA

Jonathan P. Smuts
Department of Chemistry and Biochemistry, University of Texas at Arlington, Arlington, TX, USA

Patrick A. Tishmack
Analytical Chemistry, SSCI, A Division of Aptuit, West Lafayette, IN, USA

PART 1

Introduction

CHAPTER

Introduction

1

Christopher M. Riley*, Bradford J. Mueller†, Thomas W. Rosanske**, Shelley R. Rabel Riley*,‡

*Riley and Rabel Consulting Services, Maryville, MO, USA,
†Incyte Corporation, Experimental Station, Wilmington, DE, USA, ** T.W. Rosanske Consulting, Overland Park, KS, USA, ‡Department of Natural Sciences, Northwest Missouri State University, Maryville, MO, USA

CHAPTER OUTLINE

When the first version of this book was published in 1996,[1] The International Conference on Harmonization (ICH), which is an effort by the USA, the EU and Japan to harmonize new drug applications, was still in its infancy. Since then, all the key ICH Quality Guidelines[2–26] covering specification setting (e.g. ICH Q1,[2–8] Q3–Q6)[9–21] and method validation (ICH Q2)[6] have been published, and some have been revised at least once. The ICH Guidelines, together with some of the more recent changes in regional guidelines and compendial requirements will form the general framework for this book. Where the Quality (Q1–Q11) ICH Guidelines fit into the general drug development framework is shown in Fig. 1.1.

The introduction of the earlier ICH Quality Guidelines (Q1–Q6),[2–21] which describe most of the general requirements for the analytical content of the Common Technical Document (CTD, ICH M4Q(R1))[22] and its electronic counterpart (eCTD), was followed by a series of guidelines (Q7–Q10) addressing some of the key approaches to drug development that are also to be included in the CTD. Although there are some regional differences, the CTD is the generally harmonized document used in the ICH regions for marketing authorization applications. The general framework of the CTD is also used, with appropriate modifications, for clinical trials applications. The CTD is also accepted in many non-ICH countries, such as Canada and Australia.

According to the ICH definition, the specification(s) for a new drug substance or a drug product (Q6A and Q6B) contain three elements: (1) the quality attributes (or tests), (2) references to the associated methods and (3) the acceptance criteria. The primary objective of this book is to provide a critical and comprehensive assessment of the approaches used to identify what are the key quality attributes that impact safety, efficacy, and manufacturability, select appropriate analytical methods based on the accuracy and precision needed to adequately measure and control the identified quality attributes and determine how the analytical methods are developed and validated for their intended use. The general principles of the specification-setting process are surveyed in Chapter 2 and explored in greater detail in Chapters 5–15. Chapter 16 deals with the development and validation of bioanalytical methods.

Specification of Drug Substances and Products. http://dx.doi.org/10.1016/B978-0-08-098350-9.00001-1

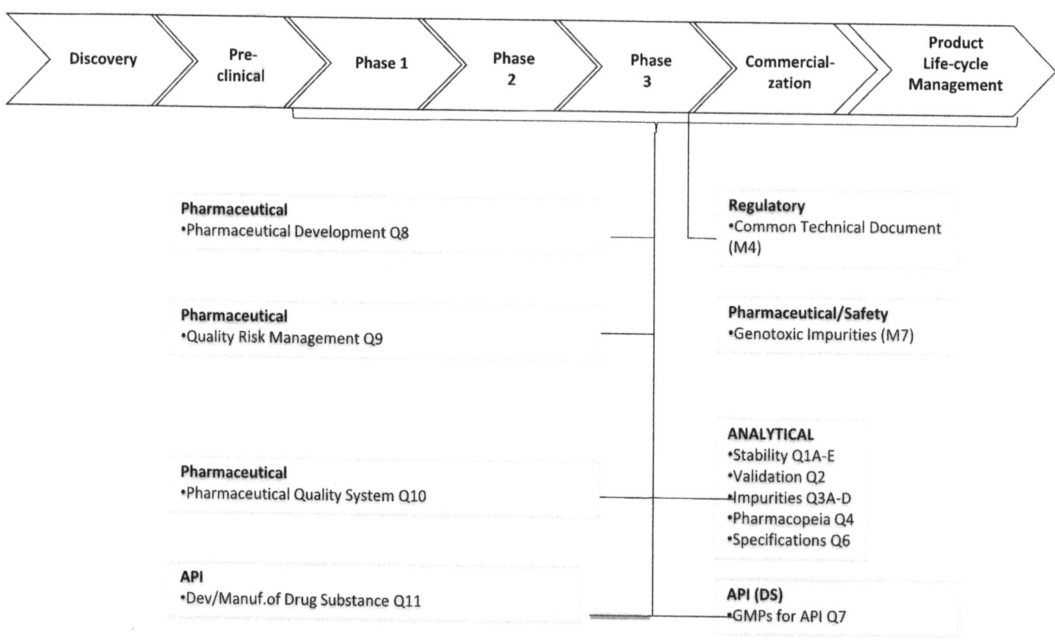

FIGURE 1.1

Summary of the ICH Guidelines applicable to pharmaceutical analysis (see also Refs 2–25) and where they fit into the drug development process.

The concept of Quality by Design (QbD) was introduced into the drug development process through the more recent ICH Guidelines (Q8–Q10)[23–25] with the primary aim of increasing the understanding and the knowledge base of the processes for the manufacturing of drug substances and products. However, the principles of QbD are equally applicable to pharmaceutical analysis. Therefore, the concept of Analytical Quality by Design (AQbD) is introduced in Chapter 3 and expanded in later chapters. Since the publication of the first version of this book, the key ICH Quality Guidelines have matured and now form the general framework for the application of worldwide marketing approvals of new drug products.

Whereas the guidelines dealing with specification setting (most notably ICH Q6A and Q6B) and Method Validation (Q2) describe <u>what</u> information regulators expect to see in a new drug application, they provide very little detail on <u>how</u> the guidelines are to be implemented at the technical level. The absence of specific direction on the implementation of the ICH Quality Guidelines allows for the application of new and improved analytical technologies targeted to the critical quality attributes which impact product performance. The use of statistical approaches to better correlate method performance with respect to control limits for critical quality attributes and to monitor long-term analytical method performance is an area which is not discussed within the guidances, but is critical to the development and maintenance of analytical methods. Whereas Chapters 2 and 3 survey the general principles of specification setting and QbD, respectively, Chapter 4 discusses conventional approaches to method validation. The ICH Guideline on Method

Validation (Q2(R1)) was primarily developed with separation techniques in mind and the following tests in particular:

- Identification tests
- Quantitative tests for impurities content
- Limit tests for the control of impurities
- Quantitative tests of the active moiety in samples of drug substance and drug product or other selected components in the drug product (e.g. preservatives, antioxidants)

Subsequent chapters will discuss how the principles of method validation set forth in Q2(R1) have been adapted to techniques as diverse as solid characterization and microbiological methods.

In keeping with the spirit of the first version of this book, this version is not intended as merely a review of existing regulatory guidance and industry practices. Rather, in addition to discussing conventional approaches, each chapter will address critical issues and novel approaches. The authors have been carefully selected as being former members of the ICH Expert Working Groups charged with developing the ICH guidelines, and/or subject-matter experts in the industry, academia and government laboratories. Thus, the book will provide the reader with not only an understanding of industry best practices and future directions, but also an insight into how international guidelines were developed and the rationale behind them.

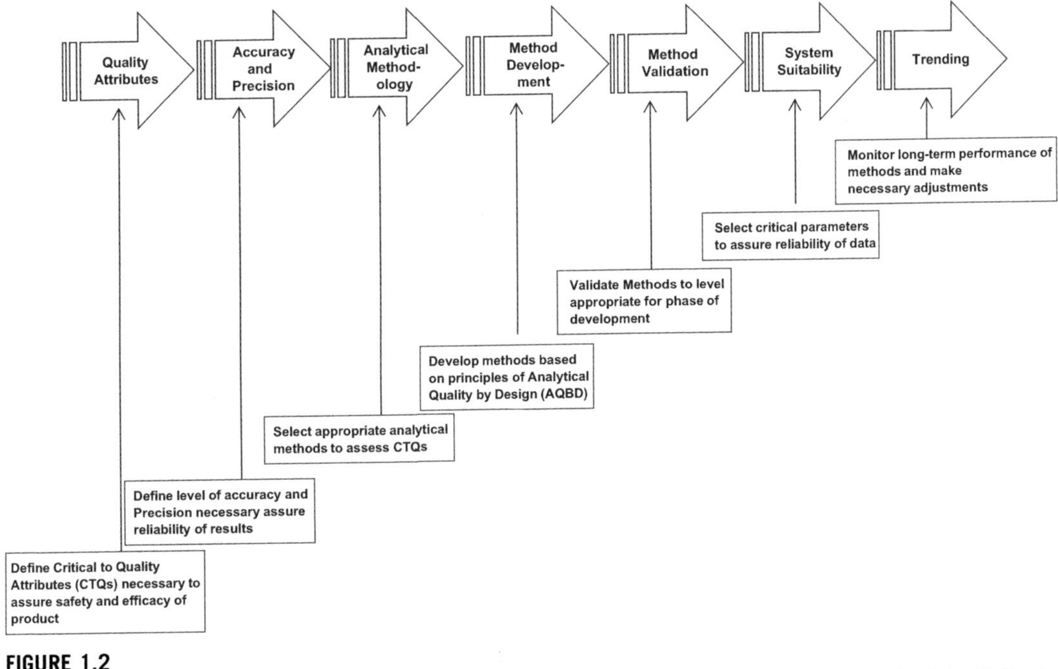

FIGURE 1.2

The evolution of analytical technology for the control of pharmaceuticals through the life cycle of the development process.

In addition to providing the "what" but not the "how" to set specifications and validate analytical methods, the ICH Quality Guidelines (Q1–Q6)[2–20] only define what is to be provided in a new drug application. They expressly exclude what is expected in the clinical stages of drug development (i.e. in an Investigational New Drug Application, IND). Therefore, a common theme throughout the book is how the methods are validated and specifications evolve over the drug development life cycle (Fig. 1.2). The intention in writing the second version of the book is to capture the many regulatory and technical advances that have occurred in the field since publication of the first version in 1996.

The "how" of the earlier Q1–Q6 Guidelines are to be applied is described in large part in subsequent guidelines (Q7–Q11). For example,[27] 16 attributes were identified for a polymeric excipient, derived from a natural product, and used in sustained release product to control the potentially variable performance of the excipient in the product. The only way to manage the 16 attributes and achieve acceptable product performance was to understand the contributions of the various attributes and the interactions between them—between each physical and chemical characteristic. By analytically measuring each of the attributes and then using statistical/chemometric approaches, it was possible to define a "design space" of all parameters which could deliver the overall desired effect of drug release.

Thus, this version is intended to be not only a review of the ICH Guidelines relating to the specification and method validation of new drugs, but also to provide a critical analysis of the regulatory guidelines and a comprehensive treatment of how those guidelines are applied to the development of new drugs. It is intended to be an educational tool and a reference source for those involved in the development and regulation of new drug products.

References

1. Development and Validation of Analytical Methods. In Riley, C. M., Rosanske, T. W., Eds.; *Progress in Pharmaceutical and Biomedical Analysis,* Vol. 3, Elsevier, 1996.
2. Stability Testing of New Substances and Products (Q1A(R2)). *The International Conference on Harmonization of Technical Requirements for Registration of Pharmaceuticals for Human Use,* Second Revision, 2003.
3. Stability Testing: Photostability Testing of New Drug Substances and Products (Q1B). *The International Conference on Harmonization of Technical Requirements for Registration of Pharmaceuticals for Human Use,* 1997.
4. Stability Testing: New Dosage Forms (Q1C). *The International Conference on Harmonization of Technical Requirements for Registration of Pharmaceuticals for Human Use,* 1997.
5. Bracketing and Matrixing Design for Stability Testing: New Drug Substances and Products (Q1D). *The International Conference on Harmonization of Technical Requirements for Registration of Pharmaceuticals for Human Use,* 2002.
6. Evaluation of Stability Data (Q1E). *The International Conference on Harmonization of Technical Requirements for Registration of Pharmaceuticals for Human Use,* 2003.
7. Stability Data Package for Registration Applications in Climatic Zones III and IV (Q1F). *The International Conference on Harmonization of Technical Requirements for Registration of Pharmaceuticals for Human Use,* 2006.
8. Validation of Analytical Procedures: Text and Methodology (Q2(R1)). *The International Conference on Harmonization of Technical Requirements for Registration of Pharmaceuticals for Human Use,* First Revision, 1995.
9. Impurities in New Drugs Substances (Q3A(R2)). *The International Conference on Harmonization of Technical Requirements for Registration of Pharmaceuticals for Human Use,* Second Revision, 2002.

10. Impurities in New Drugs Products (Q3B(R2)). *The International Conference on Harmonization of Technical Requirements for Registration of Pharmaceuticals for Human Use,* Second Revision, 2002.
11. Impurities: Guideline for Residual Solvents (Q3A(R5)). *The International Conference on Harmonization of Technical Requirements for Registration of Pharmaceuticals for Human Use,* Fifth Revision, 2010.
12. Impurities: Guideline for Metal Impurities (Q3D). Concept Paper. *The International Conference on Harmonization of Technical Requirements for Registration of Pharmaceuticals for Human Use*, 2009.
13. Pharmacopeial Harmonization (Q4A). *The International Conference on Harmonization of Technical Requirements for Registration of Pharmaceuticals for Human Use, in development*, Unpublished.
14. Evaluation and Recommendation of Pharmacopeial Texts for Use in the ICH Regions (Q4B). *The International Conference on Harmonization of Technical Requirements for Registration of Pharmaceuticals for Human Use,* 2007.
15. Viral Safety Evaluation of Biotechnology Products Derived from Cell Lines of Human or Animal Origin (Q5A(R1)). *The International Conference on Harmonization of Technical Requirements for Registration of Pharmaceuticals for Human Use,* First Revision, 1997.
16. Analysis of the Expression Construct in Cells Used for Production of r-DNA Derived Protein Products (Q5B). *The International Conference on Harmonization of Technical Requirements for Registration of Pharmaceuticals for Human Use,* 1995.
17. Stability Testing of Biotechnology/Biological Products (Q5C). *The International Conference on Harmonization of Technical Requirements for Registration of Pharmaceuticals for Human Use*, 1995.
18. Derivation and Characterization of Cell Substrates Used for Production of Biotechnology/Biological Products (Q5D). *The International Conference on Harmonization of Technical Requirements for Registration of Pharmaceuticals for Human Use,* 1997.
19. Comparability of Biotechnology/Biological Products Subject to Changes in their Manufacturing Process (Q5E). *The International Conference on Harmonization of Technical Requirements for Registration of Pharmaceuticals for Human Use,* 2004.
20. Specification: Test Procedures and Acceptance Criteria for New Drug Substances and New Drug Products: Chemical Substances (Q6A). *The International Conference on Harmonization of Technical Requirements for Registration of Pharmaceuticals for Human Use,* 1999.
21. Specification: Test Procedures and Acceptance Criteria for Biotechnology/Biological Products (Q6B). *The International Conference on Harmonization of Technical Requirements for Registration of Pharmaceuticals for Human Use,* 1999.
22. The Common Technical Document for the Registration of Pharmaceuticals for Human Use: Quality – M4Q(R1). *The International Conference on Harmonization of Technical Requirements for Registration of Pharmaceuticals for Human Use,* First Revision, 2002.
23. Pharmaceutical Development (Q8(R2)). *The International Conference on Harmonization of Technical Requirements for Registration of Pharmaceuticals for Human Use.* Second Revision, 2009.
24. Quality Risk Management (Q9). *The International Conference on Harmonization of Technical Requirements for Registration of Pharmaceuticals for Human Use.* Second Revision, 2005.
25. Pharmaceutical Quality System (Q10(R4)). *The International Conference on Harmonization of Technical Requirements for Registration of Pharmaceuticals for Human Use.* Fourth Revision, 2009.
26. Assessment and Control of DNA Reactive (Mutagenic) Impurities in Pharmaceutical to Limit Carcinogenic Risk. *The International Conference on Harmonization of Technical Requirements for Registration of Pharmaceuticals for Human Use*, Final Concept Paper, 2010.
27. Brown, B.; Caster, D.; Clarke, B.; Hopkins, S.; Llewelyn, J.; Martin, L.; Meehan, E.; Timko, R.; Yang, H. *Extended Release Formulations Comprising Quetiapine and Methods for their Manufacture,* 2011. United States Patent Application, US 2011/0319383.

CHAPTER

General principles and regulatory considerations: specifications

2

Christopher M. Riley*, Laureen E. Little**

* *Riley and Rabel Consulting Services, Maryville, MO, USA,*

** *Quality Services, Palm Desert, CA, USA*

CHAPTER OUTLINE

2.1 DEFINITIONS

2.1.1 International guidelines

The main guideline published by the International Conference on Harmonization (ICH) covering the specification of new chemical entities (NCEs) is Q6A: Specifications: Test Procedures and Acceptance Criteria for New Substances and Drug Products: Chemical Substances. The corresponding ICH Guideline covering biologicals is Q6B: Test Procedures and Acceptance Criteria for Biotechnological/ Biological Products. Additional information on specification setting can be found in other original ICH Guidelines in Q1 (Stability), Q3 (Impurities), and Q4 (Pharmacopeia). The more recent ICH

Specification of Drug Substances and Products. http://dx.doi.org/10.1016/B978-0-08-098350-9.00002-3

Guidelines, Q8, Q9 and Q10 also have important implications for specification setting, especially within the context of the application of Quality by Design (QbD) to process optimization/validation and formulation development; as well as to analytical method development and validation. The QbD concepts (in Q8, Q9 and Q10) and their application to pharmaceutical analysis are discussed in detail in Chapter 3. A full listing of all the ICH Quality Guidelines and relevant Multidisciplinary ICH Guidelines is given in Table 2.1.

According to ICH Q6A, a specification (singular) contains three elements: a list of tests (or attributes), references to test methods and acceptance criteria.[i] Both guidelines (Q6A and Q6B) distinguish between Universal Tests, which are required in any new specification for a new drug substance or drug product, and Specific Tests, which should be determined, on a case-by-case basis, depending on the nature of the drug substances or drug product. The Universal Tests for both new drug substances and drug products are:

- Description
- Identification
- Assay
- Impurities

Representative drug substance and drug product specifications of the fictitious drug S-(+) xenplifir mesylate (Exemplifi™), Exemplifi ER 200 mg tablets and Exemplifi 10 mg/mL oral solution are shown in Tables 2.2–2.4, respectively. The relevant drug substance and drug product characteristics are as follows:

2.1.1.1 Drug substance

Salt form: mesylate (methanesulfonate)
Molecular weight: 275.55
Chirality: (S)-(+): single chiral center
Solid state: form III (five known polymorphs)
Aqueous solubility: 1 mg/mL (pH 3.2)
Moisture sorption: non-hygroscopic
Recrystallization solvent: ethanol: hexane (5:95, v/v)

2.1.1.1.1 Impurities

Process-related impurities: S1, S2, S3
Degradation products of xenplivir: S4

2.1.1.2 Drug product

Daily dose: 400 mg
Impurities: Degradation products of xenplivir: S4 and P1
Dosage form 1: extended release 200-mg tablet (Exemplifi™ ER)
Dosage form 2: aqueous oral pediatric solution (Exemplifi™ Oral solution 10 mg/mL), containing 0.1% each of methyl- and propyl parabens (as preservatives), fill volume 200 mL

[i] The terms "acceptance criteria" and "specification" are often used Interchangeably. However, according to ICH Q6A and Q6B they are different. Specification refers to the entire document. Acceptance criteria refers to a specific attribute.

Table 2.1 Summary of the ICH Quality (Q) Guidelines and those Multidisciplinary Guidelines Relevant to Specification Setting

Number*	Title†
Q1A	Stability Testing of New Substances and Products
Q1B	Photostability of New Substances and Products
Q1C	Stability Testing: New Dosage Forms
Q1D	Bracketing and Matrixing Design for Stability Testing: New Substances and Products
Q1E	Evaluation of Stability Data
Q1F	Stability Data Package for Registration Applications in Climatic Zones III and IV
Q2(R1)	Validation of Analytical Procedures: Text and Methodology
Q3A(R1)	Impurities in New Drug Substances
Q3B(R1)	Impurities in New Drug Products
Q3C(R5)	Impurities: Guideline for Residual Solvents
Q3D	Impurities: Elemental Impurities
Q4A	Pharmacopeial Harmonization
Q4B	Evaluation and Recommendation of Pharmacopeial Texts for Use in ICH Regions
Q5A	Viral Safety Evaluation of Biotechnological Products from Cell Lines of Human or Animal Origin
Q5B	Analysis of the Expression Construct in Cells Used for Production of r-DNA Derived Protein Products
Q5C	Stability Testing of Biotechnological/Biological Products
Q5D	Derivation and Characterization of Cell Substrates Used for Production of Biotechnological/Biological Products
Q5E	Comparability of Biotechnological/Biological Products Subject to Changes in their Manufacturing Process
Q6A	Specifications: Test Procedures and Acceptance Criteria for New Substances and New Drug Products: Chemical Substances
Q6B	Specifications: Test Procedures and Acceptance Criteria for Biotechnological/ Biological Products
Q7	Good Manufacturing Practice Guide for Active Pharmaceutical Ingredients
Q8(R2)	Pharmaceutical Development
Q9	Quality Risk Management
Q10(R4)	Pharmaceutical Quality System
Q11	Development and Manufacture of Drug Substances (Chemical Entities and Biotechnological/Biological Entities)
M4	Common Technical Document
M7	Assessment and Control of DNA Reactive (Mutagenic) Impurities in Pharmaceuticals to Limit Carcinogenic Risk (Step 2)

*****The designation in parentheses refers to the most recent revision.
†See http://www.ich.org/products/guidelines/quality/article/quality-guidelines.html for copies of the Guidelines and other details.

Table 2.2 Suggested Drug Substance Specification of the Fictitious Drug, Xenplifir Mesylate*

Xenplifir Mesylate Specification			
Test (or attribute)	**Method**	**Acceptance Criteria**	**Chapter**
Universal tests			
Appearance	SOP 001.03	White to off-white solid	5
Identification	FTIR: TM 002.00	Conforms to reference spectrum	5
Assay	HPLC: TM 003.03[†]	98.0–102.0%	6
Impurities			
S1	HPLC: TM 003.03	NMT (≤) 0.30%	6
S2		NMT (≤) 0.25%	6
S3		NMT (≤) 0.25%	6
S4		NMT (≤) 0.20%	6
Unspecified[‡]		NMT (≤) 0.10%	6
Total		NMT (≤) 1.50%	6
Specific tests			
pH of 1% aqueous solution	SOP 005.01	2.3–3.0	
Chiral identity	Optical rotation: TM 005.03	−75.0° to +75.0°	10
Chiral impurity	HPLC: TM 007.03	NMT (≤) 1.5% R-xenplivir	10
Melting point	USP <741>		9
Polymorphic form	XRPD: TM 017.02	Conforms to reference diffractogram	9
Water content	Karl Fisher titration USP <921>	NMT (≤) 0.8%	11
Clarity of solution	SOP 002.01	Conforms to SOP 002.01	
Ethanol[¶]	GC: TM 009.01	20 ppm	6
Hexane[**]	USP <467>	290 ppm	7
Methanesulfonic acid	GC: TM 006.04	NMT (≤) 4 ppm[††]	6
Inorganic impurities[‡‡]	USP <231>	Conforms to USP[§§]	8
Microbial limits	USP <61>, <62> <1111>	Conforms to USP[***]	15

Table 2.2 Suggested Drug Substance Specification of the Fictitious Drug, Xenplifir Mesylate* *(continued)*

Xenplifir Mesylate Specification			
Test (or attribute)	Method	Acceptance Criteria	Chapter

*Assumes US submission of application for clinical trials or marketing authorization application.
†Assumes assay and impurities are measured by the same HPLC method (see Chapter 6).
‡According to ICH Q3A (R1), the limit for unspecified impurities is equal to the applicable identification threshold, which in turn is determined by the daily dose (in this case 400 mg) (see Chapter 6).
¶According to ICH Q3(R5), ethanol is a Class 3 solvent and special limits are required. However, in this fictitious example the ethanol is controlled to 20 ppm to minimize reaction with methanesulfonic acid to produce ethane methanesulfonate (a known mutagen) (see also Chapter 6).
**Hexane is a Class 2 solvent and the limit is dictated by ICH Q3C(R5).
††The limit of methanesulfonic acid is limited by the maximum daily intake of not more than (≤) 1.6 g/day; (see also Chapter 6).
‡‡Previously known as "heavy metals".
§§An acceptance criteria of "conforms to USP" is preferred to a listing of the actual limits because changes in USP general chapter will require a change to the specification.
***Assumes assay and impurities are measured by different HPLC methods (see Chapter 6).
FTIR: Fourier-transform Infrared Spectroscopy
HPLC: High Performance Liquid Chromatography

A survey of the principal methods and technologies used for the Universal and Specific Test Methods is discussed in detail in various chapters in Parts 2–6.

2.1.2 Pharmacopeial monographs and general chapters

The 1906 Pure Food and Drug Act defined the United States Pharmacopeia (USP) as the highest legal authority for the quality control of pharmaceuticals in the US. The United States Pharmacopeia and National Formulary (USP-NF) contain monographs for drug substances, drug products (USP) and excipients (NF); and general chapters (designated by the parentheses < >). Similar legal authorities exist for other regions such as the Japanese Pharmacopoeia (JP), the European Pharmacopoeia (PhEur or EP) and the British Pharmacopoeia (BP). An interesting difference between the USP and most of the other regional pharmacopeias is that, whereas the USP is independent of the Federal Government, most other pharmacopeias are governmental organizations. Another important difference between the USP and some of the other regional pharmacopeias is that the USP contains monographs on drug products and the others, with the exception of the BP, do not.

The format of monographs in the pharmacopeias is somewhat different from the information in regulatory documents such as the Common Technical Document[ii] (CTD, ICH M4). For example, the monographs in the USP contain information on Specific Test Methods unique to the drug substance, drug product and excipient in question. Methods already in the general chapters are described outside the monograph. The details of all analytical methods in the CTD are listed separately from the specification elsewhere in the investigational or new drug application.

[ii] The Common Technical Document (CTD) and its counterpart the Electronic Common Technical Document are the harmonized documents used for New Drug Applications (NDAs) and Marketing Authorization Applications (MAAs) in the ICH regions: US. EU and Japan.

Table 2.3 Suggested Drug Product Specification of the Fictitious Drug, Xenplifir Mesylate* 200 mg Oral Tablets (Exemplifi ER)

Xenplifir Mesylate 200 mg Oral Tablets			
Test (or attribute)	**Method**	**Acceptance Criteria**	**Chapter**
Universal tests			
Appearance	SOP 001.03	White caplets	5
Identification	UV: TM 010.00	Conforms to reference spectrum	
	HPLC: TM 010.00	Retention time of main peak within 5% of peak in standard solution	5
Assay	HPLC: TM 003.03†	90.0–110.0%	6
Impurities			
P1	HPLC: TM 011.01	NMT (≤) 0.4%	6
P2		NMT (≤) 0.7%	6
S4‡		NMT (≤) 1.0%	6
Unspecified		NMT (≤) 0.2%	6
Total		NMT (≤) 2.5%	6
Specific tests			
Drug release	USP <724> Method: TM: 013.04	Released at 1 h = NMT (≤) 10% Released at 5 h = 45.0–55.0% Released at 10 h NLT (≥) 90.0%	12
Content uniformity	HPLC: TM 003.03	Conforms to USP	6
Microbial limits	USP <61>, <62>, <1111>§	Conforms to USP¶	15

*Assumes US submission of application for clinical trials or marketing authorization application.
†Assumes assay and impurities are measured by different HPLC methods (see Chapter 6).
‡S4 is a degradation product and should be monitored (if appropriate in the drug product and in the drug substance). S1, S2 and S3 are process impurities of the drug substance and need not be monitored in the drug product (see Chapter 6).
§Microbial attributes of non-sterile pharmaceutical products.
¶An acceptance criteria of "conforms to USP" is preferred to a listing of the actual limits because changes in USP general chapter will require a change to the specification.

Being the highest legal authority in the US, samples seized in the field by the authorities such as the Food and Drug Administration (FDA) must meet the requirements of the pharmacopeial monograph, if such a monograph exists.[iii] If such a monograph does not exist, the next highest legal instrument is the specification in the appropriate regulatory filing (e.g. New Drug Application, NDA). Further details of the pharmacopeias as they relate to specification setting, including Pharmacopeial Harmonization, are discussed in more detail in Chapter 14.

[iii] Whereas a seized sample must meet the relevant regional pharmacopeial requirements when tested by the regulatory agency, it need not have been tested by the firm using the applicable USP pharmacopeial monograph or general chapters (unless so stated in relevant regulatory filing).

Table 2.4 Suggested Drug Product Specification of the Fictitious Drug, Xenplifir Mesylate* 10 mg/mL mg Oral Solution (Exemplifi Oral Solution)

Xenplifir Mesylate 10 mg/mL Oral Solution			
Test (or attribute)	**Method**	**Acceptance Criteria**	**Chapter**
Universal tests			
Appearance	SOP 001.03	Clear, colorless solution	5
Identification	UV: TM 010.00	Conforms to reference spectrum	
	HPLC: TM 010.00	Retention time of main peak within 5% of peak in standard solution	5
Assay	HPLC: TM 003.03†	90.0–110.0%	6
Impurities			
P1	HPLC: TM 011.01	NMT (≤) 0.4%	6
P2		NMT (≤) 0.7%	6
S4‡		NMT (≤) 2.0%	6
Unspecified		NMT (≤) 0.2%	6
Total		NMT (≤) 3.5%	6
Specific tests			
Deliverable volume	USP <698>	Conforms to USP	
pH	SOP 005.01	4.5–5.2	
Methyl parabens (0.1%)	HPLC: TM 019.01	85.0–115%	15
Propyl parabens (0.1%)		85.0–115%	
Microbial limits	USP <61>, <62>, <1111>§	Conforms to USP¶	15

**Assumes US submission of application for clinical trials or marketing authorization application.*
†Assumes assay and impurities are measured by different HPLC methods (see Chapter 6).
‡S4 is a degradation product and should be monitored (if its concentration increases over time) in the drug product as well as in the drug substances. S1, S2 and S3 are process impurities of the drug substance and need not be monitored in the drug product (see Chapter 6).
§Microbial attributes of non-sterile pharmaceutical products.
¶An acceptance criteria of "conforms to USP" is preferred to a listing of the actual limits because changes in USP general chapter will require a change to the specification.

2.2 SPECIFICATION SETTING PROCESS

2.2.1 Selection of attributes and critical to quality attributes

The attributes (tests) in the specification for a new drug substance or a new drug product are selected on the basis of the regulatory requirements prescribed in ICH Q6A or Q6B for chemicals or biologicals, respectively (i.e. Universal and Specific Tests). The critical quality attributes (CQAs) are a subset of the attributes in a specification that if not tightly controlled may adversely affect the efficacy and/or safety of the product. The emphasis on identifying the CQAs arose from the more recently published ICH Guidelines, Q8, Q9 and Q10. For example, dissolution or dissolution rate can be a CQA for a sustained release or controlled release oral formulation, particularly if the active drug substance is a highly potent compound.

Once identified, the appropriate ranges for CQAs are defined (e.g. by design of experiments, DOEs) as part of the QbD development program (see also Chapter 3). The CQAs can also form the basis of a Comparability Protocol included in the CTD to define the filing requirements for any post-approval changes to the Active Pharmaceutical Ingredient (API), formulation or manufacturing processes. The selection of the regulatory (ICH) attributes and the CQAs will be discussed in more detail in subsequent parts of this book.

2.2.2 Development of quantitative acceptance criteria

Although this is not a book on statistics, knowledge of the statistical principles involved is essential to a more complete understanding of the process of specification setting and method validation. The basic principles are summarized here and in Parts 2–6. Readers wishing to learn more on the subject are referred to several useful texts.[1–4]

2.2.2.1 Population mean, sample mean and the "target"

Data come from an underlying population of possible results. The distribution of the data in the population may be known (e.g. a discrete distribution like the binomial or continuous like the normal) or unknown. The underlying population distributions most common in analytical chemistry always have an overall mean denoted by μ which is the center (first moment) of the distribution and a standard deviation denoted by σ which indicates the spread of the data about μ. In almost all situations, μ and σ are unknown so they need to be estimated from a sample taken from the distribution. Data are generally collected to make a statement about μ and/or σ. For example, one may want to know the mean of a batch (μ) or the precision of an assay (σ) or want to compare two methods, that is the two method population means (are μ_1 and μ_2 equal?).

The population mean is estimated by the sample mean, $\bar{x}$ given by:

$$\bar{x} = \frac{\sum_1^n x_i}{n} \tag{2.1}$$

where n is the sample size.

The population standard deviation is estimated by sample standard deviation, s, is given by:

$$s = \sqrt{\frac{\sum_i^{n=1}(x_i - \bar{x})^2}{n-1}} \tag{2.2}$$

Equation (2.2) uses $n-1$ instead of N so as not to derive a "biased" (overestimated) value of the population standard deviation. One distribution that is commonly assumed is the normal or Gaussian distribution. The distribution (y) of the measurements (x_i), centered at μ with standard deviation σ, is described by:

$$y = \frac{\mathrm{e} - \left(\frac{(x_i - \mu)^2}{2\sigma^2}\right)}{\sigma\sqrt{2\pi}} \tag{2.3}$$

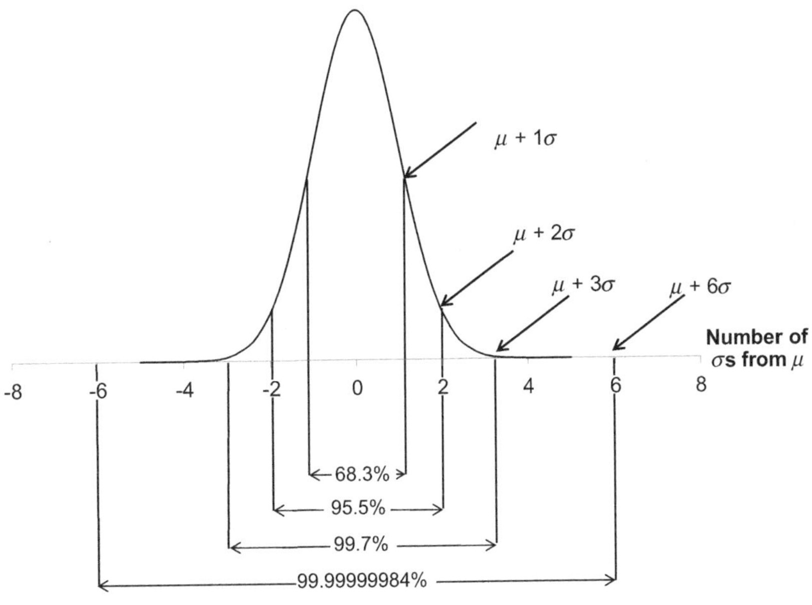

FIGURE 2.1

Distribution of data (μ–x) in units of σ.

Figure 2.1 shows the distribution of the measured values about the population mean in standard deviation units. The proportion of the data within $\mu \pm 2\sigma_1$, $\mu \pm 3\sigma_1$ and $\mu \pm 6\sigma_1$ are 95.5, 99.7 and 99.9999998%, respectively. Figure 2.1 shows that 95% of the measured values lie within the range:

$$\mu - 1.96(\sigma) \text{ to } \mu + 1.96(\sigma) \tag{2.4}$$

In many analytical methodologies a skewed or non-normal distribution is seen when calculating the reportable values. This is a common occurrence for methods for biopharmaceuticals, which are calculated using a log scale for the underlying dose–response curves. It is a common practice to perform a transformation to normalize the data. Such datasets are often said to be "log-normally distributed". Typical transformations include; log, ln and Box–Cox transformations. Biological potency assays, titer-based assays and many Enzyme-linked Immunosorbent Assay (ELISA) methodologies require transformation prior to performing any of the following calculations. Therefore the previous statistics discussed above are calculated utilizing a transformed data set.

The geometric mean (GM) for the ln is calculated by:

$$\text{GM} = e^{\text{Average}} \tag{2.5}$$

where Average equals the average of the natural log responses.

The geometric relative standard deviation (GRSD) is calculated by:

$$\%\text{GRSD} = 100 \times (e^{s} - 1)\% \tag{2.6}$$

Figure 2.1 also shows that the greater the spread in the data the (i.e. the larger the value of the standard deviation), the lesser the fraction of the data that is contained within a specific range. For example, suppose a potency assay has a mean of 100 with a standard deviation of 1. Then 95.5% of the individual results would fall within two standard deviations of 100 (98–102). If the standard deviation was doubled (2 instead of 1), then 95.5% of the individual results would lie between 96 and 104. Only 68% would lie between 98 and 102.

The statistics frequently used to describe the error associated with the repeated measurements of the mean is the standard error of the mean (sem), which is given by:

$$\text{sem} = \frac{\sigma}{\sqrt{n}} \tag{2.7}$$

It follows that 95.5% of sample means will fall between:

$$\mu - 1.96\left(\frac{s}{\sqrt{n}}\right) < \bar{x} < +1.96\left(\frac{s}{\sqrt{n}}\right) \tag{2.8}$$

As discussed previously, the population mean is the mean of all possible measurements. This population mean can always be estimated by the sample results. However, this population mean may not be equal to the "underlying" mean of what is being measured. In this case the measuring device results are biased. One way to estimate this bias is to use a reference material where the true mean is known.

Within the last 15–20 years there has been a growing interest in the application of the principles of "six sigma" originally developed by Motorola and implemented by most engineering-based companies, notably General Electric, to the pharmaceutical industry.[4,5] Six Sigma is concerned with the relationship between the rejection rate of a batch, the range of the upper and lower specification limits (USL and LSL), respectively, and the standard deviation of the process. For processes defined by two-sided specification limit (see Section 2.2.2.2), the rejection rate is 0.27% if the range of the LSL and USL is $\mu \pm 3\sigma$ (i.e. if the range of the normally distributed values are within 6σ). The rejection rate is 1.6 ppm if the range of the LSL and USL is $\mu \pm 6\sigma$.

The preceding discussion assumes that the product is manufactured at the 'target value" (T) specification, i.e. the measured mean value defined in the specification at product release is equal to T. There may be valid reasons to target a release value different than the value implied by the specification. One valid reason is manufacturing excess to allow for losses during manufacturing. However, the use of overages to allow for product degradation during storage or to allow for analytical bias is unacceptable from a regulatory perspective in many pharmaceutical products. However an exception to this is when determining the target potency for a vaccine product. In this case because of the broad safety window single use syringes are often "overfilled" to ensure sufficient potency of the dose at the end of the shelf life. These concepts of capability analysis are explored further in the following section.

2.2.2.2 Capability analysis

Capability analysis provides dimensionless (unitless) ratios or indices to allow comparison of the process distribution with the width or size of the acceptance criteria in the specification. The two most commonly used indices are C_p and C_{pk} (Eqns (2.10)–(2.14)). By convention, the capability index, C_p is given by[5]:

$$C_p = \frac{\text{USL} - \text{LSL}}{6\sigma} \tag{2.9}$$

where USL and LSL are the upper and lower acceptance criteria in the specification, respectively. In the cases of unilateral acceptance criteria (one side only), C_p values for USL and LSL are given below:

$$C_p = \frac{USL - \mu}{3\sigma} \tag{2.10}$$

$$C_p = \frac{\mu - LSL}{3\sigma} \tag{2.11}$$

It should be noted that Eqns (2.10) and (2.11) assume that μ is estimated by $\overline{x}$ and σ is estimated by s. The capability index, C_{pk}, takes into account how far the target (T) is from the center of the specification limits (μ):

$$C_{pk} = \text{minimum}((USL - \mu)/3\sigma,\ (\mu - LSL)/3\sigma) \tag{2.12}$$

It is also noted that $C_p = C_{pk}$ when the process is centered between limits in the specification (i.e. $T = \mu$). As discussed earlier, the use of a target value that is different from the mean value (i.e. the center of the specification limits) is inappropriate unless the specification limits are asymmetrical. However, the use of capability indices can be useful in calculating the acceptance criteria at product release that are necessary to give an acceptable level of assurance that the product will not go out of specification during the shelf life (see Section 2.2.4). The biggest disadvantage of the use of capability indices in this regard is the shortage of relevant data to accurately calculate the standard deviation of the process since specifications are typically established during development on relatively few key batches.

Formulating off-center will influence the "effective" shelf life of the product. For example, Fig. 2.2 shows the effect of formulating at 95% or 92% of the target on the effective shelf life of the product. The closer the assay value is to the LSL the sooner the assay value will go out of specification, reducing the effective shelf life.

2.2.2.3 "Shift happens"

Despite the best efforts of process engineers and formulators, the mean of any commercial pharmaceutical product will shift over time. A trend (gradual or immediate) to lower or higher values of the mean can, over time, reach values as high as $\mu \pm 2\sigma$. Therefore, the process must be monitored over time (e.g. by the use of control charts) and the mean recentered, if possible. Shift in the mean by such an amount can have significant effects on the "defect rate" or the rate of rejection of commercial batches. The effect of mean offset is illustrated in Fig. 2.3(a) and (b).

Process variability can also increase over time, due to lack of process control or control of key excipients, leading to an increase in the probability of the batch failing to meet the acceptance criteria (Fig. 2.3(b)). Both those cases demonstrate the importance of monitoring the process closely using control charts, (e.g. Fig. 2.3) and recentering the process and/or adjusting the process to eliminate trends and/or maintain or reduce the variance (σ^2).

2.2.3 Calculation of quantitative acceptance criteria

It is tempting to consider the use of capability analysis and the principles discussed above to set the initial acceptance criteria for in-process and final product specifications. However, the lack of a robust estimate of the standard deviation may make this approach unreliable. The lack of reliable estimate of

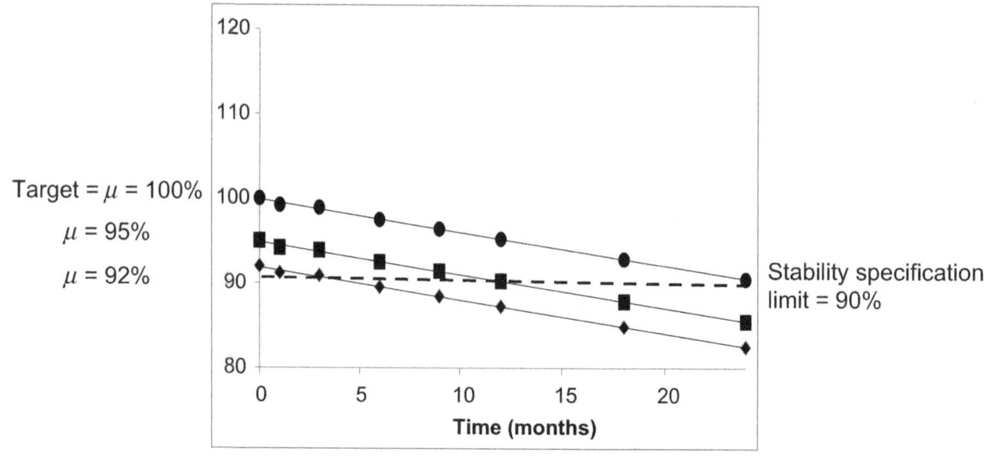

FIGURE 2.2

Effect on effective shelf life of formulating product at the mean assay value (μ) and at the lower specification limits (LSL) of either 95% or 92%, for a product with degradation rate of 0.4% per month.

the variance of the process or product arises from the fact that the standard deviation is estimated from only a few critical batches (generally pivotal clinical batches and "registration" batches). ICH Q6A and Q6B suggest that acceptance criteria be set on the basis of the three registration batches and other representative batches, which may be a very small number. Additionally, the FDA seeks to establish specifications based upon clinical experience. This represents a challenge for the industry, even with the advent of QbD approaches which will increase the data set and understanding of process and analytical variability, but will not necessarily impact the number of clinical lots. As the number of batches and amount of stability data increase during commercial production, capability analysis becomes increasingly more reliable in predicting the incidence of batch rejection and batch recall.

2.2.4 Release and stability specifications

The preceding limitations notwithstanding, capability analysis may have some utility when setting the initial acceptance criteria for the finished product (and in-process samples) and the width of the target values at product release to ensure that the product does not go out of specification prior to expiration. The width of the USL and LSL (i.e. the Acceptance Criteria in the Release Specification) can influence the potential shelf life of the product. Figure 2.4 shows that the tighter the range of USL and LSL, the shorter the potential shelf life. By convention, the Acceptance Criteria for assay (potency) in the release specification are typically 95–105% and the typical Acceptance Criteria in the stability specification are 90–110%. Thus tightening the release specification decreases the likelihood that the product will fail the stability specification during storage. However, too tight a release specification will result in more batches failing at release. Conversely, too loose a release specification can result in a greater number of batches failing the stability specification during storage.

The use of release and stability specifications is one area in ICH Q6A that is not harmonized. In the US, a product must meet a single specification at release and throughout the shelf life (this is also

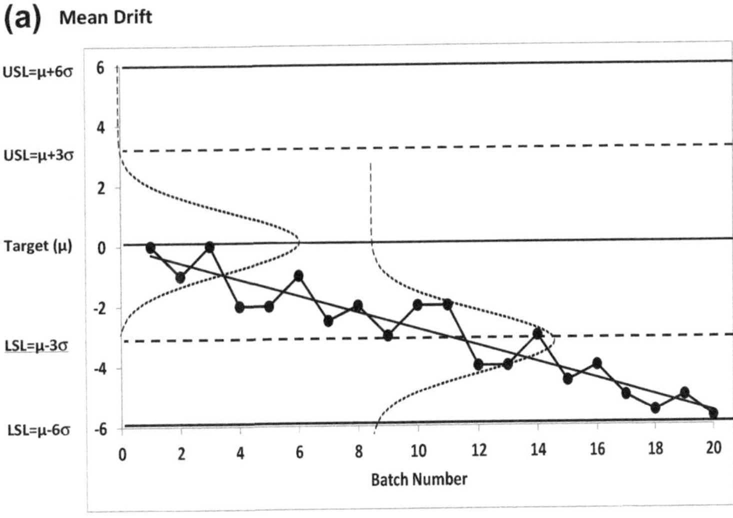

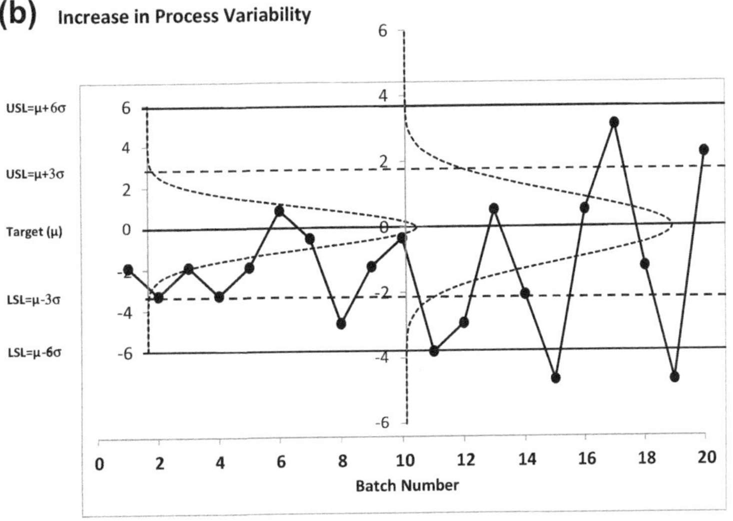

FIGURE 2.3

Control chart showing the effect of (a) mean drift with constant standard deviation and (b) increasing standard deviation (constant centered mean) during routine commercial production.

generally true for a pharmacopeial monograph). By contrast, in the EU, separate specifications are required at release and on stability. However, in the absence of a regulatory release specification, tighter "in-house" release specifications are generally used to ensure that the product will meet the regulatory specification throughout its shelf life. If this is done, it should be noted that the US

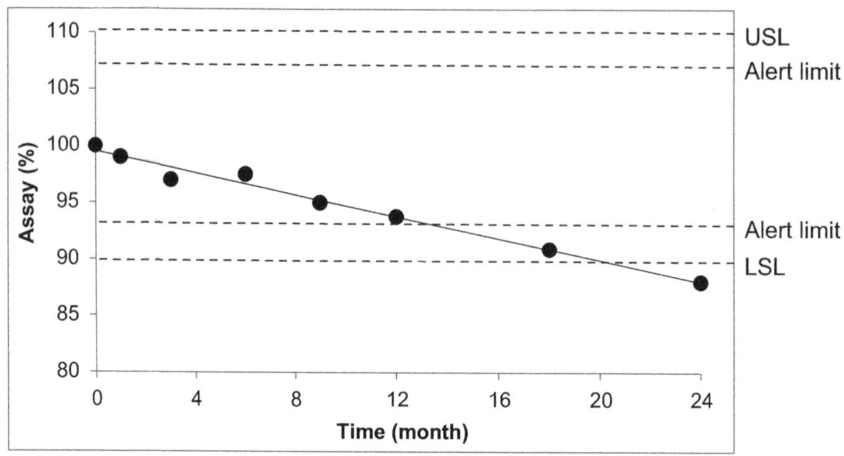

FIGURE 2.4

Alerts limits for the monitoring stability data (using assay as an example).

regulators may treat the "in-house" release specification just like any label specification when determining if a reported value is an out-of-specification (OOS) and requires a thorough, formal OOS investigation.

2.2.5 Shelf life

A shelf life between 18 and 24 months with storage at room temperature is preferred to ensure manageable inventories and supply chains for traditional (small molecule) pharmaceuticals. Many biopharmaceuticals have refrigerated or frozen storage requirements. Many of the newer biopharmaceuticals, such as cell therapies have extremely short shelf lives, often shorter than 2 weeks. Shorter shelf lives at refrigeration temperature are generally reserved for special cases or where the stability of the product does not allow room temperature storage for 18–24 months.

Figure 2.5 shows simulated data designed to demonstrate the calculation of shelf life (in this example: assay) for a product with a degradation rate of 0.5% per month and 1.5% per month at 25 °C/60%RH and 40 °C/75%RH. Data were simulated for real time at 0, 1, 3, 6 and 9 months at 25 °C/60% RH 0, 1, 3 and 6 months at 40 °C/75%RH, and analyzed according to ICH Guideline Q1A(R2). Note: this illustration uses one batch for the estimation of shelf life. In practice, several batches will be used. The data obtained at 40 °C/75%RH are also shown for illustrative purposes. The values at 12, 18 and 24 months and 25 °C/60%RH were calculated by linear extrapolation of the 0–9 month data. It is tempting to set the shelf life at 20 months (Fig. 2.5) based on the intersection of the extrapolated data and the LSL of 90%. However, based on the previous discussion there will be determinate errors in the estimation of the estimated value of the assay at the LSL. Therefore, a more statistically valid approach to the estimation of the shelf is to calculate the value of the assay at the point where the lower 95% confidence interval $CI_{0.95}$ for the predicted assay value at a given time-point intersects the LSL (e.g. Fig. 2.5) (see ICH Q1E: Evaluation of Stability Data). Using the latter approach the shelf life is estimated to be 16 months.

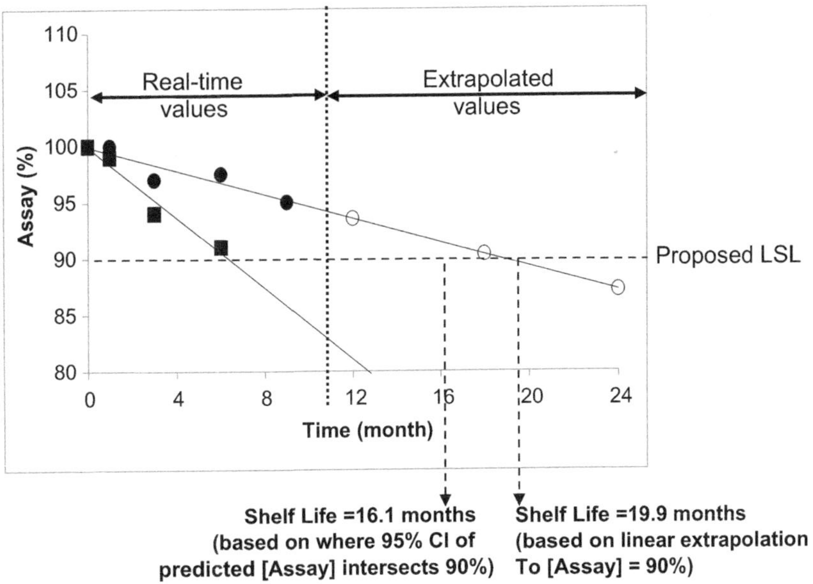

FIGURE 2.5

Calculation of shelf life (based on assay) for a product with a degradation rate of 0.5% per month and 1.5% per month at 25 °C/60%RH (circles) and 40 °C/75%RH (squares), calculated by linear extrapolation of 25 °C/60% RH data; and the lower 95% confidence interval of linear extrapolation of 25 °C/60%RH data.

In contrast to small molecules, biopharmaceutical shelf-life calculations are not usually calculated based upon accelerated temperatures. It has been shown that the complexity of the active ingredient and the requirement for a specific 3-dimensional structure is not conducive to using various models, such as the Arrhenius equation, to estimate real-time storage stability. Therefore only real-time storage data at the proposed storage temperature are typically accepted by the regulatory authorities for establishing the shelf life. It has been proposed to the World Health Organization (WHO) and other government regulatory authorities that additional real-time data points for biopharmaceutical products be collected than those recommended by the ICH Guidelines for the calculation of shelf life.

2.2.6 Contribution of analytical variability to overall process variability

There are three types of errors that contribute to analytical results: gross errors, systematic errors (determinate errors or bias) and random errors (indeterminate errors or precision). It is important that none of these errors contribute significantly to the overall variability of a process or product attribute; otherwise the method will be incapable of detecting differences between batches or changes within a batch. Gross errors (such as failing to adjust a solution to volume, or dropping a flask) should be detected during an investigation into an aberrant result and corrected.

The contributions of random or indeterminate errors have been discussed in detail in the previous sections. Determinate errors (bias) should be eliminated during method development and their absence confirmed during method validation (Chapter 4). The contribution of the random errors in the analytical measurement ($\sigma^2_{analytical}$) to overall variability (or variance) of an attribute $\sum_{i=0}^{n} \sigma^2_{other}$ can be determined from the principle of propagation of errors:

$$\sigma^2_{total} = \sigma^2_{analytical} + \sum_{i=0}^{n} \sigma^2_{other} \tag{2.13}$$

This is demonstrated by the following example. If we assume, as discussed in Section 2.2.4, the overall standard deviation of a process (e.g. assay) is equal to half the difference between the mean and the lower (or upper) specification limit (i.e. 5%) and the standard deviation (repeatability) of the assay method is 1.5%, the contribution of the remaining factors to the variability n assay is given by:

$$\sigma^2_{other} = \sigma^2_{total} - \sigma^2_{analytical} \tag{2.14}$$

or,

$$\sigma_{other} = \sqrt{5^2 - 1.5^2} = 4.72\% \tag{2.15}$$

In this example, Eqns (2.14) and (2.15) predict that although the analytical method contributes 30% of the overall variance (variability) of the measurement ($\sigma^2_{analytical}/\sigma^2_{total}$), it contributes less than 6% of the standard deviation (0.28/5.00). The contribution of analytical errors is expanded further in Chapters 3 and 4.

2.3 CERTIFICATES OF ANALYSIS, TRENDING AND OOS RESULTS

The most obvious use of drug substance and drug product specifications are in the release and stability testing of drug substances and drug products associated with studies conducted under current Good Manufacturing Practices (cGMPs, clinical trials), Good Laboratory Practices (GLP, nonclinical studies) and "unofficial" experimental studies (if required). If the samples meet the acceptance criteria, an official Certificate of Analysis (CoA), reviewed and approved by the Analytical Department and Quality Assurance (QA), is issued. Release of a batch for clinical use is always accompanied by review of the manufacturing batch record and, depending on the phase of development, partial or complete verification of the raw data. Strict interpretation of the GLP requirements reveals that review of the manufacturing batch records is not required—however, it is strongly recommended. When testing or reagent manufacture occurs in a GLP environment, the QA group often creates a standard operating procedure (SOP) and a routine study document for collecting typical manufacturing information. This can take the place of a batch record.

Although the practice is common, especially by Contract Manufacturing Organizations (CMOs), to issue a CoA after testing at each stability time-point. An equally acceptable approach is the use of stability tables, which are updated after each stability time-point. If concurrent stability testing (*cf.* testing of returned retained samples) is conducted as part of a GLP study, a CoA must be issued following testing of the first sample pulled after the in-life portion of the study for inclusion in the final (GLP) study report.

Good analytical precision, the absence of bias and gross errors, is especially critical for samples that approach the upper or lower limits of the specification. Otherwise a false OOS result may occur, which cannot be reversed by a formal investigation. An authenticated OOS verified by a formal investigation led by QA will result in batch rejection or recall of a batch from the clinic or from commercial distribution. Valid reasons for analytical errors and rejection of an OOS result are bias and gross errors: analytical imprecision is not. Examples of gross errors include: incorrect following of procedures, incorrect instrument setting, instrument failure, and failure to meet system suitability. As discussed previously, analytical bias should be detected and eliminated during method development and/or method validation. However, the OOS investigation may reveal a new source of error not previously detected during method development and validation. In this case the method should be modified and the appropriate components of the method revalidated. Graphical trending of stability data (see control charts, Figs 2.3 and 2.4) is essential to anticipate potential future batch failures and stability failures and appropriate remedial steps taken. Figure 2.2 demonstrates the importance of controlling the variance of the measurement for a particular attribute when the results approach the limit in the specification. Otherwise, an OOS may be generated on the basis of chance.

Alerts limits are also useful (Fig. 2.4). Following the well-known case of the United States vs Barr Laboratories,[6] the possibility of generating an OOS on the basis of chance is increased by the fact that any result containing a single value outside the acceptance criteria is considered OOS, even if the mean is within the allowable limits. This further emphasizes the importance of trending batch history and stability data and the application of alert limits.

It is important to note that the US vs. Barr Decision also treats biological potency assays differently than physical/chemical methods for small molecules. The USP General Chapter <111> specifically states that because of the higher variability of potency assays it is a standard procedure to calculate reportable values based upon more than a single assay run and that these averaged values may include individual values that are outside the specification. These individual values are ***not*** treated as OOS results as long as the average potency estimate is sufficiently precise as defined in the standard test method or SOP.

2.4 SPECIFICATIONS IN EARLY DEVELOPMENT

The previous discussions have focused primarily on late development specifications and specifications to be included in the marketing dossier. Very recently a working group of an industry group, the IQ Consortium[7] has published an article on "Early Development for Small-Molecule Specifications. An Industry Perspective". This article emphasizes the fact that very little information may be available to support firm specifications for the drug product and the drug substance in early development. Therefore, "*early development specifications should … focus on those tests and acceptance criteria determined to be critical for the control of product quality and supported by preclinical [safety] and early clinical safety studies*." In that paper, the authors propose standardized early phase tests and acceptance criteria for both the drug substance and the drug product. They also differentiate between test results that are to be reported to the agency at release and on stability from internal tests and acceptance criteria that are not part of the formal specifications. The paper differentiates between the early nonclinical batches where there are generally no formal specifications, and first in man batches, where the aims of the former are to:

- Ensure that the correct dosage is administered in the nonclinical studies
- Determine the correct potency value of the drug substance to ensure proper dosing of the animals
- Quantify impurities for nonclinical qualification (establish the initial impurity profile)

Table 2.5 Proposed Specifications for Clinical Drug Substance in Early Development

Attribute	Proposed Acceptance Criteria	Release Testing	Internal Testing*	Stability Testing
Description	Range of color description	+	−	+
Identification	Spectrum conforms to reference	+	−	−
Counter Ion	Report result	+	−	−
Assay	97−103% on anhydrous, solvent-free basis	+	−	+
Impurities	Individual NMT 1.0% Total NMT 3.0%	+	+	+
Chiral impurity	NMT 1.0%	+	+	+
Residual solvents	ICH Q3C or other justified limits for solvents in final synthetic step	+	+	−
Mutagenic impurities	CHMP Guideline (Ref. 8) until ICH M7 is implemented	−	+	−
Inorganic impurities	EMA limits (Ref. 9) until ICH Q3D is implemented	−	+	+
Water content	Report results	−	+	+
Solid form	Report results	−	+	+
Particle size	Report results	−	+	−
Residue on ignition	NMT 1.0%	−	+	−

**Internal testing can be performed in addition to or in replacement of release testing in the final drug substance. Internal testing may have target acceptance criteria that are tighter than the release testing criteria.*

Adapted from Ref. 7

The paper goes on to emphasize that the initial acceptance criteria for early clinical batches are targets based on the results of the initial nonclinical studies. They further emphasize the value of using the same batch of drug substance in both the early nonclinical studies and the first in man studies, in which case the impurities are inherently qualified (given the appropriate safety margins). Table 2.5 provides the standard specification for drug substances for early development as an illustration of the approach. Readers wishing to learn more about this approach including the standardized early-phase specifications for powder in a bottle, powder in a capsule, tablets and capsules are referred to Ref. 7 for more details.

Acknowledgment

The authors are grateful to James Bergum (BergumSTATS) for his thorough review of this chapter.

References

1. Miller, J. C.; Miller, J. N. *Statistics for Analytical Chemists,* 3rd ed.; Ellis Horwood: Chichester, 1993.
2. Kiemele, M. J.; Schmidt, S. R.; Berdine, R. J. *Basic Statistics: Tools for Continuous Improvement,* 4th ed.; Air Academy: Colorado Springs, 1997.
3. Riley, C. M., Rosanske, T. W., Eds. *Development and Validation of Analytical Methods;* Pergamon: Oxford, 1996.
4. Harry, M. J. *The Nature of Six Sigma Quality;* Motorola University: Schaumburg, 1997.
5. Pande, P. S.; Neuman, R. P.; Cavanagh, R. P. *The Six Sigma Way: How GE, Motorola and Other Top Companies are Honing Their Performances;* McGraw Hill: New York, 2000.
6. United States v. Barr Laboratories, Inc., 812 Fed. Supp., 458, 1993.
7. Coutant, M.; Ge, Z.; McElvain, J. S.; Miller, S. A.; O'Connor, D.; Swanek, F.; Szulc, M.; Trone, M. D.; Wong-Moon, K.; Yazdanian, M.; Yehi, P.; Zhang, S. *Pharm. Tech.* **2012,** *36* (10), 86–94.
8. CHMP. *Guideline on the Limits of Genotoxic Impurities;* EMA, 2007.
9. CHMP. *Guideline on the Specification Limits for Residual Metal Catalysts or Metal Reagents;* EMA, 2008.

CHAPTER

Application of quality by design (QbD) to the development and validation of analytical methods

3

David K. Lloyd*, James Bergum[†]

* Analytical and Bioanalytical Development, Bristol-Myers Squibb, New Brunswick, NJ, USA,
† Statistical Consultant, BergumSTATS, LLC, Howell, NJ, USA

CHAPTER OUTLINE

Specification of Drug Substances and Products. http://dx.doi.org/10.1016/B978-0-08-098350-9.00003-5

3.1 INTRODUCTION

3.1.1 Analytical quality by design

Quality by design (QbD) has been proposed as a systematic approach to product development, wherein understanding of the product is paramount.[1,2] Through a comprehensive understanding of the effects of various inputs (e.g. process parameters, materials) on the final product (active pharmaceutical ingredient or drug product), appropriate ranges of the input parameters may be defined within which the quality of the final product is guaranteed.[1,2] Achieving the appropriate level of understanding typically involves a multifactor approach to the study of a product, since changes to one input may alter the effect of another input on the final product. One-factor-at-a-time (OFAT) studies are simply not adequate to develop broad product understanding. On the other hand, given the large number of potential input parameters for any product, development of a multifactor model for all possible effects is practically impossible. Therefore, risk analysis is an important element of QbD,[3] allowing the truly impactful parameters to be identified for further investigation in studies which are of a manageable scale. These studies generally apply statistical design of experiments (DOE) or mechanistic models, and the output is a multidimensional design space in which the effects of key parameters are understood, and a product control space is defined where appropriate product quality is guaranteed (or at least highly probable!) within the parameter ranges that border the space. A further important element of QbD is the development of a control strategy, which governs how changes are assessed and implemented during the life of a product.[4]

Analogous to QbD for product, QbD may also be applied in the analytical realm.[5] If QbD for a product is defined as a full understanding of how inputs and process attributes relate to product performance, then for analytical methods it can be considered to be a full understanding of how the analytical technique attributes and operating conditions relate to analytical performance. Factors that may be considered for study include the type of analytical technique chosen, reagents used, and instrumental parameters. The method performance is defined by both the type of data that the method produces and the required quality of that data. Data quality has traditionally been determined at the *grande finale* of method development process, method validation, when method performance is determined (accuracy, precision, selectivity, robustness, etc.). By applying QbD to analytical methods, the method performance is instead largely defined and understood during the development process.

3.1.2 Why do it?

In the authors' experience, raising the topic of QbD to an analytical audience results in, at best, a mixed reaction (indeed, this may be the case for nonanalytical audiences as well[6]). Typically, concerns are raised as to what the analyst will get out of it (what benefits will be gained), and even more so about how much extra work QbD will require. One potential benefit that is sometimes proposed is that by defining a QbD design space, the method could be operated anywhere within that design space. Although this has merit in providing some flexibility in manufacturing processes, it is less obvious that this represents a major advantage in a typical analytical method. Although from the design space you may know that you can perform your measurement at a wavelength of ±10 nm from the method set point of 254 nm, it's unlikely that there would be much value in deviating from the set point; "It's Tuesday, I think I'll try 264 nm today..." is probably not a behavior typical of most analytical scientists. Instead, the great benefit comes from the identification of a robust operating region for the method. Method failure due to lack of robustness can have considerable impact, e.g. delaying a project start due to a failed technology transfer, or imperiling batch release if there is a method failure during product testing. If analytical QbD can help assure clean method transfers and routine smooth method operation, the long-term impact and value will be high. As to the question of how much extra work analytical QbD entails, the answer probably ranges from "little or nothing" to "a lot", depending on how well QbD is built into the method development and validation process. If it comes as an afterthought, it will surely result in extensive extra work. If QbD is built into the process from the beginning, good risk assessment is performed to eliminate low-value studies, and the results of systematic method development are contemporaneously documented, the impact on time and effort should be minimal while increasing method understanding and robustness.

3.2 METHOD REQUIREMENTS

The requirements for a particular analytical method are strongly tied to the product and the product attribute to which the method is being applied. The actual measured result will contain elements of method and product variability combined via their variances (see also Section 6.2.4.2 in Chapter 6):

$$\sigma^2 = \sigma^2_{\text{product}} + \sigma^2_{\text{method}} \tag{3.1}$$

It is desirable to reduce the method variability such that it becomes a relatively small contribution to the overall variability. In addition, the method may suffer from other errors such as bias, interference, etc., which reduce the data quality. The method requirements should be set such that the data generated by the method are a good reflection of the product attribute being tested, and not masked by analytical errors.

With some types of analyses, it is relatively easy to ensure that the errors of analysis are small relative to the variability of the product being measured. For example, in a chromatographic test for content uniformity (CU), it would typically be unusual for the method variance to be more than a small fraction of the product variance. With relatively cursory method development, the method requirements are likely to be met without a full-blown application of QbD. However, for other measurements the potential for analytical error may be relatively large and there is correspondingly greater justification for more extensive systematic studies.

Required method characteristics may be defined in an analytical target profile (ATP), which may be viewed as being somewhat similar to a specification for an analysis,[7] and analogous to a quality target product profile.[2] The ATP lists important method characteristics (e.g. parameters such as accuracy, precision) and describes the degree to which these must be controlled (e.g. what percentage of inaccuracy or imprecision is acceptable). Note that the ATP is essentially independent of the analytical technique used; it simply defines the characteristics that the method must have in order to adequately measure the product's critical quality attributes (CQAs). The technique-independent nature of the ATP was originally envisaged as offering a way to include required method characteristics in a regulatory filing without actually specifying exact method conditions; any method could be used to measure a product CQA so long as it was demonstrated to meet the ATP. In our experience this is not yet a concept which is broadly embraced by regulatory authorities; however, it remains a useful tool in defining what a method has to measure and how well it has to make that measurement. Given a defined ATP, one can then decide on how to fulfill the ATP's requirements, in other words, what sort of method to use. When multiple techniques offer the required analytical performance, factors not related to data quality such as cost, speed and "greenness" become important. In some cases the choice is fairly obvious, e.g. a large majority of small molecule impurity analyses are performed by reversed-phase liquid chromatography because it is a relatively routine, inexpensive technique, which is well suited to meet the ATP's requirements for typical small drug molecules.

3.3 METHOD RISK ASSESSMENT

3.3.1 Definition of risk

From ICH Q9,[3] "risk" is defined as "The combination of the probability of occurrence of harm and the severity of the harm." Risk can mean many different things in the context of pharmaceutical development, but from a viewpoint of regulatory agencies this is principally the risk of harm to the patient. In the context of analytical testing one can consider the possible harm being caused to the patient through an incorrect analytical result leading to release of a batch with undesirable characteristics. In the context of pharmaceutical analysis it has been proposed that, using a failure mode effects analysis approach (Section 3.3.2) risk = severity × occurrence × detectability, where these terms are defined by Nasr[8] as follows:

- severity = effect on patient
  - related to safety or efficacy
- likelihood of occurrence = chance of failure
  - related to the quality and extent of product and process knowledge and controls
- detectability = ability to detect a failure
  - related to suitability of the analytical methodology (sampling and testing)

It is important to note that the above expression is not a robust quantitative mathematical relationship, but a more qualitative statement that something is high risk if it has bad consequences, is likely to happen, and is not likely to be detected. The definition of "severity" may usefully be expanded; from a business perspective, harm could come about because of incorrect analytical results leading to rejection of a truly good batch of product. Thus we may choose to include business risks in the calculation (although the regulatory perspective is likely to be narrower, excluding these from consideration).

The process of risk assessment for analytical methods is thus a determination (qualitative or quantitative) of the effects of variation in factors such as method operating parameters or sample characteristics on method performance. Assessment of risk includes identification of potential risks, and analysis of these risks leading to an evaluation of the importance of that risk. The risk-assessment process brings an important benefit to method development, because a good analysis of what are truly important parameters allows a more focused systematic study of only those parameters.

3.3.2 Risk assessment toolbox

An element of risk assessment is present in any thoughtful method development activity. Traditionally, it may not have been performed as a separate activity, but informally (maybe just in the analyst's head!) as a consequence of the analyst's general knowledge of the technique and sample at hand. However, a more formal approach allows the risks to be documented and decisions justified. There exist many tools which aid in structuring the risk assessment process:

- Qualitative tools for parameter screening, e.g.
  - Process mapping
  - Ishikawa or fishbone diagrams

Such qualitative tools help define risks in a process or method by systematically laying out the various method steps and identifying the associated risks. For example, the process of a drug product analysis may be mapped as involving an automated sample preparation followed by a chromatographic analysis. These two steps in the process can each be further broken down into sub-operations, sub-sub-operations, and so on. Possible elements of risk can be associated with each operation. Fishbone diagrams illustrate the process somewhat differently (see Section 3.3.3 for an example). Many risk factors could be identified with each process, but combining one's general analytical knowledge about the technique in use and specific knowledge about the particular analyte, many hypothetical risks can quickly be discounted, leaving relatively few potentially critical parameters for further consideration.

- Semiquantitative tools for risk ranking, e.g.
  - Relative ranking
  - Failure mode effects analysis/Failure mode effects and criticality analysis (FMEA/FMECA)

Semiquantitative tools for risk ranking help define which elements of the method are truly CQAs. For example, after initial qualitative triage of risk factors, the remaining parameters can be assessed for their relative criticality; a factor that has potential to result in erroneous data which cannot easily be identified as erroneous would rate as high risk, whilst one where the error can easily be spotted would rate lower. FMEA/FMECA is specifically mentioned in ICH Q9 "Quality Risk Management".[3] FMEA starts by evaluating potential failure modes for each step in the analytical method, and correlates each failure mode with a likely effect either for patient safety or as a business risk. The root cause for each failure mode is postulated based on previous knowledge or general scientific principles. The assessment may be extended to include a consideration of the criticality of a particular risk, and hence may allow identification of method steps where additional preventive actions may be appropriate to minimize risks. It should be emphasized that although a risk probability number (RPN) may be assigned as the product of severity, likelihood and detectability (see Section 3.1 above), the RPN is not a hard number in the sense that, for example, one expects an assay value to be. Although there are criticisms of the FMEA methodology,[9,10] it nevertheless provides a useful framework within which to attempt to systematically identify and rank risks and as such, can be useful provided one does not attempt to over-interpret the numerical output.

- Experimental tools for process understanding, e.g.
  - Statistically designed experiments
  - Mechanistic models

Statistical tools can support and facilitate risk assessment. For example, a screening factorial design can demonstrate the sensitivity of a method to different parameters, and the variability encountered within the experimental space. Mechanistic models may provide similar information—a simple univariate example would be the Henderson–Hasselbach equation relating the degree of ionization of a compound ([salt]/[acid]) to its pK_a and the solution pH:

$$\mathrm{pH} = \mathrm{p}K_\mathrm{a} + \log([\mathrm{salt}]/[\mathrm{acid}]) \quad (3.2)$$

The effects of the relationship described by this equation on chromatographic method robustness are enshrined in the chromatographer's axiom that one should operate >2 pH units away from the analyte's pK_a so that minor changes in pH will not cause a large change in the ratio [salt]/[acid] (these species typically having very different retention). In the common situation of chromatographic separation of impurities where species with varying pK_as are simultaneously analyzed, an understanding of the charge state of each component will illustrate which, if any, are potentially at risk of varying retention from pH change and thus whether pH should be considered a primary factor for further investigation.

It can be very helpful to perform risk identification as a small-group activity. Having multiple viewpoints in the risk assessment reduces the possibility that potential risks will be overlooked or dismissed (the analyst considering Eqn (3.2) may believe that the analyte pK_as are far from the operating pH, but hopefully in a group discussion someone would remember that the pK_as of weak acids and bases depend greatly on the solvent mixture used and ask whether the "known" values are correct under the chromatographic conditions used!). The assessment should be appropriately comprehensive, and not just limited to the method conditions. For example, the sample itself is an important risk factor in many analyses; if there are variations in tablet properties such as particle size or hardness, these may well affect a spectroscopic calibration model or an extraction. It may well be that potential risks are identified that cannot be fully addressed at a given point in development; although analyst-to-analyst variability may be studied in a single lab, an investigation of lab-to-lab variability may not be possible (or warranted) in early development (see also Chapters 4 and 6). In later development, a wide variety of samples from product development QbD studies may become available, which cover the extremes of the process operating ranges. These may not be ready for inclusion in method development studies, but the potential risk can be identified early, and the effect (or, hopefully, lack of effect) confirmed later, e.g. in a separate ruggedness study.

3.3.3 Risk assessment example

Rather than starting with an example from a specific analytical technique, consider an example which almost any analytical chemist can relate to: the commute to work. Figure 3.1 shows a fishbone diagram

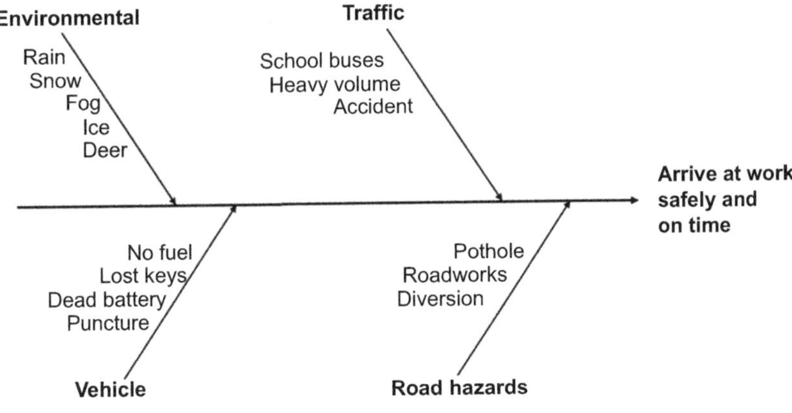

FIGURE 3.1

Fishbone diagram illustrating various risks involved in the daily commute to work. The goals are identified to the right: arriving at work safely and on time.

where four major categories of risk have been identified, with several specific risks shown in each category. Note that to the right, the goals of the commute are defined—to arrive at work safely and on time.

A qualitative assessment of the risks indicates that several can be avoided by appropriate planning. For example, a good standard operating procedure (SOP) on vehicle maintenance would likely eliminate the risks of being delayed due to a dead battery or a puncture, while another related to vehicle operation should ensure that the commuter appropriately fuels their vehicle and systematically stores their keys in a place they will not be lost. Thus, systems can be put in place to greatly minimize the identified vehicle risks. The traffic risks can also be mitigated to a significant extent, e.g. an early commute avoids volume and school buses, while forward planning can avoid major roadworks. In Table 3.1, a more quantitative analysis of these risks is presented.

The risks are rated in terms of their possible effect on achieving the stated goals. These goals need not be equally weighted when considering the risks; hopefully safety will have a greater weight in the risk analysis than timeliness. Thus, although losing one's keys in the morning may delay arrival, this is a relatively minor inconvenience compared to an unplanned close encounter with a deer (rated high severity) and was given a lesser weight for severity in the risk assessment. Similarly, judgments are made on the likelihood of a hazard occurring. If vehicle-related SOPs are in place, vehicle-related problems should have a low probability. Finally, there is the question of detectability. A hazard which can easily be detected can be avoided, and hence is given a low rating. On the other hand, black ice or a deer hiding in the roadside woods carries a higher detectability rating (i.e. poor detectability). The overall risk rating is the product of the severity, likelihood and detectability ratings. From the above analysis, it is estimated that the environmental hazards of ice and deer are the critical hazards which still need to be addressed for this commute as a result of their severity and poor detectability. These are identified as primary factors, while five other factors have a more modest impact and may optionally be assessed further.

Table 3.1 FMEA analysis of a daily commute

Risk	**Severity** High = 3 Medium = 2 Low = 1	**Likelihood** High = 3 Medium = 2 Low = 1	**Detectability** High = 1 Medium = 2 Low = 3	**Numerical Rating** Detectability × Likelihood × Severity	**Primary Factor?**
Environmental					
Rain	1	2	1	2	N
Snow	2	1	1	2	N
Fog	2	1	1	2	N
Ice	3	1	3	9	Y
Deer	3	1	3	9	Y
Vehicle					
No fuel	1	1	1	1	N
Lost keys	1	1	3	3	N
Dead battery	1	1	2	2	N
Puncture	2	1	2	4	?
Traffic					
School buses	1	2	1	2	N
Heavy volume	1	2	2	4	?
Accident	2	1	2	4	?
Road hazards					
Pothole	1	2	2	4	?
Roadworks	1	2	2	4	?
Diversion	1	2	1	2	N

The quality of the risk analysis will impact the factors studied going forward, and thus the extent of work that will be done. In this case, the hazards were identified in a brainstorming session, and the ratings were made subjectively. One can see that this process could be improved upon; for example, real statistics for road accidents or traffic flow could have been sought to better quantify the hazards and potentially to identify ones which were not recognized.

3.4 METHOD DEVELOPMENT AND OPTIMIZATION: UNDERSTANDING THE METHOD OPERATING SPACE

With many analytical techniques, one can achieve useful results with practically no method development or optimization. If rather standard approaches and method parameters give an acceptable result for the large majority of samples, and sources of error with the technique are well understood, extensive method development and optimization studies are probably not a good use of resources. Tests such as Karl Fischer (KF), simple UV measurements, or some compendial methods may fall into this category. On the other hand, there are many analyses where extensive development and optimization experiments are the norm, e.g. chromatographic impurity analyses, dissolution, or particle-size measurements. Even with these techniques, one can be fortunate and achieve a reasonable result after a few experiments based on very generic conditions; a broad acetonitrile–water gradient on a C18 column is a standard starting point for small molecule pharmaceutical analysis and it is not so unusual for many components in a sample to be resolved on the first attempt. However, this is not the whole story since the ATP will contain a variety of quality attributes such as accuracy and precision, and the choice should also include business requirements related to analytical speed or greenness of the method. Further experimentation is required to determine appropriate conditions where the ATP is met with good method robustness. The goal of these experiments is to understand the effect of the previously identified primary factors affecting the method.

There are a variety of ways in which experimental data can be transformed into useful method knowledge. Trial-and-error and more systematic OFAT approaches are limited because they provide information about points or lines in experimental space, but cannot be interpreted to understand method behavior across large regions of the experimental space. However, empirical models of the method space can be built using appropriately designed multifactor experiments, which are amenable to interpretation in a way that OFAT experiments are not. DOE approaches such as factorial designs or response surface designs fit responses to empirical functions, e.g. a quadratic function including cross terms. Such models are not intended to be expressions of the physico-chemical processes underlying the analytical method, but they do allow a result to be predicted based on a combination of input factors. Because they are not built around a method-specific model, they are universally applicable, albeit at the cost of requiring a significant number of experiments to build the model. A variety of DOE approaches useful for analytical QbD are described in Section 3.5.

In some cases, an explicit mechanistic model is available, which describes the analytical process based on a fundamental understanding of the technique. For example, in chromatography, a number of commercial software packages are available that are built around theoretical descriptions of the chromatographic separation. The advantage of this sort of approach is that the analytical response can in many cases be accurately modeled based on a very few, carefully chosen experiments. Such approaches are discussed in Section 3.6.

3.5 EMPIRICAL MODELS: DOE (SCREENING, MODELING, ROBUSTNESS)

3.5.1 Introduction

Use of an empirical model founded on statistically based DOE is a powerful tool in QbD method development. There are various software packages such as JMP, SAS, Design-Expert and Minitab that can generate a design and/or analyze the results. A DOE not only provides an organized approach to problem solving but also enables efficient and clear estimation of the effects that factors have on responses. Factors (independent variables) are chosen and controlled by the experimenter. Factors can be qualitative such as column type or solvent that are generally called "class" factors and are not on a continuous scale, or quantitative such as time or speed that are generally called "continuous" factors. Each factor is studied at one or more levels. For example, the factor may be solvent but there may be four different solvents studied. The four solvents are the levels of the solvent factor. The factor could be quantitative such as speed or time with levels of 4, 6, and 8 rpm or 1, 3, and 5 min, respectively. Responses (dependent variables) such as %recovery, %residual solvent, potency (mg/tablet) are the measured results that are generated from application of the factors to experimental material. Once the factors and levels are chosen, there are two primary components in constructing the design: (1) the combination of factor levels to include in the design and (2) the number of times to replicate each combination of factor levels. Of course, there are many details to consider prior to addressing these questions, such as: What is the question which will be answered by conducting the study? What are the available resources (materials, machines, and people)? What are the constraints on the factor levels? How will the design be carried out? What is the current available knowledge? What are the known issues about the factors and/or responses? These and other questions are addressed by definition of the ATP and risk assessment to identify factors for systematic study.

Typical uses of DOE in QbD are as follows:

1. Estimate effects of factors on responses
2. Study interactions between factors and their effects on responses
3. Estimate the precision of a measurement
4. Identify factors that have a significant effect on responses
5. Select optimum operating conditions and/or ranges
6. Identify factors that have little effect on responses (robustness studies)
7. Identify regions of failure
8. Reduce the number of factors (screening studies)
9. Reduce the number of factor levels
10. Identify factor ranges
11. Build empirical models to predict responses over the experimental range (response surfaces)
12. Estimate coefficients of known models

A general strategy for applying DOE to QbD is to perform a screening design (e.g. Plackett/Burman) to reduce the number of levels and/or factors so that a second study (e.g. fractional factorial) can be performed to further investigate the more important factors and to evaluate any possible interactions between the factors. Then, if desired, the final step is to use a response surface design so that an empirical model can be fit to establish a relationship between each response and the factors.

A DOE is used to evaluate the relationship between the factors and the responses. There can be multiple responses and/or factors in an experiment. In most cases the analysis consists of evaluating the effect of the factors on each response separately. This type of analysis is called a multifactor analysis. If the analysis is evaluating the effect of the factors on multiple responses in the same analysis (e.g. several different impurities), then the analysis would be considered multivariate. The advantage of a multivariate analysis is that it takes into account the correlations between responses. For example, several impurities may be related to each other—when impurity A is high, impurity B may tend to be low. The multivariate analysis would take this into account. At the present time, multivariate analysis is not commonly used for QbD since the analysis is much more complicated than multifactor analysis. Multivariate analysis using principal components or partial least squares is commonly used in building chemometric models.[11,12] Since multivariate analysis is beyond the scope of this chapter, only multifactor experiments will be discussed in the remainder of this section.

3.5.2 Multifactor designs

There are many types of multifactor designs that could be used in a QbD strategy, such as Full and Fractional Factorials, Nested, Split Plot, Mixture, and Response Surface designs. A number of texts have been written on the design and analysis of experiments,[13–18] which describe these designs in detail.

A design commonly used in the pharmaceutical industry consists of only one factor at several levels. For example, the OFAT strategy would be to perform a one-way experiment for one factor holding the other factors constant, then pick another factor holding the other factors (including the first factor) constant. For example, in a chromatographic experiment, all settings may be held constant except for flow rate. Flow rate could be set at specific values and several runs performed at each of these flow rate settings. This would be called a one-way experiment since there is only one factor (flow rate). This may be repeated, changing another factor while holding the rest constant. In the event of interactions between factors (see below), this is *not* an optimal strategy for multifactor experiments.

3.5.2.1 Factorial designs

Factorial designs are the most common designs used in QbD. These are used to identify important factors as well as any interactions that may exist between factors (see Chapter 5 in Ref. 13). These designs are used for method development as well as for showing the ruggedness or robustness of a method over a region. They consist of two or more factors with each factor set at two or more levels. The total number of combinations that could be tested is the product of all the levels. Each combination of factors and levels is called a treatment combination. So if there are two factors at two levels and one factor at four levels, there would be $2 \times 2 \times 4 = 16$ treatment combinations. The design before randomization would look like the following (Table 3.2):

Table 3.2 Three-factor design (two at two levels, low, L, and high, H, and one at four levels, L1-4) prior to randomization

Run	Factor A (Two Levels: L and H)	Factor B (Two Levels: L and H)	Factor C (Four Levels: L1, L2, L3, L4)
1	L	L	L1
2	L	L	L2
3	L	L	L3
4	L	L	L4
5	L	H	L1
6	L	H	L2
7	L	H	L3
8	L	H	L4
9	H	L	L1
10	H	L	L2
11	H	L	L3
12	H	L	L4
13	H	H	L1
14	H	H	L2
15	H	H	L3
16	H	H	L4

It is important to run the 16 experiments in a random order to eliminate any systematic errors. Performing a full factorial design allows estimation of the effects that each factor has on the response as well as the possible interactions between the factors. In the example above, there are three factors, so the analysis would include estimation of the main effects, 2-way and 3-way interactions. Main effects of a factor are computed by determining the difference between the average of one level of the factor averaged over all the other factors to the average of another level of the factor averaged over all the other factors. So the main effect of A is the average of the responses corresponding to runs 1–8 to the computed average response from runs 9–16. The 2-way interaction between factors A and B would compare the four combinations of the A and B levels as shown in Table 3.3A. An example of an AB interaction is shown in Table 3.3B and Fig. 3.2.

In this example, the effect of factor A depends on the level of factor B. At the low level of factor B, increasing factor A from low to high increases the response by 13 but at the high level of B, increasing factor A from low to high increases the response by only 0.5. If a factor is involved in an interaction, then interpreting the factor's main effect can be very misleading. Notice that if an OFAT strategy was performed and the scientist held factor B at the high level first and performed a run at the low and high level of factor A, there would be no difference since they both would result in a measured value around 82. If the scientist followed up holding the factor A at the low level and performed a run at the low and high level of B, then the low level of B would indicate a lower response than the high level of B. The scientist would decide that the high level of B is optimum and factor A has little effect, completely missing the fact that the low level of factor B and high level of factor A would result in a response of over 90.

In running a factorial design, replication of points is highly recommended. In the above example, suppose the experimenter only obtained one result from each of the four combinations of factors A and B of 78.3, 91.1, 80.8, and 81.3 as shown in the table. It is possible that these four results could have been obtained by performing the same combination of factors A and B four times. The variability in the

Table 3.3A Two-way interactions are determined by comparing the average AB levels

		Factor A	
		L	**H**
Factor B	**L**	Average runs (1–4)	Average runs (9–12)
	H	Average runs (5–8)	Average runs (13–16)

Table 3.3B An example of the AB interaction

		Factor A	
		L	**H**
Factor B	**L**	78.3	91.1
	H	80.8	81.3

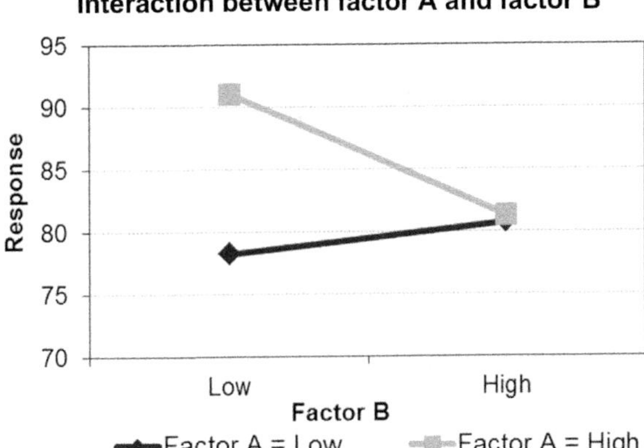

FIGURE 3.2

Graphical representation of the AB interaction shown in Table 3.3B. (For color version of this figure, the reader is referred to the online version of this book.)

results may not be due to changing the factor levels but rather just the natural variation in the method upon repeating the same treatment combinations four times. If the factors are all quantitative, then replication can be accomplished in a factorial design by adding center points at the mean of each quantitative factor. If the design also contains qualitative factors, then there is no "true" center. For example, if the design consists of the factors time (quantitative) at 5 and 10 min and two solvents (qualitative), then the replicates would be performed at 7.5 min for each solvent. Replication is used to test the significance of factor effects on the response and to provide an estimate of the reproducibility of the treatment combination. Another benefit of center points is that if the factors are quantitative and the center is the average of the high and low levels, then it is possible to obtain an estimate of curvature over the experimental region. If there is a significant curvature, then predicting the response within the factor ranges cannot be done accurately because the design cannot determine which factor is causing the curvature. Additional design points are needed to determine what factor(s) are causing the curvature. Response surface designs (Section 3.5.2.4) are often used to estimate curvature.

Performing a full factorial design with several factors each at several levels becomes large very quickly. For example, seven factors each at two levels would require 128 separate experiments! In this case, fractional factorial designs, which are subsets of full factorial designs, are generally used since they require fewer treatment combinations (see Chapter 6 in Ref. 13). These subsets are chosen in a special way so that the maximum information can be gained from the experiment. Fractional factorials generally provide less information on higher order interactions. For example, a full factorial design with five factors each at two levels would require $2^5 = 32$ treatment combinations. But a half fraction (expressed as 2^{5-1}) would require testing only 16 treatment combinations as shown in Table 3.4. The 2 represents the number of levels, the 5 represents the number of factors, and the -1 represents the fraction of the full factorial (a -2 would mean a quarter fraction).

There is some loss of information because the entire 32 run design is not performed. This is called confounding, meaning that certain terms are not separable from each other. In this design, the loss of information arises from the fact that the main effects are confounded with the four-way interactions

Table 3.4 2^{5-1} Fractional factorial design

	Factors				
Run	**A**	**B**	**C**	**D**	**E**
1	L	L	L	L	H
2	L	L	L	H	L
3	L	L	H	L	L
4	L	L	H	H	H
5	L	H	L	L	L
6	L	H	L	H	H
7	L	H	H	L	H
8	L	H	H	H	L
9	H	L	L	L	L
10	H	L	L	H	H
11	H	L	H	L	H
12	H	L	H	H	L
13	H	H	L	L	H
14	H	H	L	H	L
15	H	H	H	L	L
16	H	H	H	H	H

and the two-way interactions are confounded with the three-way interactions. For example, the interaction between A and B is confounded with the three-way interaction of C and D and E. If the AB interaction is significant in the analysis, the experimenter will not know whether the AB interaction is causing significance or the CDE interaction because they are indistinguishable. If the experimenter believes that the only possible effects are the main effects and the two-way interactions, and that the three- and four-way interactions do not exist or are very small, then not much is lost by running the ½ fraction.

3.5.2.2 *Nested designs*

These are often used to partition the total method variability into its contributing parts. For example, in an assay method validation, one could make three preparations on each of two days from the same batch and perform two injections for each preparation. The injections are nested in preparation and the preparations are nested in day. The design with results is shown in Table 3.5A.

The analysis of this design would separate the total variability into three parts (called components): between day, between preparations within day, and between injections within preparation. The analysis would provide the separation of variability as shown in Table 3.5B. As can be seen from the table, most of the variation is due to day-to-day variation. In QbD, this type of design and analysis can help to identify where the greatest sources of variability lie so that the experimenter knows where to put efforts to improve the method.

There is often confusion as to whether a design is a factorial or a nested design. For example, suppose that there are two factors, "method" and "batch". If there are only three batches, A, B, and C,

Table 3.5A Example of a nested design, with three preparations on each of two days from the same batch, with two injections for each preparation

Day	Preparation	Injection number	Results
1	1	1	40.2
		2	41.8
	2	1	43.9
		2	44.2
	3	1	39.9
		2	38.8
2	1	1	43.4
		2	45.5
	2	1	46.0
		2	47.2
	3	1	46.2
		2	46.8

Table 3.5B Variability assigned to different factors

Source	Standard Deviation	RSD(%)
Day	2.90	6.64
Preparation within day	1.78	4.07
Injections within preparation	0.92	2.10
Total	3.52	8.07

and each batch is tested by both methods, then the design is a factorial design. However, if the batches A, B, C tested by method A are totally different from the batches D, E, F tested with method B, then the batches are nested in method. In the nested design, there are six batches but in the factorial design, there are only three batches.

3.5.2.3 Split-plot design

Suppose the first four runs after randomization in a design with factors A and B are as shown in Table 3.6. The experimenter notices that factor A is at the low level for runs 1 and 2. Therefore instead of resetting A to low again, the experimenter just leaves the setting alone. In an experiment, each run should be performed as though it is the first run. All levels should be reset. However, there may be practical reasons why it is difficult to reset the factor (sometimes called "hard to change" factor). For example, the experimenter may be studying different mobile phases, flow rates, and temperatures. Remaking the mobile phase for each run may not be easy so the experimenter may want to use one

Table 3.6 The first four runs after randomization in a design with factors A and B

Run	Factor A (Two Levels: L and H)	Factor B (Two Levels: L and H)
1	L	L
2	L	H
3	H	L
4	H	H

preparation of mobile phase for several combinations of flow rate and temperature before switching mobile phase. To the analyst, this seems perfectly reasonable; they are interested in the question of whether mobile phases A or B affect their separation, so why remake the mobile phase each time? On the other hand, a statistician would consider the preparation of the mobile phase to contribute its own variability to the method and would accordingly analyze the data differently, using a split-plot design (hence, the importance of analyst and statistician discussing how the experiment is designed before it is performed!). A split-plot design for this example (before randomization) would be as shown in Table 3.7.

The randomization occurs in two steps for a split-plot design since the four flow rate by solvent combinations have to occur in each of the four mobile phase preparations. In the example, the four

Table 3.7 Split-Plot design (before randomization)

Main Plot	Mobile Phase	Split Plot	Flow Rate	Temperature
1	A	1	L	L
		2	L	H
		3	H	L
		4	H	H
2	B	1	L	L
		2	L	H
		3	H	L
		4	H	H
3	A	1	L	L
		2	L	H
		3	H	L
		4	H	H
4	B	1	L	L
		2	L	H
		3	H	L
		4	H	H

mobile phase preparations would be randomized first. Then within each of the four mobile phase preparations, the four flow rate by solvent combinations would be randomized. If the first preparation of mobile phase A is first after randomization, then all four combinations of flow rate and solvent would be run before switching to the second mobile-phase preparation in the randomization. Split-plot designs were commonly used in agricultural experiments where one factor was used for a large plot (called main plot) of land and then other factors were applied to subplots of the main plot.[18] In the example above, the mobile phase is the main plot and the four combinations of flow rate and solvent are applied to portions of the same mobile phase and are called the split plots. If this experiment was a factorial design, then mobile phase would have to be prepared 16 times whereas in the split-plot design, mobile phase is only prepared four times (2 A's and 2 B's). The analysis of a split plot takes into account that mobile phase was only applied four times but flow rate by solvent combinations were applied 16 times. The analysis allows for two sources of variability—one for the main plots and one for the split plots. Main plot variability is used to evaluate the main-plot factors and the split-plot variability is used to evaluate the split-plot factors as well as the interactions between the main-plot and split-plot factors. Split-plot designs are discussed in detail in Chapters 10 and 11 of Ref. 19.

3.5.2.4 Response surface

The designs discussed above are generally used to estimate the "true" mean response at the specific combinations in the study rather than interpolate or extrapolate outside the ranges used in the study. Response surfaces are very helpful in determining what factors are important, what the effect of changing the factor levels have on the response, estimating experimental error, and evaluating interactions between factors. However, if only two levels are used for quantitative factors, then interpolation or extrapolation can be very risky since a linear relationship must be assumed. Adding center points can allow an estimation of the overall curvature but cannot identify which factor is causing the curvature. In order to interpolate or extrapolate, designs should use an adequate number of levels to allow a reliable prediction equation. The designs discussed above can be used for this purpose by adding additional factor levels.

Response surface designs are used to develop a function that will relate the responses to the factor levels. The factors are generally quantitative. These designs help the experimenter to visualize the effects of the factors on the response. There are several texts on response surface designs.[20,21]

The empirical model that is generally used for response surface designs is a full quadratic model that includes the linear, cross product, and quadratic terms. An example of a full quadratic model in two factors is as follows:

$$\text{Response} = B_0 + B_1{}^*X_1 + B_2{}^*X_2 + B_{12}{}^*X_1{}^*X_2 + B_{11}{}^*X_1^2 + B_{22}{}^*X_2^2 \quad (3.3)$$

where

B_0 = intercept
B_1, B_2 = linear term coefficients
B_{12} = cross product coefficient
B_{11}, B_{22} = quadratic coefficients.

Since the "true" relationship between the response and the factors may be complicated, these models are approximations to the "true" model (see Chapter 10 in Ref. 13). Therefore, the model should be evaluated as part of the analysis to ensure that the model is fitting the data adequately. The range over which the model is used is also important. A full quadratic model may be adequate for a limited region of experimental space even if it cannot be used for a larger region.

Least squares are used to fit a model to the data with no assumptions to fit the model. However, to make any statistical statements such as determining significant terms, constructing confidence intervals, or lack of fit, the following assumptions are made:

1. The model is correct. There are several ways that a model can be incorrect. One incorrect model is overfitting the data. For example, a full quadratic model could be fit to a response but the true model just contains the linear term. Alternatively, two factors may be included in the model but only one factor has an effect on the response. The other incorrect model is underfitting the data. In this case, there are not enough terms in the model. An example of this is fitting a line to show linearity of reference response when the true model is quadratic. One approach is to fit the quadratic model and test for curvature by testing that the coefficient associated with quadratic term is significant. R^2 is not a good measure of linearity since the R^2 measures the combination of curvature as well as random variation about the regression line.
2. The observations about the fitted model are independent of one another.
3. The underlying distribution of the residuals about the fitted line is normal.
4. Variability of the residuals is similar at each combination of factors in the experiment. If this assumption is not satisfied, there are several strategies to correct the problem. For example, the data could be transformed so the residuals are normal or a weighted regression could be used. Weighted regression requires knowledge about the relationship between the factor settings and the variability or enough data at each treatment combination so that an estimate of the variability at each treatment combination can be made. This variability can be used as weights in the weighted regression.

A factorial design that has three levels for each factor can be used to fit a full quadratic model. The design for a 3^2 factorial (two factors each at three levels for nine treatment combinations) for the factors X_1 and X_2 is shown in Fig. 3.3.

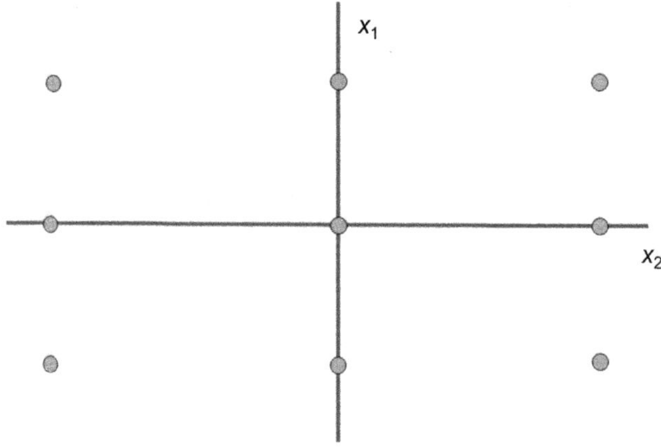

FIGURE 3.3

Three level factorial. (For color version of this figure, the reader is referred to the online version of this book.)

Table 3.8A An example of a 3^2 factorial to evaluate the effect of concentration and time on a response (recovery) before randomization

Concentration	Time	Response
1	10	73
1	20	85
1	30	80
3	10	80
3	20	90
3	30	84
5	10	85
5	20	96
5	30	90

An example of a 3^2 factorial is a study to evaluate the effect of concentration and time on a response such as recovery. The design before randomization is given in Table 3.8A. The least squares full quadratic model fit to the data provides estimates of the coefficients and the *p*-value associated with each term (Table 3.8B).

In Table 3.8B, the *p*-value is the probability of finding a coefficient as large as or larger than found in the study assuming that the "true" coefficient is zero. *p*-values less than 0.05 are considered significant since they would occur rarely if the "true" coefficient was zero. Therefore the conclusion is that the "true" coefficient is not zero. There is a significant quadratic effect of time but not concentration. Both time and concentration are involved with significant effects. In an attempt to simplify the model, one approach would be to eliminate the concentration-squared term since it is not significant. However, this approach should be performed with some caution since the design is small with low power to find significant effects. At a minimum, the fit of the full model against the actual results with the model not including the concentration-squared term should be examined to evaluate the effect of removing the term. The response surface and contour plots based on the full model are shown in Fig. 3.4. The surface plot shows the curvature in time and a rising predicted response as concentration increases. If the goal is to maximize the response, then the optimum is on the edge of the experimental region with time at around 20 and concentration of 5. However, estimating the response at given

Table 3.8B 3 × 3 factorial quadratic model coefficients

Term	Estimate	*p*-Value
Intercept	42.18	<0.01
Concentration	3.00	0.04
Time	3.68	<0.01
Concentration × concentration	0.042	0.76
Concentration × time	−0.025	0.25
Time × time	−0.083	<0.01

combinations of time and concentration is difficult to do visually from the response surface plot (Fig. 3.4(a)); the contour plot (Fig. 3.4(b)) is much better for this purpose. Any combination of time and concentration on the same contour line predicts the same response.

The number of runs required to conduct a full 3^k factorial can be problematic. Two factors require nine runs, three requires 27, and for four factors, 81 runs are necessary. Even with three factors, the number of runs may be too large in practice. However, there are other designs that allow fitting a quadratic model and do not require as many runs. The most common design is the central composite (CCD), which consists of a 2-level factorial design, center points, and axial (or "star") points. For three factors, the design is shown in Fig. 3.5.

(a)

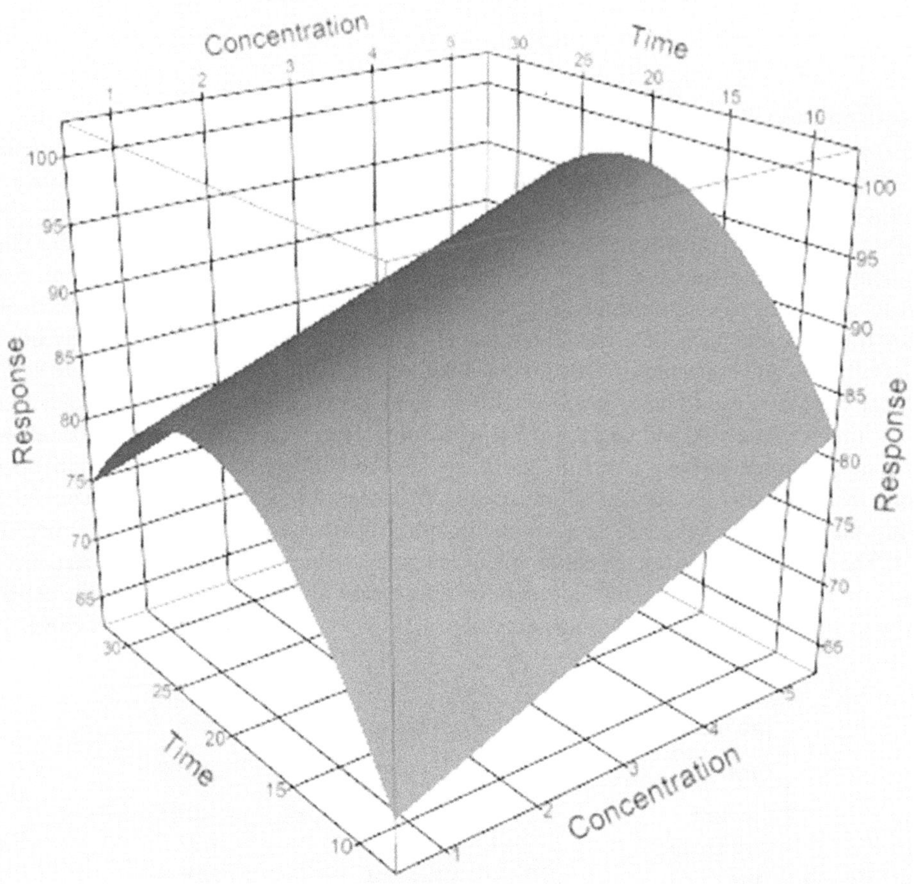

FIGURE 3.4

(a) Response surface. (b) Contour plot. (For color version of this figure, the reader is referred to the online version of this book.)

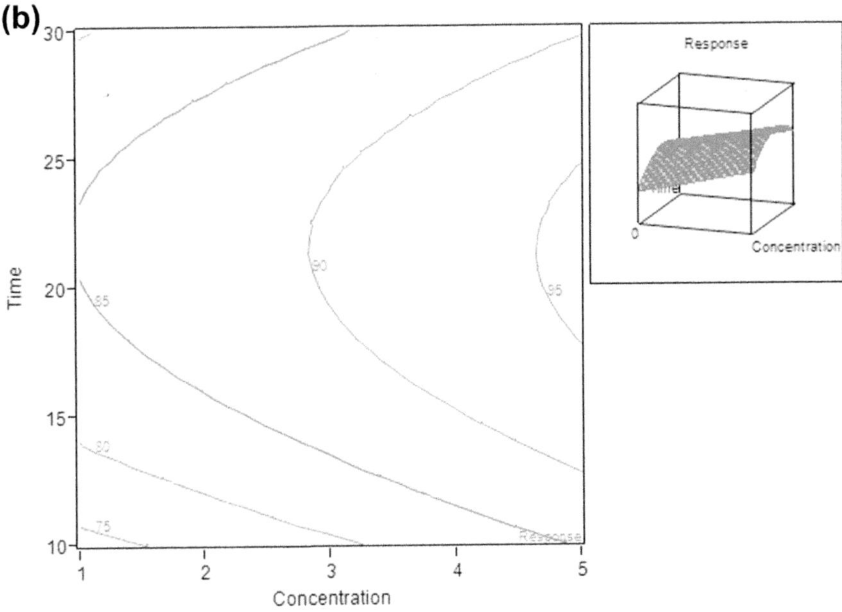

FIGURE 3.4

(*continued*).

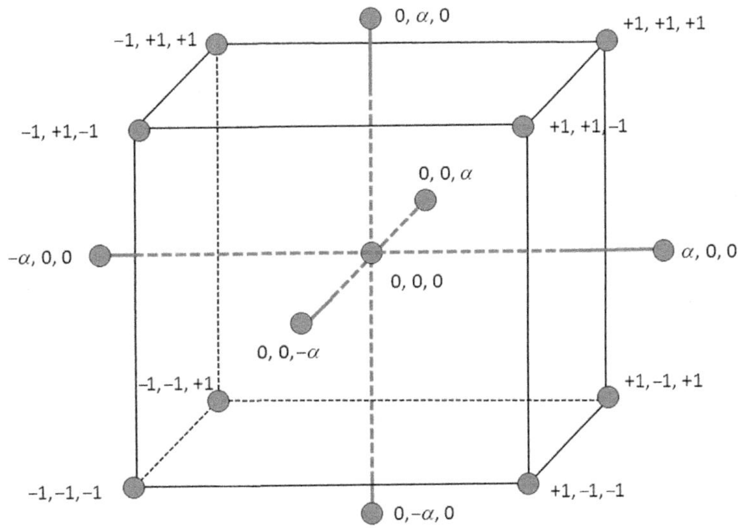

FIGURE 3.5

Central composite design for three factors. The corners of the box are the factorial portion of the design, with additional center points (0,0,0) and axial points (points with α in their coordinates). (For color version of this figure, the reader is referred to the online version of this book.)

Table 3.9 Number of runs (factorial + star + center + additional centers) required to run a central composite design

# Factors	Central Composite
k	$2^k + 2k + 1 + AC$
2	9 + AC
3	15 + AC
4	25 + AC
5	27 + AC

Table 3.9 shows the number of runs (factorial + star + center + additional centers (AC)) required to run a central composite design. Note that the number of points in the central composite is much smaller than in a 3^k factorial design.

Placement of the star points and/or number of center points can give the design different properties. The distance from the factorial points on each axis to the center point is considered one unit. The multiple of this distance is usually denoted by α. So if $\alpha = 1$, the star points would align with the factorial points. In the case of two factors, this would result in a 3 × 3 factorial. An $\alpha = 1.414$ would result in a design that is called rotatable, meaning that for any concentric circle about the center, any predicted response on the circle would have the same precision. Other values are possible, resulting in different properties.

A case study, which illustrates the use of a response surface design in sample extraction is discussed in Section 3.8.

3.5.2.4 Mixture designs

Mixture designs are used when the factor levels are proportions of a total amount. For example, a solution such as a mobile phase or an extraction solvent may consist of three components with each component representing a percentage of the total, for example 20% component A, 30% component B, and 50% component C. The sum of the proportions adds up to 100%. The goal of the study may be to find the optimum combination of factor percentages. For example, the experimenter may be interested in finding the best combination of surfactant, solvent, and oil to increase recovery. The design can be described in the diagram shown in Figure 3.6, and is also given in Table 3.10. This particular mixture design is called a lattice with each component ranging from 0 to 1 by thirds. Points are denoted by (water, surfactant, oil) giving the proportion of each component in the mixture (Fig. 3.6, Table 3.10). Note that a factorial or central composite design (CCD) cannot be used in these studies since the sum of the factor levels in some treatment combinations could add up to more than 100%. For example, if component A is to be studied between 10% and 30%, B between 20% and 40%, and C between 40% and 60%, a factorial would require a combination with A = 30, B = 40, and C = 60%, which is impossible since the components add to 130%!

Note that the interaction terms are not significant. One could investigate whether or not to reduce the model by eliminating these terms (one at a time). The corresponding contour plot using the full model is given in Fig. 3.7, showing the combinations of water and surfactant that predict the value on the contour line. The oil content would be 1 minus the sum of the water and surfactant levels.

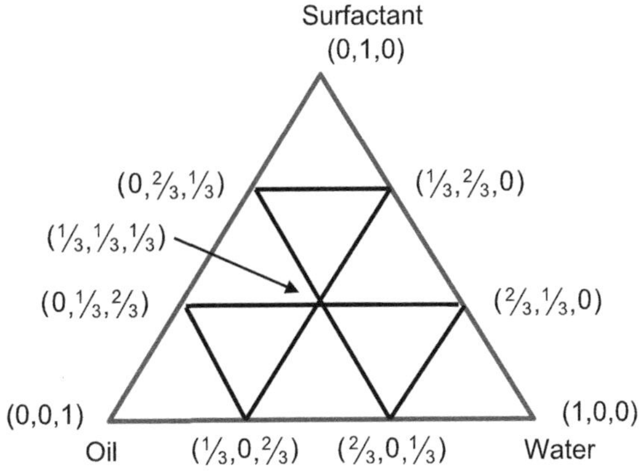

FIGURE 3.6

Lattice mixture design for three components. (For color version of this figure, the reader is referred to the online version of this book.)

Table 3.10A Input values and responses for the mixture design shown in Fig. 3.6

Water	Surfactant	Oil	Response
1	0	0	70.8
1/3	0	2/3	79.7
2/3	0	1/3	81.2
0	0	1	78.7
1/3	1/3	1/3	68.2
2/3	1/3	0	60.7
0	1/3	2/3	68.7
1/3	2/3	0	75.2
0	2/3	1/3	79.3
0	1	0	70.3

Table 3.10B The resulting model

Term	Estimate	*p*-Value
Water	69.913	0.0005
Surfactant	75.007	0.0003
Oil	76.776	0.0003
Water × Surfactant	−24.733	0.50
Water × Oil	27.802	0.40
Surfactant × Oil	−12.890	0.69

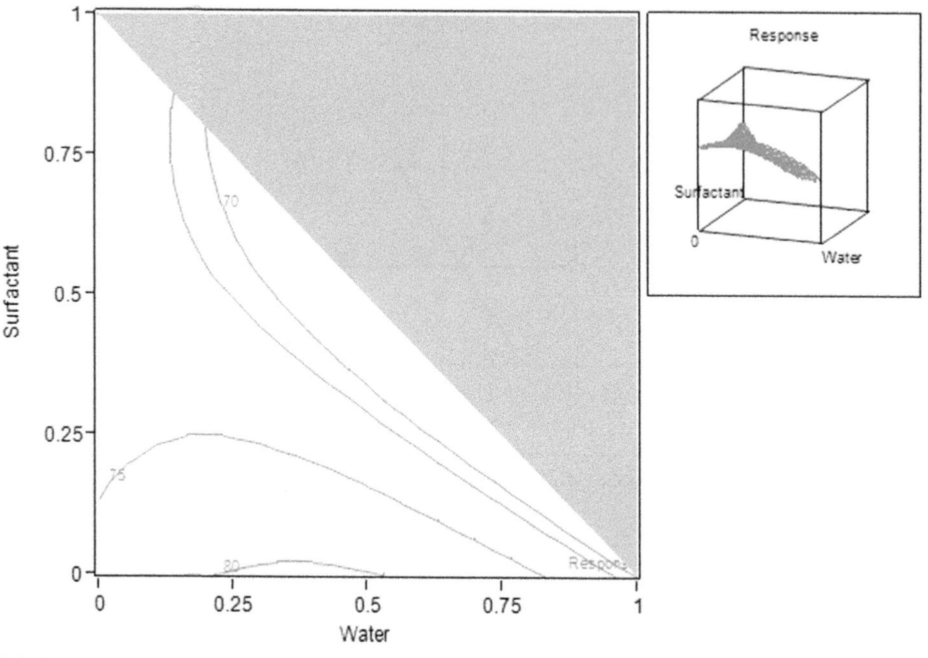

FIGURE 3.7

Contour plot for the mixture design data in Table 3.10. (For color version of this figure, the reader is referred to the online version of this book.)

To maximize the response (greater than 80) would require a surfactant level less than 5%, a water level between 25% and 50%, and depending on the choices of surfactant and water, an oil level between 45% and 75%.

3.5.2.5 Optimal designs

Prior to selecting a design, the scientist should think about the goals of the experiment. In addition to the goal, the experimental region of interest should be selected. In many experiments such as those described above, the region is rectangular by default since each single factor has a range and when all factors are combined, the result is a multidimensional rectangle. However, there may be problems where the experimental region of interest is not rectangular due to physical constraints. One example is the mixture design described above where the experimental region is triangular or a prism. If two factors are flow rate and temperature, there may be a different range of usable flow rates depending on the temperature. Another question to ask prior to creating a design is whether or not the goal is to find the best model by selecting the important main effects, interactions, or quadratic effects or to assume a specific known model but try to find the best estimates of the coefficients (or obtain the most precise predicted response). Optimal designs[19] are often used in cases where the experimental region is non-rectangular but can be defined and/or when the model can be specified but the goal is to obtain the "best" model. The "best" model could mean finding the model with the most precise coefficients or the most precise predicted value or many other statistical criteria. The most common criteria when the goal is to obtain the most precise estimates of the coefficients in the model is D-optimality, whereas the most common criteria to obtain the most precise predicted value over the experimental region are either G- or I-optimality[19].

Once the experimental region is defined (which can be all of the possible combinations or factor levels in the experiment), the model specified, number of runs are chosen and optimality criteria selected, then the optimal design needs to decide what treatment combinations to run and how many replications to perform at each treatment combination. There are various software packages such as JMP that will generate an optimal design. If the goal of the experiment is to find the best model, then the scientist should be careful when performing the analysis since the coefficients are correlated (partially confounded) with one another. Therefore, the coefficient for a term in the equation depends on whether or not other terms are included in the model. The correlation also affects the significance of terms in the model. Optimal designs can result in a large reduction in the number of runs to perform in an experiment.

3.6 APPROACHES USING EXPLICIT MODELS

The previous section focused on using empirical models in QbD. However, it may be possible to build an accurate mechanistic model, which describes the system being studied. The application of empirical and mechanistic modeling to a QbD study of chemical reactions has been compared[22]; the authors found both approaches could describe their reaction. Building the mechanistic model had the advantage of developing greater process understanding, greater ability to explore transient conditions, and aided in risk assessment by the ability to rapidly conduct simulations to test the sensitivity of the reaction to various factors. On the other hand, the empirical model offers an approach where an adequate mechanistic model cannot be developed, e.g. due to the complexity of the system. For analytical applications, several commercial software packages are available for chromatographic method development and optimization that are based around well-established models of chromatographic retention.[23] Arguably, the best known of these is DryLab, which has been available in increasingly sophisticated versions for a quarter of a century.[24] Chromatographic retention is modeled using a variety of expressions which describe the effects on retention of parameters such as mobile-phase composition, temperature, pH and additive concentration[25]:

$$\text{Solvent strength (\%B): } \log k = \log k_{\mathrm{w}} - S\varphi \tag{3.4}$$

where k is the retention factor of the analyte in the aqueous–organic mobile phase (k being the ratio of the amount of analyte in the stationary phase to that in the mobile phase, a value which can be related to the retention time of the analyte), k_{w} is the retention factor of the analyte using a mobile phase comprised only of water, S is the solvent strength parameter for this analyte, and φ is the volume fraction of organic solvent in the mobile phase.

$$\text{Temperature: } \log k = A + B/T \tag{3.5}$$

where A and B are constants for a given system, and T is the temperature (in K); this is essentially a simplified expression of the Van't Hoff equation.

$$\text{Mobile phase pH: } k = k^{0}(1 - F) + k^{\mathrm{i}}F \tag{3.6}$$

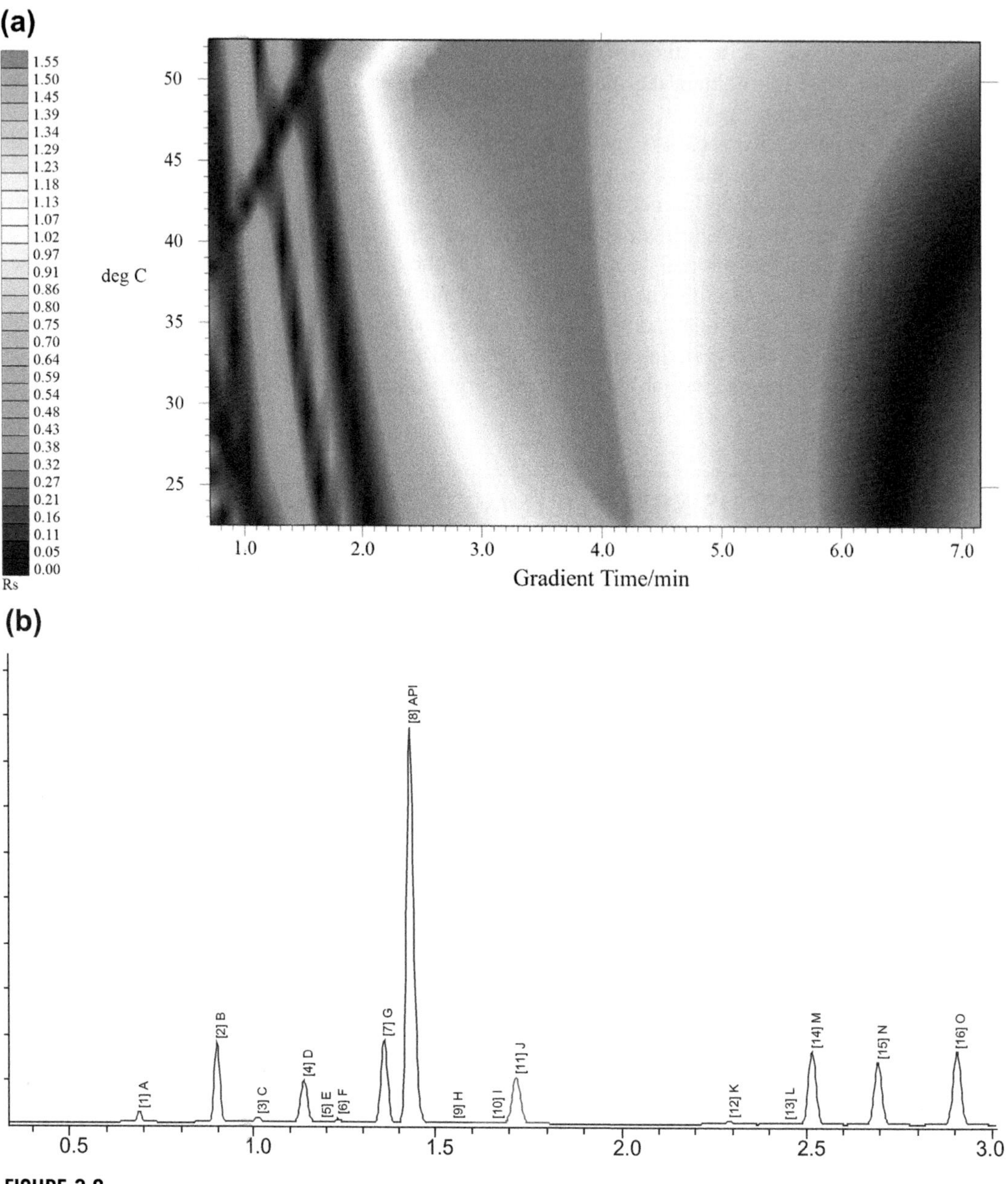

FIGURE 3.8

(a) Resolution map generated using Drylab. The scale indicates values of R_s achieved under the various separation conditions used. Optimum resolution occurs at higher temperature over a range of gradient times from approx. 2.5–4.0 min. (b) Predicted chromatogram at a gradient time of 2.5 min, at 50 °C. Peaks I (very small impurity, not visible on this scale) and J are the critical pair under these conditions. (c) Actual chromatogram. Only main components are identified with retention times. (For color version of this figure, the reader is referred to the online version of this book.)

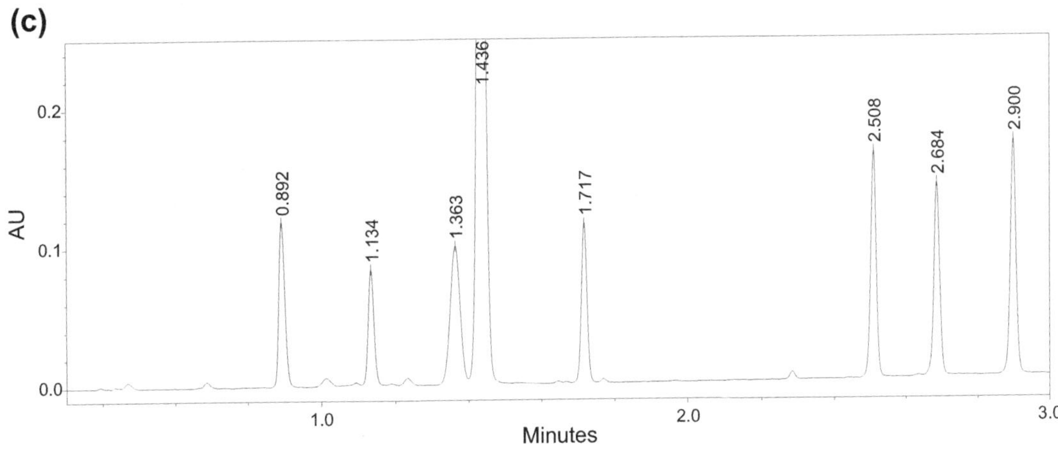

FIGURE 3.8

(*continued*).

where k^0 is the retention factor of the analyte in its neutral form, while k^i is the retention factor of the ionized species, and F is the fraction of the analyte that is ionized, which can be determined via the Henderson–Hasselbach equation (Eqn (3.2)).

$$\text{Buffer/additive concentration:} \ \log k \approx C + D \log [X] \tag{3.7}$$

where C and D are constants for a given system, and $[X]$ is the concentration of the interacting additive such as an ion-pairing agent.

Overall retention is determined as a combination of the effects of the individual parameters. Although these are a mix of empirical and more fundamental expressions, they are well established as reasonably accurate descriptions of the effects of key chromatographic variables on retention. The values of the coefficients are determined in a small number of experiments; not all variables need to be studied in each case, e.g. if there are no additives or ionizable solutes, these additive concentration and pH effects are not studied. Thus, a useful model can be obtained *with a very limited number of input experiments* (a considerable practical advantage for explicit models, when available). Although it is understood that there are some interactions when multiple parameters are changed, the effects are quite limited and, generally, accurate predictions are achieved.[25,26] As well as retention, peak width and shape are modeled, and the output is a resolution map, illustrating critical resolution between peaks as a function of parameters such as analysis time and temperature, for a given combination of solvent, column dimensions, flow, etc. An example is shown in Fig. 3.8(a), illustrating a map of resolution for a gradient very high pressure liquid chromatographic separation (resolution, R_s, is a measure of the separation of two peaks in a chromatogram based on the width of the peaks and their separation; $R_s > 1$ and preferably >1.5). The input data were from just four chromatographic runs, at two temperatures and two gradient times. From this limited input data the method was optimized. In the response surface (Fig. 3.8(a)), warmer colors indicate greater resolution between the critical pair, and an optimum region can be seen to exist in the central region of the map, with gradient times of around 2.5–4 min and temperatures in the range 40–50 °C. The predicted and actual chromatograms are shown in Fig. 3.8(b and c), illustrating the high degree of accuracy that is achieved (predicted and actual retention times agree within 2 s in this example). It should be noted that although the 2-D map in Fig. 3.8(a) plots resolution as a function of column temperature and gradient time, the effect of variables such as column length and diameter, flow rate, and gradient profile can be determined from the same data since these are accounted for in the underlying models (other parameters such as pH may also be varied if these are part of the model used and appropriate data are collected). Recent versions of DryLab are particularly focused on QbD applications, and offer 3-D visualization of the design space.[27,28]

3.7 GENERAL ADVICE ON DESIGN/ANALYSIS OF EXPERIMENTS

This section contains advice on the design and analysis strategy of experiments. As Eleanor Roosevelt said, "Learn from the mistakes of others. You can't live long enough to make them all yourself."

3.7.1 Design strategy

1. Talk to other scientists: if you are performing your first designed experiment, talk to other scientists who have already completed a design. They can provide valuable information on setting up the equipment, obtaining appropriate materials, problems encountered in setting up and running the experiment, collecting the data, formatting the data for analysis, and lessons learned.
2. Ask whether the design will answer the right question: be sure of the question before designing the experiment; think about the question that you are trying to answer. You don't want to complete the experiment, analyze the data and find out that it is not addressing the right question. One strategy once the design is determined is to enter simulated data using values which are realistic for the proposed experiment. Then analyze it and review the results.
3. Include relevant players: prior to designing an experiment, think about the scientists who will be affected by the conclusions. Include all relevant players in planning the study. The method is often transferred to another department that may have constraints that do not allow the method to be run in the same way as was optimized. If you have access to statisticians, do not wait until the data are collected before getting them involved. Most statisticians are trained to design experiments as well as analyze them.
4. Pick meaningful factor levels: after performing the risk analysis, most factors will be determined for study in the designed experiment. However, one must still pick the levels for each factor. This can be the hardest part of designing the experiment. If the levels are too close together, it will be difficult to find any effects while if the levels are too far apart, it is possible that a large number of treatment combinations will fail to provide meaningful results. It is also possible that the underlying relationship between the response and the factors has "cliffs" or nonlinear areas that are not fit well by the statistical model. An example would be an acid–base titration; this could be modeled by a linear or quadratic equation over a short range, but not for a wide range of factor levels. On the other hand, a good fit over a wide range could be made from limited data if the correct equation describing the titration process was used. In the early stages of method

development, it is desirable to allow the levels to be more spread out, but when finalizing the method for robustness or to develop a "design space", the levels should be picked over a range that provides flexibility and keeps the responses within specifications or internal limits.

5. Record data to the appropriate number of significant figures: data should be recorded with enough digits as to make the analysis reliable. Suppose a degradant is the measurement and all of the data are recorded to one place past the decimal point ranging from 0.5 to 0.8. This causes the precision estimate to be inaccurate, which in turn makes the analysis less accurate in terms of significance. It should be noted that the ICH Q3A and B Guidelines on Impurities in the Drug Substance and Drug Product, respectively, describe the number of significant figures that should be used in *reported* data. This is not meant to imply that these significant figures are appropriate for calculation of secondary data.
6. Record results and observations: it is important to keep detailed written notes during the performance of the study. This can be very helpful when the analysis shows some "outliers" or unusual results. Knowing that something different occurred during that particular run may explain the problem. This can also be useful when transferring a method. Sometimes a little change in technique may not be captured in the method (e.g. the way a vessel is shaken, or the position of a flask within an ultrasonic bath).
7. Replicate: as noted in the previous section, replication is an important part of a designed experiment. Usually the replicates are performed at the center but can also be obtained by replicating the design. However, replicating the design can result in expending greater resources. Suppose that an experiment has two factors each at two levels. Then replication could be accomplished by running a 2^2 factorial with four center points or by replicating the whole 2^2. The advantage of replicating the center points is that a measure of curvature can be obtained. Another advantage is that since the center point may be the desired settings for the method, additional data at this point may be helpful. If the whole 2^2 is run, then one gains additional precision information on each factorial point precision. The greater advantage is that the effects are estimated more accurately since they are averages of replicates. Adding centers does not have this property since center points do not increase the number of results at each factorial point.
8. Perform pilot runs: it is possible to run DOE too early in the development process. The designed experiment is not the place to still be learning how to run the equipment or learning the basics of the method. Also, pilot runs are useful to help establish levels for the factors. One strategy in picking levels is to run the "worst" case prior to starting the designed experiment. This may not be an easy decision because the "worst" case may not be all factors high or all factors low. So the decision of "worst" case prior to performing many runs may need to be based on the science.
9. Consider running designs in sequence: during the development process, it is common to perform more than one design. Based on the analysis of the first design, a scientist may decide to run a second design that may use the same factors but different factor levels or may add/eliminate factors. Planning of the second design should include thought as to what factors and levels were used in the first design. It is common that each design is analyzed completely separately. However, if the second design is well planned, the designs can be combined into a single analysis that provides much more information. One example is called the fold-over design. Suppose a design contains four factors each at two levels in a half fraction of a 2^4 factorial. This design uses eight of the possible 16 treatment combinations. The design confounds two-way interactions with each other. So if a two-way interaction is significant, one cannot tell

which of the two confounded two-way interactions is affecting the responses. A second design could be run by using the eight treatment combinations that were left out in the first design. This is a fold-over and allows all two-way interactions to be estimated, thus eliminating the confounding problem. Another example is using central composite designs. As discussed previously, the central composite design consists of a factorial, center points, and axial points. Instead of running the entire design before analyzing, one could run just the factorial part and some center points. If the results indicate that the factor levels were chosen so that the design is in the area of interest, then the second design would include the axial points as well as additional center points. When running designs in sequence, you should always use common points in both designs (usually the center) so that you can detect if a shift has occurred from the first design to the second design. This could indicate that something has changed and an investigation may be needed.

10. Consider blocking: blocking can be very useful to evaluate factor effects. Blocking is done by grouping the treatment combinations within a homogeneous set. For example, suppose that an experiment consists of two factors, A at two levels and B at three levels for a total of six treatment combinations. Each combination is used to prepare tablets that will be tested for dissolution. Since the dissolution apparatus usually consists of six vessels, there are two ways to perform the dissolution testing: (1) For each of the six treatment combinations, test six tablets with all six of the same treatment combination tested in the same dissolution apparatus or (2) test one tablet from each of the six treatment combinations in the same dissolution apparatus (the block) and run each set six times. The total number of tablets tested is 36 for each possibility but the second method is much better since all six treatment combinations are tested within the same dissolution apparatus making a much better comparison of the two factors since the apparatus run to run variation is eliminated. Another example would be comparing two potency assays using multiple batches (the block) of tablets. Instead of using one potency method on some batches and the other method on other batches, both assays would be used on each batch. Then the difference between the two assays has lower variability since the batch-to-batch variability has been removed. Blocking is discussed in Chapter 2 of Ref. 13.

3.7.2 Analysis strategy

Once the results of the study are available, a recommended strategy for analysis is as follows:

1. Review raw results: look for extreme (unexpected) observations or entry mistakes.
2. Review center points if available: since the center points were all performed at the same combination of factor levels, they should reflect the reproducibility of the factor combination. If this result is much higher than expected, this may indicate a problem with the experiment. The center point variability is used for testing the effects. If the variability is very low, then smaller differences between factor levels and interactions are more likely to be found significant. Similarly, high variability would require larger differences in the effects to be found significant. If the variability is higher than expected, it is possible that there are other sources of variation that are not being accounted for in the study.
3. Evaluate assumptions: as stated above, certain assumptions are made when analyzing data from an experiment. If these assumptions are not satisfied, then the p-values, which indicate significant effects, are affected. In many of the experiments described above that were not response surface

or mixture designs, it is often the case that only one result is available for each treatment combination, so checking assumptions can be difficult. Randomization is important to obtain independence of the results. Checking for normality and equal variance is also difficult to do in these situations. However, with response surface designs, there are plots that can help to check assumptions, such as a plot of the residuals against predicted results or against each factor. The plots should not show any patterns in the residuals of the factorial experiments described in the previous section.

4. Examine highest order interactions first: in most experiments, main effects and two-way interactions are of most interest. In this case, the two-way interactions should be examined first.
5. Examine main effects/interactions not involved in higher order interactions: as stated in List 4, if the experiment only contains main effects and two-way interactions, then main effects that are not involved in a two-way interaction should be examined. The reason for this is that if a main effect is involved in a two-way interaction, then the effect of that factor depends on the level of another factor.
6. Examine results of the curvature test.

3.7.3 Finding the best operating point

The goal of many experiments is to find the best combination of factors to either maximize (e.g. recovery) or minimize (e.g. impurity) the response or find the combination closest to a target value (e.g. label claim). The experimenter runs a DOE and obtains the responses. One option (not the best one) is to find the best result among the responses in the experiment. The results could be sorted from high to low and then just chose the "best" one. Then the combination of factors associated with that response is chosen at the optimum condition. The better option is to analyze the data either by estimating effects or fitting a response surface and determining which effects are significant based on the p-value.

Factorial designs use statistical significance to find the "BEST". If the experiment used a fractional factorial design, then not all combinations of factors were used in the experiment (e.g. a half fraction of a factor design would only use 16 of the possible 32 treatment combinations). Therefore, the best combination may be one of the treatment combinations that were not run in the experiment. The statistical analysis can be used to find the best combination even though it was not in the experiment.

3.7.4 Causes of nonstatistical significance

After running an experiment, it can be frustrating when the found effects are not significant. For example the main effect of a factor on potency is 8% but was not significant. This is usually due to the study not having enough power to detect the difference. As part of the planning for an experiment, the number of replicates of the center and treatment combinations should be considered so that there is an assurance that if a meaningful difference really exists, the design will find the difference significant in the analysis. It is dangerous to make decisive conclusions on effects that were not significant. The statistical analysis determines if the difference could have happened by chance. So if the effect is not significant, then it is possible that there is no difference and making a decision based on this inconclusive result could result in a bad decision. High variation in the center points is a sign that small differences will not be considered significant. Adding more than 4 or 5 center points loses the ability to find significant differences. Instead of adding center points, additional factorial points should be added to the experiment. This results in a better estimation of the effects since the number of points used to calculate means is increased. Another cause of nonsignificance is outliers. This could be due to a high-order interaction or an error that occurred during the experiment. Another possibility when no significant effects are found is that the factors have no effect within the experimental region. The goal of robustness studies may be to show no effects over the region so no significant effects can be a desirable result (as long as the study was large enough to detect significant effects).

3.8 CASE STUDY—SAMPLE EXTRACTION METHOD DEVELOPMENT USING A RESPONSE SURFACE DESIGN

3.8.1 Problem statement

A potency and impurities assay was being developed for a solid oral dosage form. Extraction was required prior to chromatographic analysis.

The compound is a small molecule which is poorly soluble and not very stable in water. It is highly soluble in a variety of organic solvents, and is more stable in aprotic solvents such as acetonitrile. The experimental formulation consisted of small sugar beads coated with active drug and a protective polymer to prevent drug degradation in the acidic stomach. These enteric-coated beads were then filled into a capsule.

3.8.2 Analytical target profile

The extraction objectives, and desired performance characteristics, are shown in Table 3.11.

Initially, an attempt was made to develop manual sample preparation methods; however, significant degradation was observed as the drug was exposed to water during this procedure. The drug is relatively stable in acetonitrile, but since the protective polymer coating is not soluble in acetonitrile, directly placing the intact beads in this solvent was not an option. As an alternative, beads were manually ground into a powder, followed by extraction of the drug from the powder using acetonitrile. However, this procedure was lengthy, irreproducible, and raised significant safety concerns in handling of this highly potent compound. Therefore, an alternative approach was required, and automation was chosen, using a Tablet Processing Workstation II (TPWII).

Table 3.11 Analytical target profile for extraction method

Extraction objectives	• Complete and reproducible extraction of drug from the capsules • Minimal degradation during the extraction process • Safe process
Performance criterion 1	Method accuracy better than $\pm 2\%$
Performance criterion 2	Method precision better than $\pm 2\%$
Performance criterion 3	Degradation during sample preparation $<0.1\%$
Performance criterion 4	Minimize analyst exposure to drug

3.8.3 Extraction method description

The TPWII has a robotic arm for transfer of sample capsules from input test tubes to a vessel containing a homogenizer probe. Solvent is added to the vessel and the dosage forms under test are broken up by the homogenizer in a series of pulses where each pulse is a period of time in which the homogenizer probe spins rapidly. The vessel is made to cycle up and down during part of the homogenization to ensure that capsules are drawn up into the homogenizer blades. A "mixing time" step was added, during which there is a low-speed rotation of the homogenizer probe. Under these low-energy conditions, the capsules are not broken up but the mixture in the vessel is gently stirred. This step allows extra time for the drug to dissolve after the beads have been broken up in the initial vigorous homogenization. Part of the vessel contents is then pumped through a filter and into a high-performance liquid chromatography sample vial.

3.8.4 Risk analysis

Since the TPWII removes the majority of the analyst's contact with the samples, the automated extraction approach effectively addresses the ATP's performance criterion 4—safety. Therefore, the risk analysis focused on the extraction performance in terms of criteria 1–3. Several operating parameters can significantly affect the sample preparation process, including solvent type, solvent volume, speed of homogenization, number and duration of homogenization pulses, mixing time, filter type, flush volumes used to rinse the instrument tubing, and vessel washing parameters. Potential critical parameters were identified in a brainstorming session, guided by existing knowledge of the product and of the automation platform used. A fishbone diagram illustrating the risks considered is shown in Fig. 3.9. The risks are grouped according to different parts of the TPWII sample preparation process:

Homogenization: It is intuitively obvious that the homogenization speed (measured in rotations per minute of the homogenizer probe) and the time for which homogenization takes place will

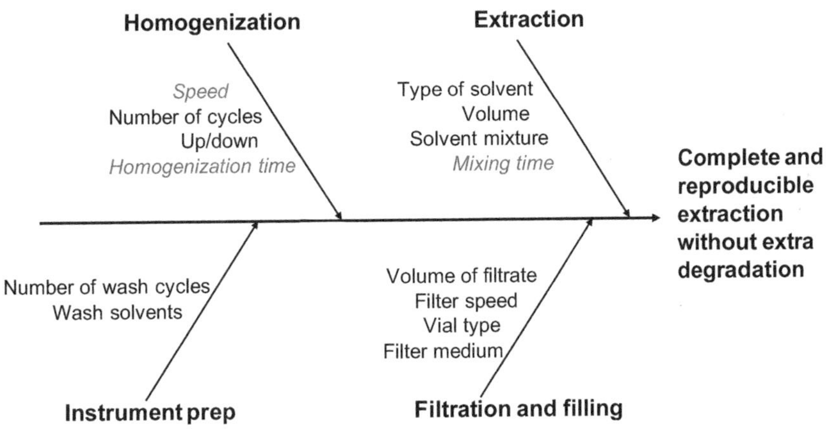

FIGURE 3.9

Risk assessment for the automated extraction method. (For color version of this figure, the reader is referred to the online version of this book.)

affect the disruption of the capsules. In addition, it was known from preliminary experiments that with more vigorous homogenization, increased degradation could occur. Thus, homogenization time and speed were considered critical for further study. The number of homogenization pulses was considered less critical—the total time being considered more important than the way the homogenization is delivered. Similarly, the number of up/down cycles may affect whether all capsules actually engage the blades and break up; clearly an intact capsule would lead to lack of extraction, but by observation after a few cycles all capsules were broken and so this was not considered a critical parameter for systematic investigation.

Extraction: As well as the physical elements of homogenization, the chemical process of drug dissolution had to be considered. Based on prior knowledge of the analyte, its stability was inadequate in water or alcohols, but good in acetonitrile. Solubility was also very high in acetonitrile. Thus the extraction solvent was fixed as acetonitrile without further study, and the volume was not considered critical because of the high analyte solubility. However, the mixing time was chosen for further study, since it was reasonable to believe that the length of time that the drug was in solution in contact with the excipients may affect its stability.

Instrument preparation was considered. The TPWII flow paths can be washed with solvents before and after use. Because of the known lability of the analyte to water, only pure acetonitrile was chosen as wash solvent; by procedurally eliminating water from the system this factor was adequately controlled and further study was not needed.

Filtration/filling: Based on prior knowledge of the compound, the filter and vial type were not considered critical. Factors such as the volume of filtrate and filter speed were not considered likely to interact with other experimental parameters, and were thus optimized separately in a univariate fashion.

3.8.5 Experimental design

A two-level factorial design was initially performed, which confirmed both the significance of the three factors chosen in the risk analysis, and that there were interactions between them. A CCD was then used to generate response surfaces for measured potency and degradation as a function of the factors homogenization time, homogenization speed and mixing time (t_h, s_h, and t_m). The CCD consisted of 15 points (2^3 = eight full factorial points, one center point, and six star points). Lower and upper limits for the three factors used in the factorial part of the design were $t_h = 100$ and 600 s, $s_h = 12{,}000$ and 18,000 rpm, and $t_m = 0$ and 300 s. The center point of the design was at $t_h = 360$ s, $s_h = 15{,}000$ rpm, and $t_m = 150$ s. The star points were chosen due to instrumental constraints: $t_h = 60$ and 850 s, $s_h = 10{,}000$ and 20,000 rpm, and $t_m = 0$ and 400 s. In addition to the three factors described above, batch-to-batch and day-to-day effects were also evaluated by running the 15 point CCD each day on several days for two different batches. One batch was tested on three days while the other batch was tested on one day. Details of the design and the associated potency and degradation results are presented in Table 3.12.

Note that the potencies are consistently higher for batch B, and the impurities are lower when compared to batch A; this reflects the actual characteristics of the batches rather than any effect of the extraction process. The data for potency and degradation were fit to each batch separately using JMP software to a full quadratic regression model that included linear, quadratic, and cross

Table 3.12 Experimental design and measured data. One batch was run on three days, a second batch on one day. The order of experiments was randomized on each day. Experimental conditions and potency and degradant results are listed in sorted factor order

Data Display										
			Potency				Degradant			
			Batch A			Batch B	Batch A			Batch B
Homogenization Time/s	**Homogenization Speed/1000 rpm**	**Mixing Time/s**	**Day 1**	**Day 2**	**Day 3**	**Day 1**	**Day 1**	**Day 2**	**Day 3**	**Day 1**
60	**15**	**150**	94.74	96.27	94.18	96.01	0.27	0.28	0.25	0.12
100	**12**	**0**	87.70	88.91	89.51	84.85	0.24	0.27	0.28	0.12
100	**12**	**300**	97.39	97.43	98.95	100.83	0.26	0.27	0.27	0.12
100	**18**	**0**	92.05	94.11	94.31	95.29	0.26	0.26	0.25	0.12
100	**18**	**300**	98.32	100.19	97.86	101.12	0.27	0.27	0.28	0.11
360	**10**	**150**	99.17	98.06	98.47	102.62	0.25	0.29	0.27	0.14
360	**15**	**0**	99.44	99.51	99.13	103.38	0.28	0.28	0.28	0.12
360	**15**	**150**	99.36	97.12	97.93	103.67	0.28	0.27	0.29	0.13
360	**15**	**400**	99.57	98.78	98.14	104.10	0.30	0.29	0.30	0.14
360	**20**	**150**	99.20	100.11	96.90	100.89	0.33	0.33	0.31	0.17
600	**12**	**0**	99.45	98.98	99.06	102.64	0.28	0.29	0.29	0.13
600	**12**	**300**	99.62	98.75	98.90	102.18	0.29	0.28	0.32	0.16
600	**18**	**0**	98.48	99.03	99.14	101.85	0.39	0.34	0.38	0.21
600	**18**	**300**	99.29	98.61	97.81	101.96	0.43	0.44	0.47	0.30
850	**15**	**150**	98.03	98.87	99.30	100.21	0.41	0.46	0.39	0.26

product terms for the quantitative continuous factors t_h, s_h, and t_m. The qualitative factor day was included in the model for batch A to estimate the day-to-day variation and determine whether or not day was a significant factor in the model. The estimated day-to-day standard deviation for the potency was 0.90% (relative to the label claim) with a *p*-value of 0.76 (not significant). Similarly, for the amount of degradant, the estimated day-to-day standard deviation was 0.017% degradant, with a *p*-value of 0.66 (not significant). Therefore, the day term was eliminated from the model. The remaining quadratic model was then reduced by eliminating terms that were not significant (*p*-values $\geq$0.10) starting with the quadratic terms, followed by the cross product terms, and finally linear terms. If a higher order term was significant, then any lower order term contained in that factor was kept in the model. For example, if mixing time squared was significant, then the linear mixing time term was kept no matter whether it was significant or not since it is a factor in the squared term.

The final models for potency and degradation are shown in Tables 3.13 and 3.14.

Since there was replication for batch A, an F-test was performed to test if the model adequately fit the potency and degradant data (lack of fit), which indicated that there was no significant lack of fit for either response (*p*-values $\geq$ 0.12).

It can be seen that the same factors can be used in models describing both the measured potency and degradation.

Table 3.13 Batch A: Factors included in the final model for potency and degradant, estimates of their coefficients and significance of each term

	Batch A			
	Parameter Estimates			
	Potency		Degradant	
Term	**Estimate**	***p*-Value**	**Estimate**	***p*-Value**
Intercept	78.65	<0.0001	0.397	<0.0001
Homogenization time/s	0.0451	<0.0001	−0.0006	<0.0001
Homogenization speed/ 1000 rpm	0.6841	0.0002	−0.0091	0.0009
Mixing time/s	0.0529	<0.0001	−0.0003	0.0612
(Homogenization time/s)^2	−1.828e−5	<0.0001	2.1499e−7	0.0003
Homogenization time/s × Homogenization speed/ 1000 rpm	−0.0011	0.0031	3.8576e−5	<0.0001
Homogenization time/s × Mixing time	−0.00005	<0.0001	2.2009e−7	0.0435
Homogenization speed/ 1000 rpm × Mixing time	−0.0012	0.0508	2.2222e−5	0.0157
(Mixing time)^2	−2.939e−5	0.0291	−1.269e−7	0.5262

Table 3.14 Batch B: Factors included in the model for potency and degradation and significance of each term

Batch B				
Parameter Estimates				
	Potency		**Degradant**	
Term	**Estimate**	***p*-Value**	**Estimate**	***p*-Value**
Intercept	73.45	<0.0001	0.231	0.0061
Homogenization time/s	0.0810	0.0068	−0.0006	0.0068
Homogenization speed/ 1000 rpm	0.8739	0.1302	−0.0070	0.1154
Mixing time/s	0.0397	0.0055	−7.97e−5	0.3537
(Homogenization time/s)^2	−3.768e−5	0.0187	2.0361e−7	0.0728
Homogenization time/s × Homogenization speed/ 1000 rpm	−0.0020	0.1598*	3.7931e−5	0.0050
Homogenization time/s × Mixing time	−7.456e−5	0.0200	4.2999e−7	0.0615

**Although this p-value was greater than 0.10, the term was kept in the model to keep the models consistent for plotting purposes. This term for potency could be deleted and the model fit again, if desired.*

In Fig. 3.10, contour plots are shown which illustrate the response surfaces for potency and degradation obtained as a function of t_h and s_h at t_m values of 0, 150 and 300 s. Any point on the same contour has the same predicted potency (or degradation). For example, in Fig. 3.10(a), any combination of homogenization time and speed associated with the blue line labeled 0.28 has a predicted degradant level of 0.28%. A homogenization time of 500 s with a homogenization speed of 13,000 rpm or a homogenization time of 300 s with a homogenization speed of 17,800 rpm have a predicted degradant level of 0.28%.

It can be seen from Fig. 3.10(a), when $t_m = 0$, maximum extraction is only achieved at large values of t_h. This is somewhat improved by increasing s_h only when t_h is relatively low. On the other hand, the minimum of degradation only occurs at $t_h < 300$ s. Increasing t_m to 150 s brings the optimum regions closer together, with a clear plateau for potency seen at lower values of t_h in Fig. 3.10(b). This trend continues as t_m is increased to 300 s, and in Fig. 3.10(c), the region of maximum potency is seen to closely approach the area of minimum degradation. A global optimum of method performance exists at the intersection of the individual optimum regions of the contour profiles for recovery and degradation. From Fig. 3.10(c), it can be seen that optimum conditions are approximately $s_h = 12{,}000$ rpm, $t_h = 400$ s and $t_m = 300$ s. Although this does not correspond to the absolute minimum for degradation, a difference of 0.01% is not significant and so these extraction parameters represent a good compromise. Greater t_h or s_h would place the method on the rapidly rising part of the degradation surface, which is not considered acceptable.

The forms of the response surfaces are in agreement with the interpretation that more vigorous conditions (longer extraction, higher homogenization speed) lead to more complete extraction of the

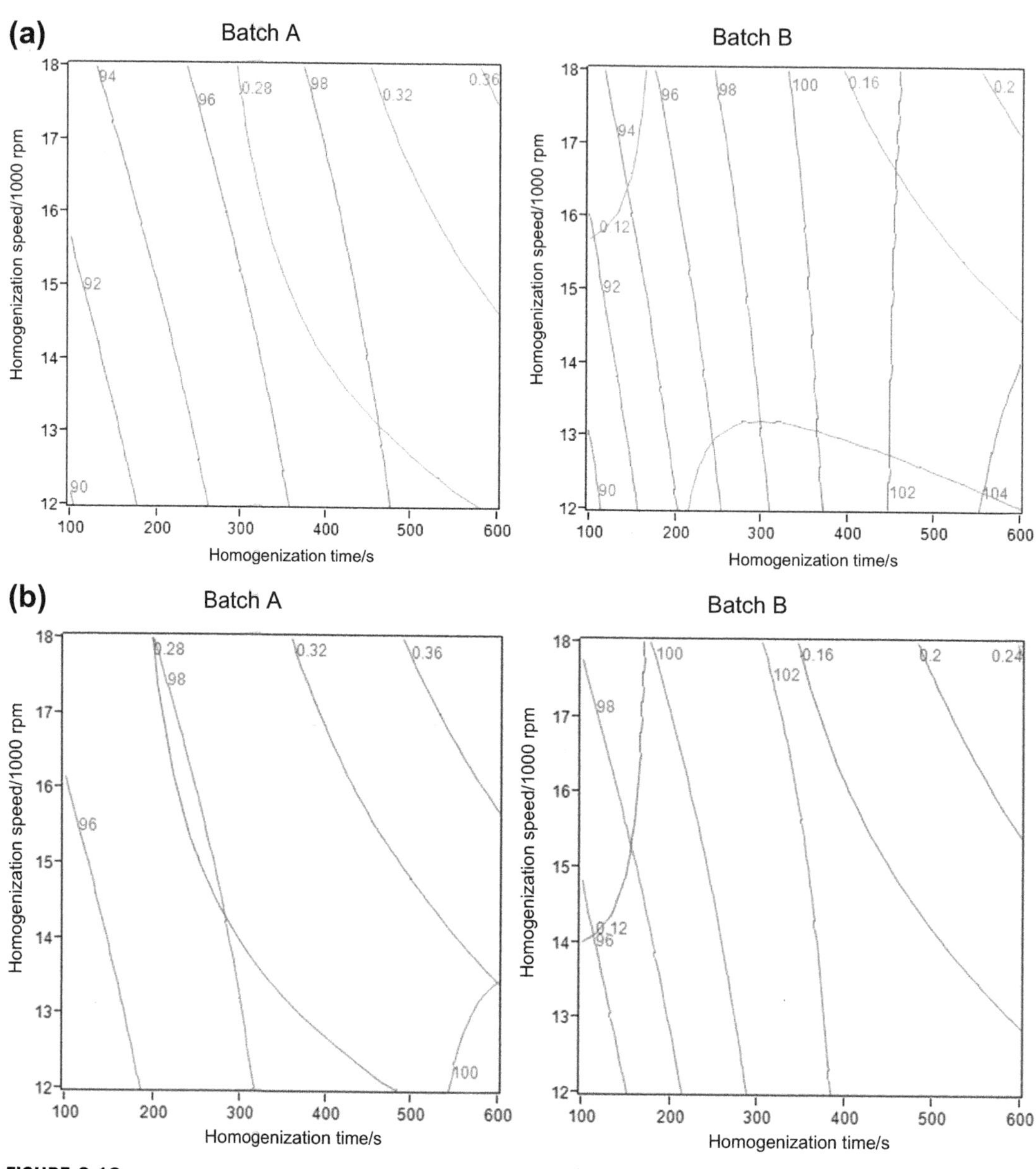

FIGURE 3.10

Contour plots for potency and degradant as a function of t_h and s_h at different values of t_m. The optimum region exists at the point where potency is maximized, and amount of degradant is minimized. (a) $t_m = 0$ s, (b) $t_m = 150$ s, (c) $t_m = 300$ s. (For color version of this figure, the reader is referred to the online version of this book.)

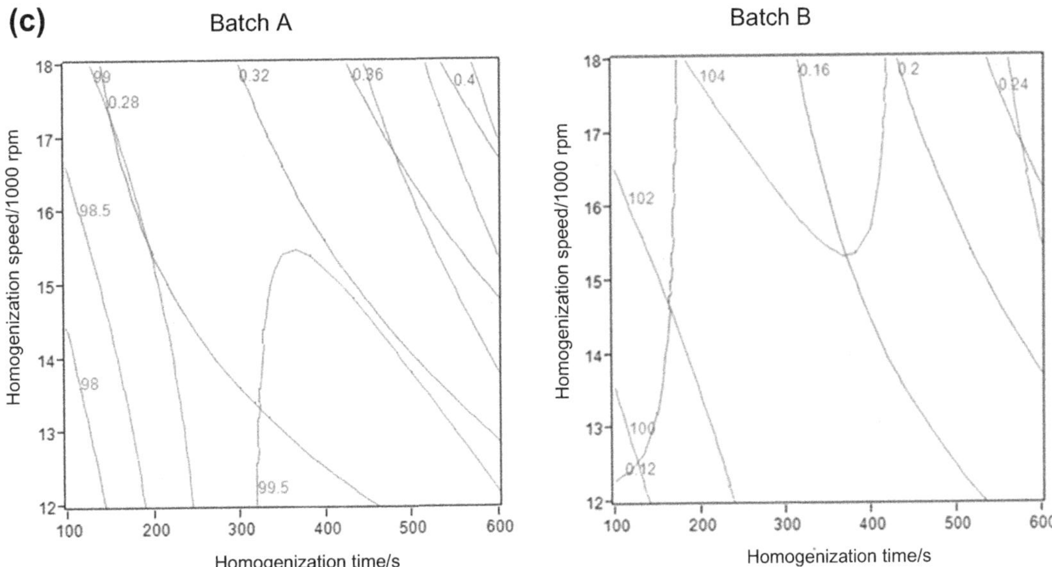

FIGURE 3.10

(continued).

drug from the capsules, whilst promoting degradation. It is interesting to note that the potency response surface begins to curve down at the most vigorous extraction conditions, e.g. dropping below 99% for $t_h > 500$ s and $s_h > 16{,}000$ rpm (Fig. 3.10(c)). This may to some degree reflect true changes in the amount of drug in solution, since the measured potency will decrease as the amount of degradation increases. However, with a predicted potency loss of greater than 1% under the most vigorous extraction conditions, the increase in degradant is only around 0.2%. Since the response factors of both compounds are similar, this could be interpreted as predicting a mass balance deficit. However, this apparent deficit is due to the limitations of the quadratic fit in modeling the sigmoidal relationship between the extraction conditions and the measured potency; the model describes the response surface as being a symmetrical hill, when in reality it is more like an asymmetrical plateau.

3.9 DEVELOPMENT TO VALIDATION

Regulatory guidance exists on the required elements of validation (see Ref. 29 and Chapter 4 of this book), and validation is typically performed as a discrete activity at the end of the development process. This guidance has proved extremely useful in standardizing expectations for method validation, but a consequence has been that validation tends to follow a rigid, procedure-driven path for determination of accuracy, precision, etc. Following a QbD approach to method development, the systematic studies performed should result in extensive knowledge of the primary factors which are critical to successful method operation, demonstrated operating ranges, and data on method performance within those ranges. Appropriate controls will also have been identified, either as elements of the method itself, or as qualification requirements for instruments, SOPs for operators or facilities, etc. Therefore, it can be argued that in a QbD approach a final validation in its traditional format is not required; much of the method performance is defined during method design (in studies which are scientifically justified for the method under study, but which may differ greatly from method to method). Consequently, it has been proposed that in analytical QbD a life cycle approach be adopted, comprising method design, method qualification (involving a modest degree of experimentation to demonstrate the method meets the requirements laid out in the ATP under routine operating conditions; perhaps such studies can simply be documented from the development phase), and by continued verification of method performance during the method lifetime.[30] It will be interesting to see how validation guidance evolves in the future to incorporate such concepts.

ICH guidelines[29] include some well-defined approaches for validation of a variety of method performance parameters. However, for assessment of method robustness the guidance is less specific: "The evaluation of robustness should be considered during the development phase and depends on the type of procedure under study. It should show the reliability of an analysis with respect to deliberate variations in method parameters." More generally, a method should be rugged, i.e. insensitive to factors external to the method, such as where it is run and by whom. A QbD approach to method development facilitates achieving these goals. The risk assessment process identifies primary factors (those expected to have a significant effect on the experiment), and the modeling process demonstrates the range within which adequate performance is achieved as these key parameters are varied. So, in the chromatographic example described above, the response surface in Fig. 3.8(a) illustrates the sensitivity of the method to changes in gradient time and temperature. Similar maps may be generated after making small variations in the mobile phase composition, flow rate, etc., to determine whether the method is sensitive to these factors. Thus, a region may be defined where adequate resolution is achieved for any operating parameter setting. Within this, the method operating space may be defined (likely to be smaller than the absolute maximum ranges determined from the model). Running the method at the extremes of the range defined can verify the predicted performance. In the case of a chromatographic method, where resolution is typically modeled, verification runs may be useful in that other attributes such as accuracy and sensitivity may be checked. If a factorial or response surface design has been performed with replication, this will demonstrate the method performance across the studied space.

Although considerable understanding is gained during method development, the more sophisticated models created will typically include a small subset of possible method parameters (primary factors) for thorough investigation, as defined via risk analysis. Secondary factors identified during risk assessment (those not expected to have a significant effect on the method) will likely not be extensively studied during method development. However, secondary factors should still be evaluated to make sure that they do not have an unanticipated effect. So should both primary and secondary factors be studied in a screening study first in a highly fractionated design as part of risk analysis followed by selection of the most important factors for follow up experiments so that interactions can be studied? Or should the primary factors be evaluated first holding the secondary factors constant, then after optimizing the responses, perform a second study showing that the secondary factors originally left out have no or little effect on the responses? The advantage of the first approach is that one finds out early if any of the factors that were not expected to have an effect really do have an effect. If this happens, then follow up experiments can include those factors. The advantage of the second approach is that the first step may result in changing the factor levels to optimize the method. Then the secondary factors can be evaluated against the optimum factor levels. However, if there are secondary factors that have a significant effect on the response or interact with the factors already studied, then additional work would be required. The first approach may be better as long as there are not too many factors and the design does not require too many runs. On the other hand, if there is a high confidence that the risk analysis really has identified key parameters (e.g. through extensive knowledge of the technique employed) the second approach may be justified. A separate ruggedness study at the end of the process may in any case be required to encompass a broader range of factors that were not known or available when the original method was developed, e.g. to test the method against formulations prepared at the limits of the product design space, or to include testing at multiple sites. To limit the amount of work involved, very sparse designs may be employed for robustness studies.[31,32] If a factor is identified as important at this stage, then it will be necessary to add further controls and/or redefine the method operating space to ensure robust operation.

3.10 KNOWLEDGE MANAGEMENT

If QbD involves developing a full understanding of how method attributes and operating conditions relate to method performance, there needs to be a suitable mechanism for gathering all this knowledge together throughout the life of the method (indeed, knowledge management is an expectation outlined in ICH guidance[4]). A fundamental first step is to ensure that method development experiments are adequately documented with the reasons for performing the experiment and a conclusion based on the results gained. Exercises such as risk assessment or choice of study design should also be appropriately documented, such that the rationale for the decisions made is not lost. The approach taken to systematically collect the knowledge gained will depend greatly on questions such as the infrastructure available within a given organization. For example:

- Paper-based records, e.g. paper lab notebooks. Indexing and retrieval of data from QbD experiments is a considerable challenge. This may be aided by generation of a contemporaneous method development report, listing, for example, experiments performed and critical conclusions.
- Electronic notebooks (ELNs) offer much better search, indexing and retrieval capabilities than in the paper world. However, some systems are better than others and so it will likely be helpful if the analyst is systematic in using appropriate identifiers such as keywords so that method development records can be linked together. There is still an argument for creating an overview record which identifies key experiments and conclusions. Integration of data collection, modeling packages and corporate documentation systems with ELNs may allow a comprehensive solution to QbD data and knowledge management.
- Specialized data management systems have been proposed for analytical QbD. They may include the option to import data into shared, standard tools for analysis and report generation.

The final output may include method history, development and performance reports as well as the method description including validated operating ranges if desired. This package of knowledge becomes part of the transfer to receiving laboratories where the method will be put into use. Furthermore, it provides the basis for any future revisions.

3.11 QBD THROUGHOUT THE METHOD LIFETIME

Once the method is put into routine use at one or more quality control laboratories, a wealth of data will be generated which will indicate how it is performing, including:

- Simple observations by operators. Does the method continue to perform "as advertised" or are adjustments needed, e.g. a factor such as instrument equilibration time was identified as noncritical during development, but now extra equilibration time is needed.
- Is method performance changing, e.g. if there are system suitability criteria, is system suitability routinely met? Is there other evidence of the method misbehaving, e.g. out-of-specification (OOS) results (ones where the analysis is found to be the root cause, but also ones where the root cause is indeterminate and thus may be related to the analysis)?
- Systematic data collection and analysis, e.g. analyses of reference material data or system suitability data interpreted using control charts[33] to monitor method performance over time.

Observation of a pattern such as repeated OOS results related to the method, or a drift in quantitative performance is cause for further investigation and remediation. It should be rare that a new risk factor is identified at this stage, but this is not impossible. For example, an unannounced modification to the manufacture of a chromatographic column could result in a change which simply falls outside of the experimental space investigated in developmental studies, and may not become apparent until after aberrant results are generated and an investigation performed. Changes to site facilities and personnel may present similar challenges, although such changes are typically planned and thus can be prepared for. A change to the product which moves the product outside of the range of samples studied during method development may require the method to be reassessed, possibly even to the point of reevaluating the ATP. Such modifications to the analytical methodology should be planned in conjunction with the process modification, within the context of the firm's change management procedures.[4]

3.12 CONCLUSIONS

Although analytical QbD is not as well established as the application of QbD to product development, there is the potential for significant benefit in terms of robust performance of a method throughout its life. Definition of the ATP allows the method goals to be clearly stated, and risk analysis allows the development effort expended to be focused in the most important areas. Systematic studies allow the definition of a method operating space, extensive study of primary factors, and screening of a broader range of factors to determine robustness. If the comprehensive knowledge acquired during development becomes part of the package transferred to labs which will actually run the method, this will form the basis for understanding method performance and for continuous improvement.

The above description includes a variety of elements which, individually, are valuable. Hopefully, the whole is greater than the sum of the parts, and a revolutionary approach to the full implementation of QbD in the analytical laboratory could involve considerable upheaval, albeit with the maximum potential benefit. Alternatively, a step-by-step approach to implementing analytical QbD may be advocated, first incorporating elements of the analytical QbD toolkit where they make most sense in terms of existing workflows, and then looking for the greatest gaps in existing practices where most benefit can be gained. Within an organization, many individual elements of QbD may already be practiced, but perhaps in an informal way, or with less consistency than desirable. These are areas where gains can be made for a relatively modest effort, potentially acting as stepping-stones on the path to a more comprehensive application of analytical QbD.

Acknowledgments

The authors would like to acknowledge the work of Katie Wu who performed the chromatographic work described in Section 3.6 and the extraction experiments described in the case study in Section 3.8, and Emil Friedman who performed the initial analysis of the extraction data. The authors also acknowledge members of the QbD team in BMS Analytical and Bioanalytical Development for many discussions related to analytical QbD, in particular, Mario Hubert, Ruben Lozano, Adele Patterson, Yueer Shi, Liya Tang and Peter Tattersall.

References

1. *ICH Q8 (R2) – Guidance for Industry, Pharmaceutical Development,* International Conference on Harmonization, 2009.
2. McCurdy, V. Quality by Design. In *Process Understanding: For Scale Up and Manufacturing of Active Ingredients;* Houston, I., Ed.; Wiley, 2011.
3. *Quality Risk Management (Q9),* The International Conference on Harmonization of Technical Requirements for Registration of Pharmaceuticals for Human Use, Second Revision 2005.
4. *Pharmaceutical Quality System (Q10(R4)),* The International Conference on Harmonization of Technical Requirements for Registration of Pharmaceuticals for Human Use, 2009.
5. Borman, P.; Chatfield, M.; Nethercote, P.; Thompson, D.; Truman, K. The Application of Quality by Design to Analytical Methods. *Pharm. Technol.* **2007,** *31,* 142–152.
6. Torbeck, L.; Branning, R. QbD: Convincing the Skeptics. *BioPharm Int.* **2009,** *22,* 52–58.
7. Schweitzer, M.; Pohl, M.; Hanna-Brown, M.; Nethercote, P.; Borman, P.; Hansen, G.; Smith, K.; Larew, J.; Carolan, J.; Ermer, J.; Faulkner, P.; Finkler, C.; Gill, I.; Grosche, O.; Hoffmann, J.; Lenhart, A.; Rignall, A.; Sokoliess, T.; Wegener, G. Implications and Opportunities of Applying QbD Principles to Analytical Measurements. *Pharm. Technol.* **2010,** *34* (2), 52–59.
8. Nasr, M. Quality by Design (QbD): Analytical Aspects. In *32nd International Symposium on High Performance Liquid Phase Separations and Related Techniques;* Baltimore: MD, 2008.
9. Franklin, B. D.; Shebl, N. A.; Barber, N. Failure Mode and Effects Analysis: Too Little for Too Much? *BMJ Qual. Saf.* **2012,** *21,* 607–611.
10. Bowles, J. B. An Assessment of RPN Prioritization in a Failure Modes Effects and Criticality Analysis. *J. IEST* **2004,** *47,* 51–56.
11. Frank, I. E.; Friedman, J. H. A Statistical View of Some Chemometric Regression Tools. *Technometrics* **1993,** *35,* 109–148.
12. Wold, S.; Esbensen, K.; Geladi, P. Principal Component Analysis. *Chemom. Intell. Lab. Syst.* **1987,** *2,* 37–52.
13. Box, G. E. P.; Hunter, W. G.; Hunter, J. S. *Statistics for Experimenters: Design, Innovation and Discovery,* 2nd ed., Wiley, 2005.
14. Montgomery, D. C. *Design and Analysis of Experiments;* Wiley, 1997.
15. Hicks, C. R.; Turner, A. V. *Fundamental Concepts in the Design of Experiments;* Oxford University Press, 1999.

16. Wu, C. F.; Hamada, M. *Experiments: Planning, Analysis and Parameter Design Optimization;* Wiley, 2000.
17. Cornell, J. A. *Experiments with Mixtures;* Wiley, 1990.
18. Milliken, G. A.; Johnson, D. E. *Analysis of Messy Data* In: *Designed Experiments,* Vol. 1; Chapman and Hall, 1992.
19. Goos, P.; Jones, B. *Optimal Design of Experiments: A Case Study Approach;* Wiley, 2011.
20. Myers, R. H.; Montgomery, D. C. *Response Surface Methodology;* Wiley, 2002.
21. Box, G. E. P.; Draper, N. R. *Empirical Model-Building and Response Surfaces;* Wiley, 1987.
22. Hallow, D. M.; Mudryk, B. M.; Braem, A. D.; Tabora, J. E.; Lyngberg, O. K.; Bergum, J. S.; Rossano, L. T.; Tummala, S. An Example of Utilizing Mechanistic and Empirical Modeling in Quality by Design. *J. Pharm. Innov.* **2010,** *5,* 193–203.
23. Garcıa-Alvarez-Coque, M. C.; Torres-Lapasio, J. R.; Baeza-Baeza, J. J. Models and Objective Functions for the Optimisation of Selectivity in Reversed-Phase Liquid Chromatography. *Anal. Chim. Acta* **2006,** *579,* 125–145.
24. Molnar, I. Computerized Design of Separation Strategies by Reversed-Phase Liquid Chromatography: Development of DryLab Software. *J. Chromatogr. A* **2002,** *965,* 175–194.
25. Dolan, J. W.; Lommen, D. C.; Snyder, L. R. High-Performance Liquid Chromatographic Computer Simulation Based on a Restricted Multi-Parameter Approach: I. Theory and Verification. *J. Chromatogr. A* **1990,** *535,* 55–74.
26. Snyder, L. R.; Dolan, J. W.; Lommen, D. C. High-Performance Liquid Chromatographic Computer Simulation Based on a Restricted Multi-parameter Approach: II. Applications. *J. Chromatogr. A* **1990,** *535,* 75–92.
27. Molnar, I.; Rieger, H. J.; Monks, K. E. Aspects of the "Design Space" in High Pressure Liquid Chromatography Method Development. *J. Chromatogr. A* **2010,** *1217,* 3193–3200.
28. Monks, K.; Molnar, I.; Rieger, H. J.; Bogati, B.; Szabo, E. Quality by Design: Multidimensional Exploration of the Design Space in High Performance Liquid Chromatography Method Development for Better Robustness Before Validation. *J. Chromatogr. A* **2012,** *1232,* 218–230.
29. *Validation of Analytical Procedures: Text and Methodology (Q2(R1)),* The International Conference on Harmonization of Technical Requirements for Registration of Pharmaceuticals for Human Use, 1995.
30. Nethercote, P.; Ermer, J. Quality by Design for Analytical Methods: Implications for Method Validation and Transfer. *Pharm. Technol.* **2012,** *36,* 74–79.
31. Box, G. E. P.; Hunter, J. S.; Hunter, W. G. Additional Fractionals and Analysis. In *Statistics for Experimenters: Design, Innovation and Discovery;* Wiley: Hoboken, 2005; pp 281–316.
32. Torbeck, L. D. Ruggedness and Robustness with Designed Experiments. *Pharm. Technol.* **1996,** *21,* 169–172.
33. Massart, D. L.; Vandeginste, B. G. M.; Buydens, L. M. C.; De Jong, S.; Lewi, P. J.; Smeyers-Verbeke, J. *Control Charts, in Handbook of Chemometrics and Qualimetrics: Part A;* Elsevier: Amsterdam, 1997. pp 151–170 (chapter 7).

CHAPTER

4 General principles and regulatory considerations: method validation

Thomas W. Rosanske*, Bradford J. Mueller†

* *T.W. Rosanske Consulting, Overland Park, KS, USA,* † *Incyte Corporation, Experimental Station, Wilmington, DE, USA*

CHAPTER OUTLINE

Specification of Drug Substances and Products. http://dx.doi.org/10.1016/B978-0-08-098350-9.00004-7

4.1 INTRODUCTION

4.1.1 Design, proof of concept, and life cycle management of analytical methods

If one views an analytical method as a "product", the process by which it is developed, validated, and maintained can be put into terms familiar to development teams charged with bringing a new product to the market. As such, the evolution of analytical methods can be seen as progressing through three stages: design, proof of concept, and life cycle management.

4.1.1.1 Design

Once the quality parameters and acceptance criteria that must be met to achieve the desired product or process performance have been defined, the analytical chemist will identify approaches to measure the quality parameters with appropriate accuracy, precision, specificity, and sensitivity to achieve what is required for control. By establishing the analytical method "product" characteristics and assessing "design" options, the analyst can develop a strategy and approach to selecting the appropriate analytical methodology to achieve what is required to measure the quality parameter.

Chapter 3 of this book describes how the principles of Quality by Design and Experimental Design can be used in the Design stage of method development. The data and knowledge gained at this stage of an analytical method will provide a high level of assurance that the analytical procedure has the desired characteristics. This chapter discusses core technical attributes or validation parameters which should be evaluated to support the applicability of the analytical method "product". One aspect that is often overlooked when designing an analytical method is the needs of the customer. The analytical method "product" should be customer friendly with respect to ease and efficiency of use, cost effectiveness, and resources (personnel and equipment) required to implement and maintain. Therefore, an understanding of the customer needs is essential during the design phase.

The changing needs of the customer during the life cycle of a product often serves as the basis for setting analytical method requirements and the level of method validation required. This chapter discusses how a phased approach to method validation can be implemented to ensure that the key customer analytical data needs are met at various stages of product development.

4.1.1.2 Proof of concept

The next step of the process is verifying that the method is able to achieve its design goals. This is the "Proof of Concept", verifying that the method is capable of achieving predetermined performance criteria. In conventional thinking, this constitutes traditional method validation. The validation parameters normally employed to demonstrate the Proof of Concept are described in the International Conference on Harmonization (ICH) Quality Guidance Q2 (R1), Validation of Analytical

Procedures: Text and Methodology.[1] While this guidance document defines the typical method quality parameters and assessment approaches for an analytical method, it is often augmented by learnings in the Design phase of the method development. The guidance, while applicable to most analytical techniques, may not adequately cover key performance aspects of some methodologies which are more complex in nature.

Traditional validation parameters such as specificity, linearity, precision, accuracy, quantitation and detection limits, and robustness are well suited for chromatographic methods using external standard calibration, e.g. see Chapter 6 for Assay and Impurities. Spectroscopic methods, such as Near Infrared (NIR) spectroscopy might be sensitive to density, hardness, particle size, and polymorphic form. Extensive calibration training sets may need to be generated for NIR methods which are based on multivariate analytical signals.[2] The mathematical algorithms contained in the analytical data to enable accurate quantitative determinations will most likely need to be maintained and updated. As a result, the calibration model used to extract the key analytical information becomes a critical validation parameter for the method. In addition, calibration training sets may not be directly transferable to other spectroscopic systems.

Similarly, traditional method validation largely focuses on the analytical determinant step. Robustness studies typically target variations in instrumental settings or variations in mobile phase composition. The suitability of sample preparation techniques used to generate the analytical sample is often inadequately studied. Specificity for the analyte in the sample matrix can routinely be ensured, but the impact of the sample matrix on the accuracy and precision of the method is often evaluated by artificially fortifying the sample matrix with the analyte of interest through spiking studies and determining the recovery of the spiked analyte. These fortification studies often do not reflect the true sample matrix. Recovery studies during validation often show a quantitative recovery of the analyte, but in day-to-day use on real samples, low recoveries may routinely be observed due to the inadequate extraction of the analyte from the true sample matrix. This bias is often uncovered when comparing congruent analyses (i.e. assay and dissolution of tablets). Therefore, a thorough evaluation of the robustness and accuracy of the sample preparation step is essential. A recent text, *Sample Preparation of Pharmaceutical Dosage Forms: Challenges and Strategies for Sample Preparation and Extraction, B. Nickerson, Editor*[3] provides an overview of the dosage form and diluent properties that impact sample preparation of pharmaceutical dosage forms and the importance of sampling considerations.

Ultimately, the decision of what to validate and how to conduct suitable validation studies to demonstrate the "proof of concept" of the method depends on the parameter being measured, the target acceptance criteria, and the analytical method quality attributes that impact these. In addition the inherent properties of the test article being analyzed and the relationship between these properties and the analytical test method strengths and weaknesses need to be considered. This chapter reviews the key validation parameters to be considered for all analytical methods. In general these parameters will apply to any analytical method, but the approach to their evaluation will be unique to each situation and analytical technique.

4.1.1.3 Life cycle management

Despite the best efforts in the design and development of an analytical method and following verification of the suitability of that method through appropriate validation studies, the true reliability and sustainability of the analytical method need to be assessed throughout its life cycle. The application of meaningful system suitability tests is the first step in ensuring method performance during the analytical method life cycle. When properly chosen, the system suitability tests and acceptance criteria will demonstrate on the day of analysis that data obtained through testing are reliable and meet the requirements of the quality control test. It is an affirmation of the "proof of concept" validation studies targeting the analytical method equivalent of critical process parameters. The frequency of system suitability failures can also be indicative of fundamental issues with the analytical method that need to be addressed.

In addition to system suitability testing, longer term analytical method performance is valuable to assess. The use of stable quality control samples, which challenge the critical quality attributes of the analytical method, can be useful in trending method performance over time as well as on the day of analysis. Data trending is also useful in identifying potential issues with an analytical method. The use of quality control charts, often used to monitor manufacturing performance, is also useful to identify trends in analytical method performance. The application of process capability indices (i.e. *Cpk*, see Chapter 2) can similarly be used to determine long-term performance of analytical methods.[4] The use of these assessment tools during the life cycle of an analytical method is critical in a world where product and method transfers become more routine. This chapter discusses in more detail the use of these tools in identifying when method remediation may be required to continue to meet the design criteria of the analytical method.

4.2 DEFINITIONS

The purpose of validating an analytical method is to ensure that the method provides results that can be considered true and reliable for the intended use, whether the method is run in the laboratory where it was developed or any other laboratory as may be necessary. The extent of validation necessary for a given method will in large part depend on the product development stage of a drug candidate. The guidance provided by ICH Q2 (R1)[1] sets forth method parameters considered critical in validating certain, primarily separation, methods at a registration level. These parameters provide a sound basis, however, for validating methods at any stage of development. The parameters outlined by ICH Q2(R1) and their definitions per ICH Q2 (R1) are described below (see also Chapter 6 for the application of the guideline to assay and impurities).

4.2.1 Accuracy

Accuracy is a measure of the "trueness" of a result, and is represented by the closeness of a value obtained to a true or accepted reference value. The term "recovery" is also used as a measure of accuracy when evaluating the ability to quantitate an analyte of interest in the presence of other sample components such as a formulation matrix.

4.2.2 Linearity

Linearity is a measure of the ability of a method to provide a detector response directly (linearly) proportional to the concentration of the analyte injected into a system. Although it is acceptable to validate methods that do not show a linear response, e.g. some bioanalytical methods, it is more convenient to work in analytical ranges where responses are linear.

4.2.3 Range

The range of a validated method is the analyte concentration interval over which accuracy, linearity, and precision have been adequately demonstrated for the intent of the method.

4.2.4 Specificity

Method specificity is the ability of a method to identify and/or quantitate, unequivocally, an analyte of interest in the presence of other matrix components. Matrix components could range from synthetic process impurities and degradation products to formulation excipients.

4.2.5 Precision

The precision of a method is a measure of the variability of a series of measurements made on a homogeneous sample or set of samples. The precision is usually expressed in terms of the standard deviation of a series of measurements. Other statistical values used to measure precision include variance or coefficient of variation. There are three levels of precision to consider, depending on the intended use of the method.

4.2.5.1 Repeatability

Repeatability refers to the variability of a method under the same operating conditions over a short period. Repeatability is generally used if the intention is to routinely run the method in a given laboratory with a single analyst. Another term used for repeatability is intra-assay precision.

4.2.5.2 Intermediate precision

Intermediate precision is used to assess the variability within a laboratory but under different conditions. The condition variations may include analysts, days, and equipment.

4.2.5.3 Reproducibility

Reproducibility is a measure of the variability between laboratories and is used in method transfers or collaborative studies between laboratories.

4.2.6 Detection limit

The detection limit is the lowest sample concentration at which an analyte of interest can be detected but not quantitated. As such, in a quality control setting, detection limit applies only to limit tests and never to quantitative tests (see Chapter 6 for more details).

4.2.7 Quantitation limit

The quantitation limit is the lowest sample concentration at which an analyte of interest can be quantitated with the necessary accuracy and precision.

4.2.8 Robustness

Method robustness is a measure of the ability of a method to withstand variations in method operation conditions. Robustness is typically evaluated during the development stage of a method and includes variations in mobile phase, column, flow rate, and column temperature.

4.2.9 System suitability

System suitability refers to the ability of a given system to perform the method on a given day under a given set of conditions. Typical system suitability parameters for chromatographic methods include precision, tailing factor, and peak separation factor.

4.3 GUIDELINES

Validation of analytical methods is an essential part of the product development process on a global basis. While the directives of the various regulatory agencies require that analytical methods be validated, there are no specific regulations on the requirements of a validation process. There is, however, an abundance of guidance documents that one can refer to when considering the validation of a method for a specific application. Some of the various guidance documents may be referred to throughout this book in reference to specific applications, but this chapter presents an overview of some of the more widely used guidance documents. It should be noted that most guidances on validation focus on what would be suitable for the final registration method. Method validation, by its definition, is generally a phase-dependent process and the amount of effort expended on a method validation is generally less at an early development phase than at registration.

Perhaps the most widely used guidance for validation is that established by ICH Q2 (R1)—"Validation of Analytical Procedures: Text and Methodology".[1] This guidance was established by representatives from industry and regulatory agencies from the United States, Europe, and Japan in an effort to harmonize the processes used to validate methods for registration. While the principles of method validation covered by the ICH have general applicability, the focus of the guidance is for application to identification methods and quantitative methods for the drug substance and impurities in either the drug substance itself or in a drug product matrix. The "Text" section of this guidance discusses the types of methods that are covered in the guidance along with a discussion of the parameters considered critical in demonstrating that a method is suitable for inclusion in a regulatory filing submission. The "Methodology" section provides guidance on testing protocols for the various validation parameters.

The US Food and Drug Administration (FDA) has issued a number of guidance documents and policies relating to validation.[5–9] A draft guidance on "Analytical Procedures and Method Validation"[5] makes reference to and is based substantially on the ICH Q2 (R1). Other FDA guidances or draft guidances cover bioanalytical method validation,[6] submitting samples and analytical data for method validation,[7] and the use of mass spectrometry for confirmation of the identity of animal drug residues.[8] In addition, the FDA also has a policy guide for requesting method validation in support of Abbreviated New Drug Applications.[9] The United States Pharmacopeia (USP) addresses method validation and method verification in General Chapters <1225> and <1226>, respectively.[10,11] As with the

FDA, the USP discussions are closely aligned with the ICH. More detailed discussions on the verification of compendial methods are given later in this chapter and in Chapter 14.

There are two European guidance documents of note. EURACHEM has published a guidance document entitled "The Fitness for Purpose of Analytical Methods".[12] This document is one of the more detailed official documents on method validation. The European Medicines Agency (EMA) has also issued a guidance document on generating and reporting methods of analysis in support of pre-registration data requirements.[13] The Australian Therapeutic Goods Administration (TGA) has published a guide on "Starting Material Analytical Procedure Validation for Complementary Medicines".[14]

4.4 PHASE APPROPRIATE METHOD VALIDATION

The requirements for analytical method validation for marketing applications are described in ICH Quality Guidance Q2 (R1), "Validation of Analytical Procedures: Text and Methodology",[1] and other regulatory guidances and pharmacopeias.[5,10,15] Requirements regarding analytical methods and their associated validation in clinical stages of development are less defined in applicable regulatory guidances, but nevertheless support the concept of phased method validation.[16–18]

The underlying rationale supporting a phased approach to method development lies in three fundamental purposes that apply at various stages of pharmaceutical development. These fundamentals were described by Boudreau et al.[19] following a 2003 PhRMA workshop about acceptable analytical practices, and are summarized in Table 4.1.

Table 4.1 Purpose of Analytical Methods by Phase of Development

Clinical Purpose	Pharmaceutical Purpose	Purpose of Methods
Early	Early	Early
- To determine the safe dosing range and key pharmacological data (e.g. bioavailability and metabolism) in Phase I trials involving a few healthy volunteers - To study efficacy in Phase II trials in patients while continuing to test safety	- To deliver the correct bioavailable dose - To identify a stable, robust formulation for the manufacture of multiple, bioequivalent lots for Phase II and III trials	- To ensure potency, to understand the impurity and degradation product profile, and to help understand key drug characteristics - To indicate stability and begin to measure the impact of key manufacturing parameters to help ensure drug substance or product consistency
Late	Late	Late
- To prove efficacy, confirm safety, and obtain the desired label through phase III trials involving a large number of patients	- To optimize, scale-up, and transfer a robust and controlled manufacturing process for the commercial product	- To be robust, cost-effective, transferable, accurate, and precise for specification setting, stability assessment, and approval of final marketed products

Adapted from Ref. 19.

If the fundamental purposes of an analytical method are taken into consideration during drug development, then one can construct an approach to phased method validation which supports the analytical method target profile required for the critical drug substance/drug product/process attributes necessary for the method at that stage. The analytical method capabilities should be developed and validated such that a progression of the confirmation of the method's ability to ensure patient safety, confirm efficacy, and assess product/process robustness is achieved. By constructing the validation studies in such a way, a demonstration that the method is "suitable for its intended purpose" (at a given stage of development) can be established.

As a product or process is developed, information regarding those quality aspects critical to measure and control is established using quality by design and risk assessment approaches. Early on in development, analytical methodology needs to be available which will allow for trending of quality attributes and product or process performance. The analytical methodology which supports these development studies needs to have sufficient specificity, accuracy, and precision to enable the identification of these critical quality and process attributes. Optimizing specific method attributes is not normally required unless specific quality criteria for the product or process are identified that require tighter control. The relationship of the analytical method's capabilities and the specification acceptance criteria is often the key to determining if a method has been shown to be suitable for its intended purpose and will help define the degree of method validation required and what is deemed acceptable analytical method performance.

In addition to understanding the main purpose of an analytical method at a specific clinical stage of development, it is also important to consider the number of individuals or laboratories that will utilize the analytical methodology and the variability in analytical instrumentation that might be utilized when conducting analyses. Most often in the early stages of development, analyses may be conducted by a single laboratory with a limited number of analysts and instrumentation. The method is likely to be run by the same analyst who developed the method and is, therefore, very familiar with the nuances of the procedure. Consequently, the need to demonstrate intermediate precision and robustness of the analytical procedure is not as critical in these situations. In later stages of development where more laboratories and analysts may be asked to run the analytical method, these two validation parameters become more critical to evaluate to ensure consistency of the data being generated.

Too often, full ICH validation requirements are applied to an analytical method in the early stages of development, and as a result, less time is devoted to the actual development of the method. In addition, as changes to a synthetic process or formulation occur, the analyst avoids changing the original method to avoid extensive and time-consuming validation studies. As a result, the method begins to lose its efficiency and core purpose over time. A more practical approach may be for the analyst to focus on what the most critical method performance criteria are and to demonstrate through reduced validation studies that these performance criteria can be met. Including key method performance tests during system suitability assessments on the day of analysis may be more useful than performing extensive validation studies on a method, which will most likely evolve during drug development to address new analytical challenges. One might consider the approaches to validation for early stage development projects as outlined in Tables 4.2 and 4.3 for high-performance liquid chromatography (HPLC) procedures for assay and impurities, respectively.

At the time of Phase III clinical studies or before the transfer of analytical methods to other laboratories, validation studies consistent with the requirements for marketing submissions should be considered.

Table 4.2 Early Stage Validation Approach for Assay HPLC Procedures

	Drug Substance	Drug Product
Specificity	Show resolution of drug substance from most likely impurities.	Show non-interference of likely impurities and excipients with drug substance
Linearity	Include multilevel standards in early method	Include multilevel standards in early methods
Range	Supported by daily method linearity	Supported by daily method linearity
Accuracy	Supported by daily method linearity	Demonstrate recovery from sample matrix at target level
Precision	Supported by system suitability injection precision test	Demonstrate repeatability at the target level
Detection/quantitation limit	Not required	Not required
Robustness	Based on method development studies. Assess standard and sample solution stability	Based on method development studies. Assess standard and sample solution stability

Table 4.3 Early Stage Validation Approach for Impurity Procedures

	Drug Substance	Drug Product
Specificity	Include a daily system suitability test for critical peak pairs	Show non-interference of excipients and the most likely degradation products with drug substance
Linearity	Include multilevel standards of the active around the reporting limit	Include multilevel standard of the active around the reporting limit
Range	Supported by daily method linearity	Supported by daily method linearity
Accuracy	Supported by daily method linearity	Demonstrate recovery of the active from the sample matrix at the specification limit for impurities
Precision	Supported by system suitability injection precision test	Determined from variability of accuracy replicates
Detection/quantitation limit	Add a sensitivity solution of the active at the reporting limit	Add a sensitivity solution of the active at the reporting limit
Robustness	Based on method development studies. Assess standard and sample solution stability	Based on method development studies. Assess standard and impurity solution stability

4.5 VERIFICATION OF COMPENDIAL METHODS

Compendial methods are often used to evaluate pharmaceutical compounds or products. The most frequently used compendial references are the United States Pharmacopeia, European Pharmacopoeia (EP), Japanese Pharmacopoeia (JP), and the British Pharmacopoeia (BP) (Chapter 16). For materials that have a compendial monograph, it is often easier and less costly to use the compendial method for analytical determinations rather than to develop and validate a new method. It must be realized, however, that the methods provided in compendial monographs were generally established for an Active Pharmaceutical Ingredient (API) material manufactured by a specific process or for a specific, proprietary formulation of a drug product. Many of the methods provided in the current compendia were developed using old, nonspecific, and/or outdated technologies. Although validation documentation generally accompanies submissions of monographs for incorporation in the various compendia, many compendial methods were published prior to the establishment of current validation practices or guidelines.

It is an acceptable practice to use compendial methods for API or drug product release or sometimes even stability. Prior to doing so, however, it is the responsibility of the user to verify that the method is suitable for the intended use. USP General Chapter <1226>, Verification of Compendial Procedures, presents a number of considerations for ensuring that a compendial method is acceptable for use. If one is planning on using a compendial method for a purpose other than for which it was established, the first order of business is to understand the principles of the method and assess whether the principles of the method as written are applicable to the intended use. If the fundamental principles do not apply, it is best to develop and validate an alternative method. Each compendial method was established for a specific intended use. As such, the method principles are not necessarily applicable to other intended uses of the method. Another factor to be considered is the scientific soundness of the method as written. Often, compendial methods do not provide specific details such as sample preparation. This requires analyst interpretation of often critical steps or activities in the conduct of a compendial method "as written". As a result, different analysts following the same procedure can get widely varying results. The verification process would involve eliminating any ambiguity in the procedure.

Verification of a compendial method does not necessarily mean revalidation of the method. The usual validation parameters are often a key consideration in the verification plan. However, it is usually sufficient to include only those parameters that are directed at differences in a particular application relative to the original application for the method. For example, specificity would need to be evaluated if the method will be applied to a drug manufactured by a different process than that for which the compendial method was originally issued. Similarly, if a method is used to evaluate a drug product formulation that is different from that for which the compendial method was published, the absence of interference from alternative formulation ingredients must be demonstrated.

It is not uncommon to use compendial methods for stability testing. Compendial methods, as written, however, cannot be assumed to be stability indicating. As such, prior to using a compendial method for stability, the verification process must include a demonstration that the method can suitably separate and quantitate the active ingredient and degradation products. This may require a forced degradation study.

4.6 REVALIDATION OF METHODS

Throughout drug development, change is inevitable. Changes may be made to improve a product or process from both a quality and an efficiency perspective. When these changes occur, an evaluation of the changes on the suitability of the analytical methods used to control the quality of the product or process needs to be conducted. Established analytical methods should be revalidated when significant changes are made to laboratory equipment/analytical instrumentation or the operating conditions of the analysis, or when the drug substance or drug product being analyzed has undergone significant changes.

Significant changes to equipment, analytical instrumentation, or operating conditions may include:

1. Use of automation to replace manual methods (e.g. sample preparation and sampling)
2. Changes in sample preparation equipment (e.g. shakers, homogenizers)
3. Changes in principles of detection (e.g. photodiode array vs single wavelength detection systems, changes in detection cell path length)
4. Changes in instrumental operating principles (e.g. low vs high pressure gradient mixing in HPLC systems, axial vs radial plasma orientations in inductively coupled plasma optical emission spectroscopic methods)
5. Use of alternate suppliers (e.g. changes in grade and quality of reagents, changes in column supplier)
6. Modifications to sample diluent/extraction solvents (if not studied in robustness studies of original method)

The validation parameters that need to be reconsidered for the example changes listed above are dependent upon the likely impact of such changes on the quality of the analytical data generated. For methods that include system suitability tests capable of assessing the impact of such changes, little or no additional revalidation may be required.

For changes which may impact the method in a way that cannot be assessed through system suitability testing, comparative studies of the modified method and the original validated method along with some additional robustness studies may be warranted. When conducting comparative testing, appropriate statistical approaches to demonstrate equivalence should be applied.[20,21] Since comparative testing is generally conducted under set conditions in a limited time frame, additional robustness studies to assess the modified method are highly recommended. Regardless of the change or modification, an assessment of the impact on the critical validation parameters of the method should be conducted. Those parameters most likely to be impacted should be subject to revalidation.

Significant changes to the drug substance are likely to occur during development in an effort to simplify the process, increase yield, reduce cost, and drive green chemistry. Changes to the drug substance synthetic process may result in different impurities or changes to physicochemical properties of the drug substance. The introduction of new impurities, resulting from different chemistries and different reagents or starting materials, will generally require additional validation of the method. The specificity of the method with regard to new impurities will need to be demonstrated. Additional linearity, accuracy, and precision studies for newly introduced impurities will also need to be conducted. In addition, the ability of the method to detect and quantitate the new impurities at levels that support the specification requirements for these impurities will need to be verified through additional validation. The solution stability of new impurities should also be assessed.

Changes to the synthetic process can also result in changes to the physicochemical properties of the drug substance. These changes can impact properties such as particle size, surface area, and polymorphic form. As a result, solubility differences (equilibrium and kinetic) can occur that can impact the efficiency and effectiveness of analytical sample preparation for methods requiring solubilization of the sample. Changes to sample preparation techniques and sample diluents may be necessary to prepare the analytical sample. Method attributes which are most likely to be impacted by these types of changes include accuracy, precision, and robustness. Similarly, changes to the drug product are to be expected during development to address changes in clinical doses, maximize bioavailability, and optimize the manufacturing process. New dosage forms will require full method validation. Minor changes in formulation composition and excipient ratios of a dosage form can generally be supported by reduced method validation studies.

If modified formulations introduce new excipients, the impact of these excipients on the specificity of the method needs to be reassessed. New excipients could also impact the stability of the drug product and could promote different degradation mechanisms. In those cases, method revalidation studies analogous to those applied to drug substance impurity profile changes should be conducted. New excipients could also impact the accuracy and precision of the method, especially if those excipients impact the ability of the analytical sample preparation to effectively extract analytes of interest from the dosage form matrix due to drug or impurity interactions with the newly introduced excipient.

Minor changes to the ratio of excipients in a drug product formulation are generally considered to have minimal impact on the critical quality attributes of the analytical method because degradation pathways are unlikely to change, and the original method specificity is generally maintained. The accuracy and precision of the method, however, may sometimes be impacted. An example of this would be making minor changes to the ratio of polymers used in a sustained release erodible matrix formulation. The same changes that will impact the in vivo release of drug from the formulation can also impact the accuracy and precision of the analytical method.

Modifications to the manufacturing process also need to be considered as they may introduce the potential for new impurities or impact the physicochemical properties of the dosage form. Regardless of the change or modification to the manufacturing process, an assessment of the impact on the drug product and how that may influence the acceptability of the original analytical method should be conducted. Those method parameters most likely to be impacted should be subject to revalidation.

4.7 METHOD REMEDIATION

During the Design (method development) and Proof of Concept (method verification/validation) phases of an analytical method the objective is to establish a procedure which is robust, accurate and precise. During the life cycle of the method (long-term use, equipment changes, and method transfers) it may be found that methodology does not have the necessary sustaining attributes originally hoped for. Tracking of a method's performance over time and through what might be considered to be minor changes is as critical as being able to assess drift or changes to a manufacturing process. By effectively assessing method performance over time, potential issues can be identified early and continuous improvement of the method can be realized resulting in better data and the ability to trend manufacturing performance, and assess stability changes.

The use of meaningful system suitability testing not only ensures data integrity at the time of sample analysis but also provides longer term method performance information. Drifts in the system suitability test results can indicate instrument, column, reagent changes that can impact the accuracy and precision of the data obtained. By carefully assessing the results of system suitability tests, preventative actions can be taken to minimize the risk of erroneous test results which could lead to time consuming investigations or product failures. In most cases, degradation in system suitability test results can be corrected through routine maintenance of instrumentation, replacement of key analytical components of the system such as a chromatographic column, or a change in supplier or grade of reagents used. If these efforts fail to improve the situation, the method may require modifications to restore its robustness to changes which are beyond the control of the analyst, such as step change in column chemistry by the manufacturer.

While monitoring changes in system suitability test results can be useful in assessing the long-term performance of a method based on system components (instrumentation, reagents, and critical analytical components such columns), they are less useful in aspects of the method which are dependent upon the sample. Variability in sample preparation over the life cycle of the method may be indicative of a procedure that has fundamental robustness issues. The variability is often attributable to inconsistency in sample preparation techniques or extraction solvent efficiency. The effectiveness and robustness of sample preparation can be assessed over time through the use of well-characterized, homogeneous, and stable quality control (QC) samples. Quality control charts can be generated for these QC samples to help identify trends in method performance. The use of quality control charts and their interpretation is well established in the pharmaceutical production environment to identify manufacturing trends, but their use to assess analytical method performance is less common. Where appropriate, they could also be included in the cadre of system suitability tests to identify potential problems at time of sample analysis. The acceptable tolerance of the day-to-day drift of the QC samples should take into account the normal variability of the method and the specification requirements. Many statistical software packages exist (e.g. Statistica by StatSoft) which are useful in the generation and interpretation of quality control charts.

As organizations have become more focused on improving manufacturing processes and reducing product failures, statistical approaches to evaluate the capability of a process have been developed and utilized to drive continuous improvement in the manufacturing sector. The two main capability indices (Chapter 2) most commonly used are *Cp* (for processes centered at the mean of the specification limits) and *Cpk* (for processes which are not centered at the mean of the specification limit):

$$Cp = (\text{USL} - \text{LSL})/6\sigma \tag{4.1}$$

$$Cpk = \min[(\text{USL} - \mu)/3\sigma, (\mu - \text{LSL})/3\sigma] \tag{4.2}$$

where LSL and USL are the lower and upper specification limits of the process, respectively, and μ and σ are the mean and standard deviation of the process, respectively. Processes with capability indices >1.33 are generally regarded as adequate to meet specifications. Processes with indices <1.00 indicate that the process is inadequate for the product routinely meeting specification.[22]

Direct application of the capability indices to analytical methods would seem reasonable and a good way to assess the applicability of the method capabilities with respect to the specification criteria. Bouabidi et al.[4] have demonstrated that the application of the traditional process capability indices to analytical methods will overestimate the true capability of the method, especially if applied to method validation data and the associated small data set. They introduce the concept of a modified capability index, *Cpk-tol*, and the corresponding estimator of proportion of nonconforming results

($\pi_{Cpk\text{-}tol}$). Through the use of Monte-Carlo simulations, these parameters have been shown to be much better estimators of an analytical method's capability, especially when limited data sets are available.

As described above, there are a number of ways to track and monitor analytical method performance. The effective use of these tools during the life cycle of the analytical method can help identify when method remediation may be needed.

4.8 METHOD TRANSFER

Analytical methods are typically developed and validated in Research and Development laboratories, but are almost always transferred to other laboratories for more routine sample testing. These "receiving" laboratories could be, for example, quality control laboratories within the same company, or, as is often the case in today's industry, contract laboratories. Oftentimes, the methods are transferred to multiple laboratories across the globe. It is essential that it be clearly demonstrated and documented that the receiving laboratories are capable of running the methods and obtaining results with the same accuracy and precision as the developing laboratory.

4.8.1 Protocols

All method transfers that will result in the receiving laboratory generating any data that will involve testing of clinical or registration materials should be formal and protocol driven. The transfer protocols can be generated by either the development or receiving laboratories, but both parties are responsible for review and sign-off of the protocol prior to commencement of the transfer activities. The protocols must clearly delineate the responsibilities of both parties. Transfer protocols must clearly outline all pertinent details of the method, materials, equipment, sample lot numbers, and method parameters to be evaluated. Of particular importance in a method protocol are the acceptance criteria. There should be no ambiguity when testing is completed and data evaluated as to whether the transfer is successful.

4.8.2 Transfer processes

There are several ways that method transfers can be accomplished. Some of the more common practices for method transfer are described below.

4.8.2.1 Results comparison

This is the simplest of the method transfer processes. If the development laboratory has completely validated a method and system suitability can be demonstrated in the receiving laboratory, a comparison of results generated by both laboratories may be sufficient to complete a successful transfer. This comparative testing is often, and preferably, carried out by having both the sending and receiving laboratories generate data on a selected lot or lots of material within a reasonably short, but defined, time period. The results from each laboratory are then compared and the comparison evaluated against the acceptance criteria established in the protocol. Under certain circumstances, it may be acceptable to compare receiving laboratory data against a Certificate of Analysis (CoA), provided the CoA was generated within a reasonable time relative to the data generated by the receiving laboratory.

4.8.2.2 Revalidation

If a method to be transferred has not been validated, partially validated, or perhaps validated for a different intended purpose, it may be necessary for the receiving laboratory to revalidate the method, either partially or fully. This is the most involved of the method transfer processes, but may provide the highest level of assurance that the receiving laboratory can adequately perform the testing for the intended use of the method. The level of revalidation required for this transfer process will depend on the type of method and amount of documented validation data available from the receiving laboratory. The revalidation must still be protocol driven and coordinated with the development or sending laboratory.

4.8.2.3 Covalidation

Perhaps the most efficient way to effect a transfer is to have the receiving laboratory participate in the validation of the method. In this case, there would be no need for a "formal" transfer, but as the receiving laboratory has generated data used in the validation, provided the validation is successful, no additional transfer activities need take place. The validation protocol and report, co-authored by the sending and receiving laboratory can be considered as having completed a successful transfer. This process requires strategic planning and is most often used, for example when a development laboratory will be transferring a method to a quality control laboratory at the same site or within the same company. If several laboratories will be receiving the method, transfer by this process may be very cumbersome, if not impossible.

4.8.2.4 Transfer waivers

Transfer waivers can be affected in certain circumstances provided adequate justification is documented. Possible reasons for such waivers may include transfer of analysts familiar with the method to a receiving laboratory, changes in the method that do not change the ability of the analysts to run the method, or the method is very similar to a published method.[23]

4.8.3 Transfer reports

The final step in executing a formal transfer is the transfer report. This report will include all the data generated per the protocol, discussion of the results, comparison to the acceptance protocol, and a statement that the transfer has been successfully completed. The transfer report should be signed by all parties involved in the transfer.

References

1. Validation of Analytical Procedures: Text and Methodology Q2(R1), The International Conference on Harmonization of Technical Requirements for Registration of Pharmaceuticals for Human Use, First Revision, 1995.
2. Broad, N.; Graham, P.; Hardy, A.; Holland, S.; Hughes, S.; Lee, D.; Prebble, K.; Salton, N; Warren, P. Guidelines for the Development and Validation of Near-infrared Spectroscopic Methods in the Pharmaceutical Industry. In *Handbook of Vibrational Spectroscopy;* Chalmers, J. M., Griffiths, P. R., Eds.; John Wiley & Sons Ltd.: Chichester, 2002.
3. Nickerson, B., Ed. *Sample Preparation of Pharmaceutical Dosage Forms: Challenges and Strategies for Sample Preparation and Extraction;* Springer, 2011.

4. Bouabidi, A.; Ziemons, E.; Marini, R.; Hubert, C.; Talbi, M.; Bouklouze, A.; Bourichi, H.; El Karbane, M.; Boulanger, B.; Hubert, Ph.; Rozet, E. Useful Capability Indices in the Framework of Analytical Methods Validation. *Anal. Chim. Acta* **2012,** *714,* 47–56.
5. U.S. Food and Drug Administration, Draft Guidance for Industry—Analytical Procedures and Methods Validation Chemistry, Manufacturing, and Controls Documentation (August, 2000).
6. U.S. Food and Drug Administration, Guidance for Industry—Bioanalytical Method Validation (May, 2001).
7. U.S. Food and Drug Administration, Guidelines for Submitting Samples and Analytical Data for Methods Validation (February, 1987).
8. U.S. Food and Drug Administration, Draft Guidance for Industry—Mass Spectrometry for Confirmation of the Identity of Animal Drug Residues (June, 2001).
9. U.S. Food and Drug Administration, Office of Generic Drugs, Manual of Policies and Procedures MAPP5221.1, Requesting Methods Validation for Abbreviated New Drug Applications (ANDAs), November, 1998.
10. United States Pharmacopeia, 35th rev., <1225> (United States Pharmacopeial Convention, Rockville, MD, November 2011).
11. United States Pharmacopeia, 35th rev., <1226> (United States Pharmacopeial Convention, Rockville, MD, November 2011).
12. Eurachem, The Fitness for Purpose of Analytical Methods: A Laboratory Guide to Method Validation and Related Topics, 1998.
13. European Commission, Directorate General Health and Consumer Protection, SANCO/3029/99 rev.4, Residues: Guidance for generating and reporting methods of analysis in support of pre-registration data requirements for Annex II (part A, Section 4) and Annex III (part A, Section 5) of Directive 91/414, Working document, November, 2000.
14. Australian Government, Department of Health and Aging, Starting Material Analytical Procedure Validation for Complementary Medicines, March, 2006.
15. Kamberi, M., et al. Setting Acceptance Criteria for Validation of Analytical Methods of Drug Eluting Stents: Minimum Requirements for Analytical Variability. *Eur. J. Pharm. Sci.* **2011,** *42,* 230–237.
16. Japanese Pharmacopeia, 16th rev., English Translation, General Information <G1> (Japanese Pharmacopeia, Tokyo, Japan 2011), pp. 2148–2150.
17. Food and Drug Administration, INDs for Phase 2 and Phase 3 Studies, Chemistry, Manufacturing, and Controls Information (May, 2003).
18. Committee for Medicinal Products for Human Use (CHMP), Guideline on the Requirements to the Chemical and Pharmaceutical Quality Documentation Concerning Investigational Medicinal Products in Clinical Trials, CHMP QWP/185401/2004 (March, 2006).
19. Boudreau, S. P., et al. Method Validation by Phase of Development: An Acceptable Analytical Practice. *Pharm. Technol.* **2004,** 54–66.
20. Chambers, D., et al. Analytical Method Equivalency: An Acceptable Analytical Practice. *Pharm. Technol.* **2005,** 64–80.
21. Chatfield, M. J.; Borman, P. J. Acceptance Criteria for Method Equivalency Assessment. *Anal. Chem.* **2009,** *81* (24), 9841–9848.
22. United States Pharmacopeia, 35th rev., <1080> (United States Pharmacopeial Convention, Rockville, MD, November 2011).
23. Radebaugh, G. *Global Transfer of Analytical Methods.* presented at the 2nd Annual QbD Conference in Israel: Rugged and Robust Product Development in the Pharmaceutical Industry, May, 2010.

PART

2

Universal Tests

CHAPTER

Description and identification

Ernest Parente

Mallinckrodt Pharmaceuticals, St. Louis, MO, USA

CHAPTER OUTLINE

Specification of Drug Substances and Products. http://dx.doi.org/10.1016/B978-0-08-098350-9.00005-9

5.1 INTRODUCTION

In this chapter, two important quality and safety attributes for drug product and drug substance evaluation, description and identification, will be discussed. While the overall goals of these evaluations are the same for drug products and drug substances, i.e. to ensure quality, safety and compliance with standards, there are special considerations in the test method selection and validation that need to be addressed. While not specifically discussed, the methods used for the evaluation of drug substances can be extended to include pharmaceutical excipients. In order to avoid duplication, methods used for identification are mainly presented in the drug substance section. However, with modifications basically involving extraction and purification of the drug substance from the drug product, the methods described are also applicable to drug product identification. In developing the chapter, concepts from the International Conference on Harmonization (ICH) Guidance documents, requirements from major compendia, regulatory expectations and current good manufacturing practices (cGMP) have been included.

5.2 DESCRIPTION

According to ICH Guideline Q6A,[1] which addresses setting specifications for new drug substances and drug products, a description statement is a universal test that needs to be included in specifications for all new drug substances and drug products. The description should address the physical state and color for drug substances and the appearance of drug products. The Guideline further states that, "If any of these characteristics change during storage, this change should be investigated and appropriate actions taken." So, while the scope of the Guideline is specifically aimed at new drugs, in some cases, particularly for investigations of changes on storage for older drugs, it may be retroactively applied as a cGMP requirement.

With regard to compendial specifications, in the United States Pharmacopeia (USP) descriptions are no longer included in article monographs. However, they are provided in a separate section, "Description and Relative Solubility of USP and NF Articles." These descriptions are provided for information only and are not intended as material specifications. In the European Pharmacopoeia (EP) and Japanese Pharmacopoeia (JP), descriptions are provided in the monographs. However, the general notices in each pharmacopeia indicate that they are not used to evaluate material compliance with specifications. The World Health Organization (WHO) considers the description declaration as part of the identification specification, which may include other elements that define identity (i.e. melting points, eutectic temperatures, optical rotation, etc.). In United States New Drug Applications and European Union Marketing Authorizations, descriptions are considered as part of the product specifications.

Up until the early 1970s, many descriptions included statements with regard to organoleptic characteristics such as odor or taste. While these characteristics may be useful in describing the properties of a material (e.g. "odorless liquid, odor of peppermint," etc.), they generally are not used for regulatory control of materials in most major compendia. The inclusion of organoleptic characteristics in drug substance and drug product specifications is systematically being eliminated since the evaluation is subjective and may present hazards to analysts.

5.2.1 Drug substance description

As given in ICH Guideline Q6A, a drug substance description should include "a qualitative statement about the state (e.g. solid, liquid) and color of the new drug substance."[1] An example of a description statement for a solid drug substance could be given as "White, crystalline powder." This statement meets the criteria of the Guideline and addresses both the physical state and color of the material. However, while the statement seems straightforward, there are potential pitfalls. Since there are numerous shades of white, if a standard for "whiteness" has not been defined, it may present some ambiguity to the evaluation.[i] Further, there are potential issues with the requirement "crystalline" since X-ray diffraction data are often needed to definitively prove that a material is "crystalline" and there are many materials that could appear crystalline that are in fact amorphous.[ii] Clearly, it is not the intent of the description to require X-ray diffraction testing. It is suggested that a more robust description would be "White to off-white powder," which allows for some observer variation in the color interpretation and still indicates that the material is a finely divided solid.

With regard to color, an evaluation to determine compliance with the description specification is typically done visually. However, the description of color is complex since the visual perception of color depends on the material properties (e.g. particle size, surface roughness, etc.), the characteristics of the illumination source, and the ability of the observer to distinguish subtle differences in color. Taken together, these factors introduce significant subjectivity into the evaluation. Therefore, the current practice has been moving toward quantitative instrumental methods in an attempt to eliminate potential analyst-to-analyst variation. A more detailed discussion will be provided later in the section "Color", which applies to both drug substances and drug products.

The Acceptance Criteria for the evaluation of description are usually on a pass/fail basis and stated as either "conforms" or "nonconforms" to the Acceptance Criteria depending on the result of the test. Any lot of material that fails a description test at the release stage should be investigated to determine the root cause of the failure.

Finally, to address concerns that the material appearance continues to meet the description requirements throughout the expiry period of the product, limits on acceptable change need to be established. As part of a stability program evaluation, often a 3-point evaluation system is used that involves assignment of descriptions as (1) unchanged, (2) changed but acceptable and (3) unacceptable. All unacceptable changes must be investigated and can impact the product expiration date. Unacceptable changes may indicate that special packaging and storage are required.

5.2.2 Drug product description

Based on the definition in ICH Guideline Q6A,[1] "A qualitative description of the dosage form should be provided (e.g. size, shape and color)." Similar to the drug substance, the guidance indicates that investigations be undertaken to evaluate changes in color during manufacturing and storage and also suggests that "If color changes during storage, a quantitative procedure may be appropriate." Following are examples for tablets, capsules and oral liquids/suspensions.

---

[i] In some cases, a "standard" lot may be used for comparison. However, the visual difference from the standard that would constitute a nonconformance is still subjective and difficult to define.

[ii] For example, while ground glass may appear "crystalline", it is in fact amorphous.

5.2.2.1 Tablets

Tablet descriptions should include color, shape and any specific markings related to the trade dress of the marketed product. This includes printing, embossing, debossing, identification numbers, logos, film coatings, scoring, etc. A distinctive trade dress may be used to differentiate a brand from potential generic versions. It is increasingly important in detecting, and making it harder to produce counterfeit products. It is also important in identifying misuse of prescription drugs with abuse potential. An example of a tablet description statement is "Light orange, round, bi-convex, scored, film-coated tablet imprinted with '77' on one side and blank on the other." Generally, the actual dimensions of the tablet are not included in the description.

The Physicians' Desk Reference (PDR®)[2] is an authoritative source for identifying Food and Drug Administration (FDA)-approved prescription tablet and capsule products (also available online). There are a number of other good sources available online that provide authentic images of marketed tablets that are useful for identification.[iii]

5.2.2.2 Capsules

Capsule descriptions should include the capsule size, material, cap color, body color, a statement as to whether they are clear or opaque and description of distinctive trade dress identifiers (i.e. identification numbers, banding if used, logos, etc.), and a description of the contents. An example of a capsule description is "Size 0, gelatin capsule, opaque yellow cap with a blue opaque body with the number '777' printed with black ink on the body and the number '01' on the cap, containing a white to white-off powder."

5.2.2.3 Oral liquids/suspensions

Oral liquid descriptions should include the color and whether the product is a clear liquid or a suspension. Frequently, flavorings are included in the description and conformance to the description is determined by odor. Examples of typical descriptions are "White, raspberry-crème suspension" and "Light-green, clear solution with the odor of peppermint" for an oral suspension and a clear oral liquid, respectively. Again, while the odor description for an oral liquid is subjective, it is usually included to identify the product.

5.2.3 Color

The United States Pharmacopeia (USP) defines color as "the perception or subjective response by an observer to the objective stimulus of radiant energy in the visible spectrum extending over the range 400–700 nm in wavelength. Perceived color is a function of three variables: spectral properties of the object, both absorptive and reflective; spectral properties of the source of illumination; and visual characteristics of the observer".[3] Typically, the color of a drug substance is evaluated visually under ambient lighting conditions, which may include a combination of fluorescent/incandescent artificial light and natural sunlight. The evaluation is made visually and is sometimes compared to a "control" sample of the test article with the intent of establishing a color match to establish conformance to specifications.

Since visual observation is subjective and depends on the perception of the observer, various attempts have been made to standardize color evaluations. For clear liquids or solutions of solids that

[iii]One example for prescription medications is the "RxList Pill Identifier", available at www.RXList.com.

produce clear liquids, the so-called APHA[iv] color scale (also, commonly referred to as the Hazen color or more correctly the platinum–cobalt color) has been used. The test is based on comparing the color of a liquid or solution to a selected dilution of a platinum–cobalt stock solution. The scale is defined over a range of 0–500, where 0 represents "water-white" and 500 is the color of a 500 ppm platinum–cobalt solution. While it was initially developed to quantify the color of waste water (see ASTM D1209—Standard Test Method of Color of Clear Liquids (Platinum–Cobalt Scale)), it has been useful in the pharmaceutical industry since many solid drug substances are white to off-white and the color scale is a measure of "yellowness". The color evaluation is typically done visually by comparison to dilutions of the standard solution. However, the solutions can also be measured using a spectrophotometer. For example, a limit test for absorbance in the range of 400–410 nm can be used as a control for "yellowness". This would involve qualitatively establishing what absorbance limit would be acceptable.

United States Pharmacopeia General Chapter <631>, "Color and Achromicity",[3] describes the use of 15 liquid color matching standards (A–T). These color matching standard solutions are prepared by mixing three primary solutions made using cobaltous chloride, ferric chloride and cupric chloride in various ratios and dilutions. The color descriptions are given using the closest color match (e.g. "Near USP F" would be specified instead of light-yellow.)

European Pharmacopoeia Chapter 2.2.2, "Degree of Coloration of Liquids",[4] describes a similar approach to that of the USP using dilutions and mixtures of three primary standard color solutions (yellow, red and blue) to produce a set of 37 reference color solutions. Color descriptions are based on the best match of the test article to the color standard (e.g. "Near EP Y2" would specify a light, yellow solution).

The Pantone Matching System (PMS)[v] uses a printed set of Color Guides made using the CMYK process, which involves the use of four ink colors commonly used in commercial printing, cyan, magenta, yellow and key (black). However, 15 pigments are actually used (13 base pigments plus white and black) to produce the 1114 spot colors in the Color Guide. The printed guide is provided as a "fan-deck" in which each sheet gives different shades in a color family and every color is numbered. Using this system, a color specification could be given as "between PMS 100 and PMS 102" for a light, yellow color.

5.2.3.1 Color space model

The perception of color is derived from the response of the human eye to different wavelengths of visible light. To develop the sensation of color, the human eye utilizes photoreceptors, called cone cells. There are three types of cone cells, each with peak sensitivity in a different spectral region, corresponding to the colors red, green and blue. This response can be used to construct a three-dimensional RGB color space that encompasses all colors that can be made by mixing red, green and blue. In 1931, the International Commission on Illumination (CIE)[vi] developed the CIE 1931 XYZ color space standard,[5] based on human eye spectral response functions. In this color space each perceptible color is characterized by three coordinates (i.e. tristimulus values) X, Y and Z in a three-dimensional color space. Using this model and the earlier work of Hunter,[6,7] in 1976, the CIE

[iv] American Public Health Association.
[v] Available from X-Rite, Grand Rapids, MI.
[vi] Commission Internationale de l'Eclairage (CIE).

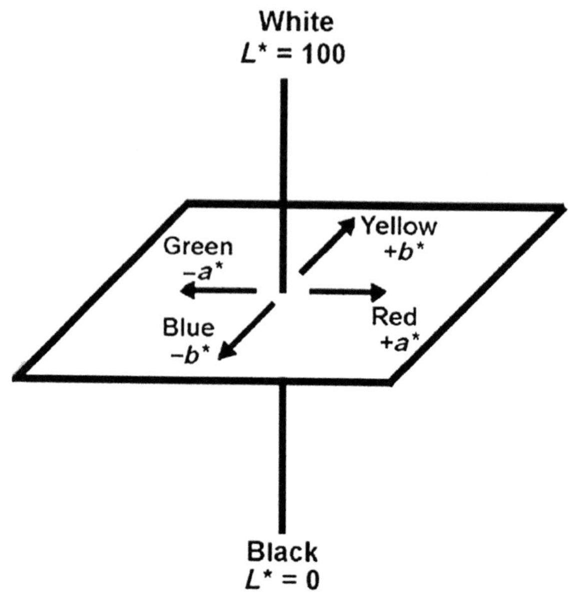

FIGURE 5.1

CIELAB color space.

Reprinted by permission of HunterLab. Copyright HunterLab 2008.

proposed the CIE $L^*a^*b^*$ (CIELAB) model for quantifying and communicating color attributes. A diagram[8] of the color space is shown in Fig. 5.1.

Using a spectrophotometer or a colorimeter,[vii] the tristimulus values for a color can quantitatively be measured and assigned values for L^* (lightness) and a^* and b^* (chromaticity). For each of the three parameters, the difference between the test article and a standard sample is calculated. Two colors are said to match if they have similar color value. The overall color difference, ΔE, can be calculated using the following equation:

$$\Delta E = [(\Delta L^*)^2 + (\Delta a^*)^2 + (\Delta b^*)^2]^{1/2} \tag{5.1}$$

where,

L^* = lightness ($L^* = 0$ for black and $L^* = 100$ for white)
a^* = chromaticity: green/red ($-a^*$ is green and $+a^*$ is red; range: -128 to 127)
b^* = chromaticity: blue/yellow ($-b^*$ is blue and $+b^*$ is yellow: range: -128 to 127).

5.2.4 Developing acceptance criteria for a color specification

In pharmaceutical manufacturing, color is an important indicator of lot-to-lot consistency and may signal the presence of impurities and degradation products. In some cases, the presence of very low

[vii]Instruments that make measurements in reflectance and transmission mode are available from Hunter Associate Laboratories, Inc., Reston VA (http://www.hunterlab.com).

amounts (e.g. in the ppm range) of highly colored materials can result in significant color changes. It is also important to monitor color change on stability throughout the shelf life of drug substances and drug products and investigate any significant changes observed. Even slight changes in color and appearance may result in a significant number of customer complaints related to the quality of the product.

While the visual method is subjective and to some extent depends on the visual characteristics of the observer, it is still the most widely used method. As mentioned earlier, qualitative visual color evaluations are generally "scored" either based on the description itself or with reference to a standard sample or color chart. For stability testing usually a three-point scale is used and the sample color is rated as (0) unchanged, (1) slight acceptable change or (2) significant unacceptable change. It is important that both the standard and test articles are viewed using the same light source to provide illumination. For example, it is possible that while two colors are judged to match using one illumination source, they may be judged to be different using a different illumination source. Colors that differ in this way are called a metameric pair and are a consequence of the fact that both samples do not have the same absorption and reflection spectra. Thus, in developing a test, it may be important to specify the properties of the illumination source. Another problem has to do with the fact that color changes may be difficult to describe since color may change in hue, value or saturation (chroma), any of which may lead to an unacceptable change.

To determine if there is a potential color change it is often useful to conduct accelerated aging studies (e.g. exposure to elevated temperature, strong visible/UV light, etc.).[viii] When a potential for change is apparent, protective packaging should be considered and cautions like "protect from light" can be stated on the product label. For drug products, light-resistant primary packaging (i.e. packaging materials that have product contact), opaque, or amber-colored containers or protective secondary packaging (e.g. cardboard carton or opaque overwrap) should be considered.

5.2.5 Acceptance criteria for instrumental methods

While visual color evaluation is most commonly used, there is a growing trend to replace the visual techniques with instrumental methods to make the tests less subjective.

When there are changes in color on stability, using a quantitative versus qualitative approach is recommended. While the color difference, ΔE (Eqn (5.1)), is measured instrumentally and eliminates subjective interpretation associated with the color measurement, there is still a subjective component to the color evaluation related to establishing how much of a change in ΔE (Eqn (5.1)) constitutes "significant" color change. One approach to establishing significance is to measure the variation in ΔE for samples with acceptable colors and set limits based on a statistical analysis of the data (i.e. mean value $\pm$ 3σ). While this method accounts for acceptable variation and gives an estimate of the measurement error, it does not address the question of what is or is not acceptable. If the original data set had little variation, narrow limits on ΔE could result in Acceptance Criteria failure without any perceptible visual change in the test article. A second and more desirable approach would be to establish the failure limits using samples with unacceptable color changes. This can be done by exposing the sample to accelerated aging conditions (e.g. heat and light) until the color change was

[viii]Some common colorants are known to be susceptible to light (e.g. FD&C Blue 2 and various purple pigments made using FD&C Blue 2).

found to be unacceptable and correlate this with the observed change in ΔE value. This method could also be combined with a statistical analysis. But, as mentioned above, a somewhat subjective judgment of what is and what is not acceptable still must be made. Finally, with regard to specifications based on setting an Acceptance Criteria for total color difference, ΔE, it is possible to have an acceptable value match for total color while the values for ΔL^*, Δa^*, and Δb^* result in unacceptable changes. It is suggested that during development, the need to set individual limits for each of the parameters should be evaluated.

5.3 DRUG SUBSTANCE IDENTIFICATION TESTING

As given in ICH Guideline Q6A,[1] "Identification tests should optimally be able to discriminate between compounds of closely related structures which are likely to be present." In general, identification using high-performance liquid chromatography (HPLC) retention times alone is not considered sufficiently specific. However, the guidance indicates that in the absence of a single specific test, a combination of nonspecific tests (i.e. two different chromatographic tests or tests such as HPLC/UV diode array, HPLC/MS or GC/MS) may be acceptable. While the guidance indicates one specific test is acceptable, commonly two tests, preferably one that is specific and one that can be nonspecific, but uses a different method of analysis (so-called, "orthogonal" method) are given in the product specifications. The use of two methods in tandem can increase the overall specificity. For example, suppose it is known that the probability of a match of the infrared spectrum of the test article to an authentic standard is 90% and the specificity of detecting a match using HPLC retention time is 70%, if both tests give a positive result, the overall specificity of establishing the correct identification is $(1 - 0.10 \times 0.30) \times 100 = 97\%$. Practically, it may be difficult to obtain a reliable estimate of the specificity of each method, but it does illustrate the power of using two independent tests for identification to increase the specificity of the identification. Including HPLC as an identification test in the specification may not require any additional testing since it can be obtained from the assay or test for impurities if these tests are done by HPLC (see Chapter 6).

To perform identification testing, three general approaches that are typically used include spectroscopic analysis, chromatographic analysis and wet-chemical/physical analysis. The following is a survey of some of methods that are commonly used for drug substance and drug product identification. The discussion is not intended to describe the theory and operation of the methods. Rather, it focuses on the strengths and weaknesses of the methods and some considerations that should be taken into account during method development and method validation.

5.3.1 Spectroscopic methods

5.3.1.1 Infrared spectroscopy

In most cases, infrared (IR) spectroscopy,[9] a form of vibrational spectroscopy, is still considered the method of choice for routine identification testing. Due to the richness of the spectral information that is characteristic of functional groups and structural elements of the drug substance of interest in this spectral region (4000–650 per cm), in some cases, solid-state IR analysis is selective enough to discriminate different salt forms and polymorphic forms of a given drug substance. In most applications, Fourier transform infrared spectroscopy (FTIR), which uses interferometer-based instrumentation, has largely replaced the traditional wavelength-dispersive, IR grating instruments. Older instruments produced a spectrum using percent transmittance (%T).[ix] However, most modern instruments provide a spectrum in absorbance versus wavelength and the trend is to use absorbance.

For solid samples, while there are a number of sample preparation choices (e.g. pellet, film or mull), the use of a compressed potassium bromide (KBr) pellet is the most commonly employed. A dispersion of ca. 0.3% of the test article of interest in KBr is prepared by mixing the sample with Spectrograde KBr using an agate mortar and pestle or a mechanical mixer (e.g. Wig-L-Bug) to form a solid–solid dispersion that can be compressed into a pellet. Since it is well known that the quality of the spectrum can be affected by particle size, it is important to grind the sample finely to prevent scattering, which results in band broadening and a drifting baseline.

With regard to hydrochloride salts of drug substances, care must be taken to ensure that the spectrum obtained is actually that of the hydrochloride salt and not that of the corresponding hydrobromide salt.[10] The reason for this is that many hydrochloride salts can exchange anions with the KBr pellet matrix, especially if a small amount of moisture is present and depending on how the sample is mixed with KBr to form the dispersion before compression (e.g. grinding time, grinding force, compression force, etc.). This usually goes undetected since the test article spectrum is usually compared to a spectrum of a standard that has been similarly prepared and has also undergone anion exchange. However, occasionally, the ruggedness of the test is brought into question by anomalous absorption bands or movement of the band position, especially, in the fingerprint region of the spectrum (ca. 1300–650 per cm), which is sensitive to these types of subtle changes. During method development, the ruggedness of the dispersion with regard to anion exchange can be tested by comparing the spectrum obtained using a KBr pellet dispersion to that obtained using split-mull liquid–solid dispersion technique using mineral oil (Nujol) and fluorinated hydrocarbon solvents (Fluorolube). While KBr is preferred for its compression characteristics and extended transmission range, if anion exchange is suspected, pellets can be prepared using KCl so that if exchange of anions occurs the spectrum will not be affected. Comparing a spectrum obtained using KBr to one obtained in KCl or one obtained using attenuated total reflectance sampling can also be used during development to detect the potential for anion exchange.

In addition to preparing a solid dispersion in KBr, a spectrum of a solution of a solid or liquid can be obtained by dissolving the material in a suitable solvent and employing a liquid cell holder. The cell holder uses two IR-transparent windows that are compatible with the liquid. Commonly, NaCl is used for nonaqueous samples and AgCl for aqueous solutions. However, other window materials are available. Neat liquid samples can also be prepared by "sandwiching" a drop of the liquid between salt plates with the use of a solvent. When solvents are used, a blank of the solvent is typically obtained and subtracted from the sample spectrum to eliminate bands due to the solvent.

A typical specification would require the sample to have absorption maxima at the same wavelengths as the standard. Since the IR spectrum is directly correlated to chemical structure and functional groups present, a spectrum with a missing peak is a clear indication of nonconformance. Since some bands may appear as shoulders or as fused peaks, the spectrum should be carefully evaluated. On the other hand, the presence of additional bands may indicate nonconformance or may be due to impurities. IR spectroscopy is generally not very sensitive to low-level impurities (i.e. impurities need

[ix]The use of transmittance, instead of absorbance, was largely due to the difficulty of electronically and/or mechanically taking the log (1/T).

to be at a level of several percent to be detected) so the ruggedness of the method should be established to support the level of any known impurities that could potentially cause interferences resulting in negative identifications.

The specification may include a requirement that the absorption bands also have the same relative intensity. While not required by the USP, this is required by the EP. Since the relative band intensity can be influenced by the sample preparation technique, the ruggedness of the sample preparation technique should be established as part of the method validation. In many cases, the comparison is between a sample and a standard similarly prepared at the time of testing. However, comparisons can also be done to using library spectra. For identification, the JP requires the comparison be done using a spectrum published in the pharmacopeia to ensure that a suitable spectrum with adequate detail is obtained.

In addition to transmission spectroscopy techniques, two other methods employing reflectance spectroscopy for IR identification are in common use, attenuated total reflectance (ATR) spectroscopy[11,12] and diffuse reflectance infrared Fourier transform spectroscopy (DRIFTS).[13,14] ATR spectroscopy employs a special cell made using a high refractive index crystal usually made of ZnSe, Ge, thallium halides, ZnS or diamond. If the cell material has a higher refractive index than the sample, the IR beam is reflected from the sample and is then used to obtain a spectrum. The spectrum is given as % reflectance or 1/(log reflectance), which are equivalent to percent transmittance and absorbance, respectively. Usually, the cell is designed in such a way to allow multiple reflections for greater sensitivity. The technique can be used for liquid and solid samples. One advantage of ATR over transmission techniques is that no sample preparation is needed. Since the penetration depth of the beam into the sample is limited, it is also useful for studying thin-surface films.

DRIFTS is another form of reflectance IR spectroscopy that utilizes a parabolic mirror to collect internally and externally reflected and transmitted light. For solid powders, typically no sample preparation is needed. The spectra are usually presented in Kubelka–Monk units versus wavelength; so to establish a positive identification the sample and standard should be both obtained using DRIFTS. To obtain good spectra, it is generally necessary to grind the sample to a particle size of 5 μm or less, so particle size and sample preparation needs to be considered during method development and validation.

5.3.1.2 *Ultraviolet and visible spectroscopy*

The comparison of the ultraviolet–visible (UV–VIS) spectrum of a sample to that of a standard can be used to support identification. While the UV–VIS spectrum (i.e. 190–780 nm) is characteristic of the material, the absorption bands, which result from electronic transitions, are usually broad and do not provide the same wealth of structural information given in the vibration spectrum obtained by IR or Raman spectroscopy. Thus, identity tests using UV are usually combined with another complementary technique. Also, since the spectrum is usually obtained in solution, solid-state structural information is lost.

Typical acceptance criteria for a positive identification would indicate that the test article had maxima and minima at the same wavelengths as a standard preparation and had the same absorptivity as a similarly prepared standard at a given wavelength, usually the absorption band maximum. Generally, the limit for an acceptable maxima match is ± 1 nm and limits for absorptivity matches should be within ca. $\pm 3\%$.

5.3.1.3 Near-infrared spectroscopy

Since spectra obtained in the near-infrared spectral region (12,000–4000 per cm) generally are the result of overtones and combination bands, they show broad absorption bands that lack the fine detail obtained by IR or Raman spectroscopy. Unlike IR or Raman, the identification is not based on similarity to a reference standard, but is based on chemometric analysis of data obtained from a library of reference spectra. Based on a multivariate model (e.g. principal component analysis, neural network analysis, partial least squares, etc.), the sample is assigned a "score". This score is compared to the computed scores in the reference library. If the distance between the two scores is within the capability of the model to discriminate closely related material that could be present, a positive identification is established.

Since identifications using near-infrared spectroscopy (NIR) are based on a specific library database, it is generally not considered as a primary identification method. However, NIR is very useful as an alternate method since it is rapid and usually requires little or no sample preparation. It is most useful for the identification of drug substances and excipients when the analysis of individual drums is needed to verify the identity of drums from a multi-drum shipment.[15] Since instruments are available that employ hand-held probes, it is also useful for identification at the point of use of a material.

5.3.1.4 Raman spectroscopy

While the Raman effect is due to scattering of radiation and arises in a fundamentally different way than IR, Raman spectroscopy, like IR, is a vibrational spectroscopic technique that gives a spectrum rich in structural information with enough specificity to provide unambiguous identification. It is a complementary technique to IR since each technique has different active, allowed vibrational modes. Unlike NIR, which requires a spectral library, identity is established by comparison to a standard spectrum. The method is rapid and portable, and hand-held devices are available. In most cases, no sample preparation is necessary. It is possible to obtain a spectrum of a drug substance stored in a polyethylene bag without opening the bag, making it attractive for nonintrusive identification of materials. Thus, it is particularly useful when drum identifications are required to verify the identity of each drum of a multi-drum material shipment, which is currently required for products for distribution in the EU. Since water has very little Raman activity, the method is also suited for the analysis of aqueous solutions.

5.3.1.5 Nuclear magnetic resonance spectroscopy

Nuclear magnetic resonance (NMR) spectroscopy is a powerful tool for establishing a positive structural identification. While it is particularly useful for proof of structure, it is not a method of choice for routine identification since it requires expensive equipment. Further, since the test is most commonly employed using a solution of a solid, information characterizing the solid-state structure (i.e. polymorphism) that can be obtained by IR identification is lost. Identification by NMR can be used in early development studies, but unless there are special situations, other alternatives should be considered.

5.3.1.6 Mass spectrometry

Like NMR, mass spectrometry (MS) is an extremely powerful tool for identification and structural elucidation. Compounds are identified by comparing the mass of the molecular ion and ions that result from its characteristic fragmentation pattern, which can be measured with very high resolution. In cases where there is little fragmentation and additional selectivity is needed, tandem mass spectrometry (known as MS/MS) can be used to obtain additional structural information to establish a positive identification. Its high sensitivity also makes it particularly suited for identification of low-level impurities. Ion mobility spectrometry has also been used and is particularly useful for the identification of low dose, high potency drugs.[16] However, like NMR, MS is generally not cost-effective for routine use. While it is considered as a powerful development tool, it should only be considered for routine use when all other options have been exhausted.

5.3.1.7 X-ray powder diffraction

In cases where a drug substance is known to exist in different polymorphic crystal forms, it may be necessary to include an identification test for a specific polymorphic form. It is well known that different polymorphic crystal forms can have different solubilities that can influence bioavailability. Since polymorphism can affect the safety and effectiveness of the drug, it is important to identify and control polymorphs. For new molecular entities, the existence of polymorphs should be investigated during early development as part of a preformulation program. Since the final crystallization step in the synthesis of the drug substance is most important in producing the final solid-state form, the solvent system used in this step should be carefully studied. This can be accomplished by varying the composition of the solvents used in order to determine if the desired polymorph can be consistently isolated.

The stability of the polymorphic form of the drug that will be used in the drug product should also be established since it is possible for a more soluble, less stable polymorphic form to convert to a less soluble, less bioavailable, more stable form with time. Stability studies for the drug substance should be conducted using accelerated storage conditions such as elevated temperature and humidity. While it is often not possible due to the low concentration of a drug substance that is diluted in a formulation matrix, the stability of the crystal form during manufacturing and on stability of the drug product should also be considered.

X-ray powder diffraction is the most definitive way to identify and characterize crystalline polymorphs. The sample is mounted in a scanning goniometer that rotates the sample as it is exposed to a collimated X-ray source. The diffraction pattern is obtained by measuring the intensity of the energy scattered as a function of the scanning angle, which is usually expressed as 2θ, where θ is the angular displacement of the rotated sample relative to the X-ray beam. The diffraction bands are characteristic of the crystalline interplanar spacing, d, which is independent of the radiation source used, as defined by the Bragg equation:

$$n\lambda = 2d \sin\theta \tag{5.2}$$

where,

n = order of the reflection
λ = the wavelength of the X-ray radiation source
d = interplanar spacing
θ = angular displacement.

A positive identification is established if the sample has the same diffraction pattern as an authentic standard similarly prepared. The specification for a positive match involves comparison of the interplanar spacing values and their relative intensities. Since the relative intensity can be influenced by preferred orientation effects of the crystal matrix, the robustness of the sample preparation (i.e. particle size, grinding technique etc.) should be studied as part of the method validation.

5.3.1.8 Melting point and differential scanning calorimetry

While it may be important to conduct X-ray powder diffraction analysis as part of a development program, it is not considered as a suitable test for routine use due to cost and complexity. However, if during development a relationship of the crystal form determined by X-ray diffraction can be correlated with thermal characteristics, the use of a melting point test or differential scanning calorimetry (DSC) may be considered as a routine control test. One problem in using tests based on melting point is that if the compound decomposes with melting, the melting point can become very dependent on the conditions of the test. So, variables like starting temperature, heating rate and particle size need to be considered to determine ruggedness during method validation. One solution is to use DSC, in which these variables can be carefully controlled. Often, the endotherm onset temperature is not sensitive to decomposition and can be used to establish a rugged Acceptance Criterion. Currently, simple melting point tests are being phased out in lieu of DSC tests.

5.3.1.9 Optical rotation

For a chiral drug substance that can exist as a pair of enantiomers, and the drug substance is supplied as a pure enantiomer, a chiral identification test should be included in the specification to distinguish it from the other enantiomer or the racemate (see also Chapter 10). One way to establish a positive identification is by using polarimetry. The degree to which each enantiomer can rotate plane polarized light is characteristic of the substance. The observed angular rotation, α, is a function of concentration and the path length of the polarimeter cell and is influenced by the wavelength of the light source used and the temperature. The Acceptance Criteria are usually expressed in terms of the specific rotation, which is given by the equation:

$$[\alpha]_{\lambda}^{T} = \frac{\alpha}{l \times c} \tag{5.3}$$

where, α, is the measured angular displacement, l is the path length in decimeters and c is the concentration in g/mL for a sample at a temperature T (given in degrees Celsius) and wavelength λ (in nanometers).

Typically, the experiment is conducted at 20 °C using the sodium-D line (589 nm) as the radiation source. Due to the relationship of the angular rotation and concentration, it is important in developing and validating a specification that the concentration is set such that the measured rotation would be sensitive enough for an unambiguous identification. It should be established during development that if other impurities are present at their limit, they would not interfere with the optical rotation test.

If optical rotation is used as an identity test for a racemate, the sensitivity of the test should also be established since a positive identification would be a specific rotation of zero. If pure samples of individual enantiomers are available, the sensitivity can be established by preparing mixtures to evaluate if there is adequate discrimination power. For example, an Acceptance Criterion can be established using a mixture with a 51:49 ratio of one enantiomer to another (i.e. a 2% enantiomeric excess) to ensure that the sample preparation is concentrated enough to give measurable and repeatable angular rotation values. If it is not sensitive enough, circular dichroism is preferred.[17]

While still common in most pharmacopeias, optical rotation identity tests are largely being replaced by chiral chromatographic methods. Like other chromatographic methods that are generally considered nonspecific, the combination of a test specific for the chemical moiety, like IR spectroscopy, that is not sensitive to enantiomeric modifications with a chiral-sensitive chromatographic test is sufficient to establish a positive identification.

5.3.2 Chromatographic methods

5.3.2.1 High-performance liquid chromatography and gas chromatography

While chromatographic methods are not considered specific enough for unambiguous identification, chromatographic retention time is often used as a second confirmatory identification test. For a positive identification, the retention time of a sample is compared to the retention time of a reference standard that is similarly prepared and analyzed. The selectivity of the method needs to be validated and the peak of interest should be well separated from all known impurities and degradation products or interferences from drug product formulation components. The specification should indicate an acceptance time window for a positive match. The acceptance window depends on the retention time of the peak and should be evaluated during method validation and should be based on repeatability. If there is doubt about the retention time match, a common method to verify the peaks that have the same retention time would be to make a 1:1 mixture of the sample and standard, reinject and examine the chromatographic peak obtained.

While the most common detection mode is UV absorbance at a selected wavelength, there are a number of detectors available (e.g. refractive index, fluorescence, electrochemical) that can be used to enhance selectivity. It is also possible to operate the detectors in series and compare retention times and responses for other detection modes for further support of identification.

5.3.2.2 Tandem chromatographic techniques

HPLC and gas chromatography (GC) can be used in combination with other in-line and/or off-line techniques to increase the probability of obtaining an identification. The most common in-line application for HPLC uses a photodiode array detector to obtain a UV spectrum in conjunction with a chromatographic retention time for a given peak. Another powerful combination is HPLC or GC with mass spectrometry (HPLC–MS and GC–MS). These techniques combine the selectivity of the chromatographic techniques with the selectivity/specificity of the mass spectral identification. As mentioned earlier, the use of two orthogonal methods greatly enhances specificity.

5.3.2.3 Chiral HPLC

Chiral HPLC, which employs special column stationary phases and/or mobile phase combinations that are capable of resolving chiral compounds, is a method for the identification of specific enantiomers. While in and of itself it is generally not considered specific enough to identify a given material, it is specific enough to establish a positive identification for a given enantiomer if the system has adequate resolution and the enantiomers to be separated are known. In combination with a more specific test, like FTIR, as mentioned earlier, positive identifications can be established. Like other chromatographic methods, a positive identification is based on retention time match to an authentic standard.

5.3.2.4 Thin-layer chromatography

Similar to retention time in HPLC, the comparison of the thin-layer chromatography (TLC) retardation factor, R_F, for a sample compared to a standard can be used to establish identity. The retardation factor[18] is defined as:

$$R_F = a/b \tag{5.4}$$

where,

a = distance traveled by the spot measured at the center
b = distance traveled by the mobile phase as measured to the solvent front.

Unlike HPLC, in which the samples are run in series, many TLC samples can be run simultaneously in parallel on one TLC plate affording faster analysis times. TLC is also relatively inexpensive and does not require expensive instrumentation. Since it frequently uses a normal-phase separation mode, compared to HPLC, which is typically done in the reversed-phase mode, it is considered to be a different and complementary technique to HPLC. However, it does not have the resolving power of HPLC. Again, like HPLC, TLC alone is not considered specific enough to establish positive identifications. While still in widespread use, TLC is gradually being replaced by HPLC. HPLC identification data can usually be derived from HPLC data used for drug substance or drug product assay methods (see Chapter 6).

5.3.3 Chemical tests

When the drug substance is provided as a salt, an identification test should be provided for the counter-ion as well as the active moiety. Often, these use wet-chemical tests and involve formation of precipitates and/or colored species as a positive endpoint. For a list of common chemical identification tests, the reader is referred to USP General Chapter <191> "Identification Tests–General",[19] which gives wet-chemical tests for over 40 metals and ions (e.g. Al, Cl, Br, citrate) to support testing of USP/NF articles. In many cases, these tests are somewhat nonspecific and can be replaced by other more specific tests, such as ion chromatography, atomic absorption spectroscopy, or inductively coupled plasma spectroscopy. In cases where a single test is not specific enough to establish the identity of a material, two or more tests may be needed to establish a positive identification.

5.4 DRUG PRODUCT IDENTIFICATION

According to the ICH Guideline Q6A,[1] "Identification testing should establish the identity of the new drug substance(s) in the new drug product and should be able to discriminate between compounds of closely related structures which are likely to be present." For identification when the drug products are formulated with a variety of inactive ingredients, it is usually necessary to first extract the active ingredient from the formulation matrix before conducting the identification test. Once isolated, the active ingredient can then be tested using similar methods as those used in the identification of drug substances. For IR identification tests, in cases were the drug substance cannot be completely extracted from formulation excipients, it is possible to establish a specification based on picking six or more absorption bands that are characteristic of the active ingredient. One drawback of the extraction method is that once in solution, information with regard to polymorphism, salt form or crystal habit is lost. Nevertheless, extraction followed by IR identification is a preferred identification procedure due to the selective nature of the spectral information obtained. This is usually combined with a second identification test, usually HPLC, to establish a positive identification. However, other spectral or orthogonal chromatography procedures, like TLC, can be used to increase the probability of obtaining a positive identification.

While generally not required in the US, the EU requires identification tests for colorants to be provided if they are used. These tests may not be routinely run for batch release. However, methods should be available so that products can be tested on a periodic basis. Since many of the common colorants approved for drug use employ iron oxides, titanium dioxide and other inorganic pigments, the colorant tests are typically colorimetric. The colorants are usually sampled by carefully scraping the coatings from the tablets without disturbing the tablet core ingredients. For capsules, the colorants are isolated by dissolution of the capsule shell in an appropriate medium, usually water. Drug product identification specifications usually do not include tests for counter-ions since many of the components of the formulation would cause interferences.

5.5 SUMMARY

Appearance and identity are two important characteristics in controlling the safety and efficacy of drugs in commerce. Methods for appearance and identity testing must be sufficiently specific to unambiguously distinguish the active ingredients and identify drug products to ensure product quality. Due to the international reach of the industry, it should also be considered that these methods will most likely be transferred to or performed in other laboratories so that the technology used needs to be widely available, robust and cost-effective to support potential global product distribution.

References

1. *Specifications: Test Procedures and Acceptance Criteria for New Drug Substances and Drug Products: Chemical Substances (Q6A),* In *International Conference on Harmonization*, October, 1999.
2. *Physicians' Desk Reference;* PDR Network LLC: Montvale, NJ.
3. General Chapter 631. Color and Achromacity. In *USP 35-NF 30;* United States Pharmacopeial Convention: Rockville, MD, 2012.
4. Chapter 2.2.2, Degree of Coloration of Liquids. In *European Pharmacopeia,* 7th ed., European Directorate for the Quality of Medicines and HealthCare (EDQM): Strasbourg, 2013 (7.6).
5. *CIE (1932),* In *Commission Internationale de l'Eclairage Proceedings*, Cambridge, 1931.
6. Hunter, R. S. Photoelectric Color Difference Meter. *J. Opt. Soc. Am.* **1948,** *38* (7), 661.
7. Hunter, R. S. Accuracy, Precision and Stability of New Photo-electric Color-Difference Meter. *J. Opt. Soc. Am.* **1948,** *38* (12), 1094.
8. HunterLab. CIE L*a*b* Color Scale. *Applications Note* **2008,** *8* (7).
9. Smith, B. C. *Fundamentals of Fourier Transform Infrared Spectroscopy,* 2nd ed., CRC Press: Boca Raton, FL, 2011.
10. Mutha, S. C.; Ludemann, W. B. Solid State Anomalies in IR Spectra of Compounds of Pharmaceutical Interest. *J. Pharmaceut. Sci.* **1976,** *65* (9), 1400–1404.
11. Harrick, N. *Internal Reflectance Spectroscopy;* John Wiley & Sons: New York, 1967; pp 342.

12. Willard, H. H.; Merritt, L. L., Jr.; Dean, J. A.; Settle, F. A. *Instrumental Methods of Analysis,* 7th ed., Wadsworth Publishing Company: Belmont, CA, 1988; pp 311–313.
13. Sherman HSU, C.-P., Ph.D. *Handbook of Instrumental Techniques for Analytical Chemistry;* Prentice-Hall: New Jersey, 1997; pp 262.
14. Mitchell, M. Fundamentals and Applications of Diffuse Reflectance Infrared Fourier Transform (DRIFTS) Spectroscopy. In *Advances in Chemistry;* ACS Publications: Vol. 236, pp 351–375.
15. Gemperline, P. J.; Webber, L. D.; Cox, F. O. Raw Materials Testing Using Soft Independent Modeling of Class Analog Analysis Near-Infrared Reflectance Spectra. *Anal. Chem.* **1989,** *61* (2), 138–144.
16. Likar, M. D.; Cheng, G.; Mahajan, N.; Zhang, Z. *J. Pharm. Biomed. Anal.* **June 2011,** *55* (3), 569–573.
17. II. Introduction: General Principles, Chiral Substances. In *European Pharmacopeia;* European Directorate for the Quality of Medicines and HealthCare (EDQM): Strasbourg, 2013.
18. General Chapter 621 Chromatography. In *USP 35-NF30;* United States Pharmacopeial Convention: Rockville, MD, 2012.
19. General Chapter 191 Identification Tests–General. In *USP 35-NF 30;* United States Pharmacopeial Convention: Rockville, MD, 2012.

CHAPTER

Assay and impurities

Christopher M. Riley[*], **Shelley R. Rabel Riley**[*,†], **Robyn L. Phelps**[‡]

[*] *Riley and Rabel Consulting Services, LLC, Maryville, MO, USA,*
[†] *Department of Natural Sciences, Northwest Missouri State University, Maryville, MO, USA,* [‡] *PharmAdvance Consulting, Inc., Sequim, WA, USA*

CHAPTER OUTLINE

Specification of Drug Substances and Products. http://dx.doi.org/10.1016/B978-0-08-098350-9.00006-0

6.1 INTRODUCTION

Collectively, assay and the analysis of the impurity profile[1–6] characterize the overall purity of the drug substance and the drug product. Consistent with the general principles outlined in ICH Q3A (R2)[7] and Q3B (R2),[8] this chapter deals with small organic molecules present as active pharmaceutical ingredients in pharmaceutical drug substances and drug products.

6.2 ASSAY

According to ICH Q6A[9] and Q6B,[10] assay measures the content of the active ingredient in the drug substance or drug product. The term strength is also commonly used to define the content of the active ingredient in a drug product.

6.2.1 Drug substance[i]

The content of the active ingredient in the drug substance (A_i) may be expressed as a percentage of the total weight (W) of the sample assayed, i.e.

$$\text{Assay} = \frac{A_i}{W} \times 100\%. \tag{6.1}$$

The total sample is composed of the active ingredient plus organic impurities (previously known as "related compounds"), residual solvents[11] (see Chapter 7), inorganic impurities[12] (Chapter 8), water (Chapter 11), and other contaminants. Although the assay value may be expressed according to Eqn (6.1), it is more common to express assay on a solvent-free, anhydrous basis. This may be achieved in one of two ways. In the first case, the sample (and the reference standard, if appropriate) is dried to remove the residual solvents and any water. If the drug substance is unstable to heat, the result obtained according to Eqn (6.1) is corrected for the calculated amounts of residual solvents (Chapter 7) and the water (moisture, Chapter 11).

6.2.2 Drug product

The assay value for solid dosage forms is generally expressed as the amount of drug substance in a unit dosage form (e.g. 2 g/tablet). For liquids or semisolids, the assay value is generally expressed as a

[i]In some ICH Guidelines the term Active Pharmaceutical Ingredient (API) is used instead of Drug Substance. The two terms can be considered interchangeable.

percentage of the sample assayed (i.e. Eqn (6.1)). For the purpose of the drug product specification, the assay value for all types of dosage forms is generally expressed as a percentage of the label claim (e.g. 90.0–110.0%).

6.2.3 Analytical techniques and calculations

The analytical techniques for assay determination may be divided into two: (1) those employing specific methods (e.g. HPLC), and (2) those employing nonspecific methods (e.g. titration). Nonspecific methods, such as titrations, may be considered absolute methods since they use calibration standards that can be traced back to National Institute of Standards and Technology (NIST), or appropriate pharmacopeia. However, nonspecific methods, by definition, are susceptible to interference from related compounds (especially impurities and excipients) that can lead to the potential for bias in the assay value. Therefore, for complete characterization of purity, nonspecific assay methods must be supported with a specific, stability indicating method, such as HPLC, for the determination of impurities.

Stability-indicating HPLC methods require the use of an external reference standard that is generally a highly purified batch of drug substance that has been fully characterized and a purity value assigned. It should be noted that if a calibrated reference standard is in short supply, an "in-house" reference standard may be used, provided its purity has been established relative to the calibrated reference standard. For neutral compounds, the calculation of assay using HPLC may be performed by comparison of the responses for the drug in the sample with a calibration curve. Alternatively, a very common approach (and the one used by the pharmacopeias) is to use a single external standard (Eqn (6.2)).

$$\text{Assay} = \frac{R_i \times C_s \times \text{DF}}{R_s \times C_i \times P} \times 100\%, \tag{6.2}$$

Where R_i is the response of the drug substance in the sample (peak height or area), C_s is the concentration of the drug in the standard solution, DF is the dilution factor, R_s is the response of the reference standard, C_i is the nominal concentration drug in the sample and P is the purity of the reference standard. Although often referred to as a "single-point calibration", strictly speaking this is a two-point calibration method since it assumes a linear relationship between the response for the standard and the origin. Therefore, it is very important during the method-validation exercise that the linearity of the relationship between the response and concentration be established, and that the intercept is not statistically different from zero.

When the assay for the drug substance is calculated according to Eqn (6.2) without correction for solvents and water, the value is sometimes referred to as the "use-as-value" since it represents the content of the drug relative to the total content of the sample. This value is particularly useful for the calculation of the amount of drug substance to add to a formulation. As discussed previously, for the purposes of the drug substance specification, the assay value is corrected for the presence of residual solvents and water and reported as the solvent free, anhydrous value. If the samples and/or the reference standards were dried prior to analysis then the values of the residual solvents and water should be subtracted from the solvent free, anhydrous assay value to give the use-as-value.

If the drug is a salt of a weak acid or a weak base, some modifications of the basic equation for the calculation of assay (Eqn (6.2)) may be necessary. No modification of the calculation is necessary if the reference standard is the same salt form as the drug substance. However, correction for the differences in molecular weights of the drug substance and reference standard are necessary if the salt forms of the drug substance and the reference standard are different. Similar modifications in the calculations are also required if the drug substance and the reference standard are different solvates or hydrates. (Such modifications to correct for solvates or hydrates may be unnecessary in the case of drug substance if the reference standard and the drug substance are heated prior to analysis to remove the residual solvents and water). In addition to modification of the calculations, an assay should also be performed on the counter-ion for drug substances that are salts, and limits for the counter-ion included in the specification. The general principles and calculations for the assay of counter-ions are the same as those for the drug substance itself.

6.2.4 Validation of methods for assay and impurities

The original ICH Guidelines on Validation of Analytical Procedures comprised two documents, Q2A Text (the parent guideline) and Q2B Methodology. In 2005, the parent guideline Q2 (R1) was renamed Validation of Analytical Procedures: Text and Methodology, and the Q2B guideline on methodology was incorporated into the parent guideline.[13] The ICH Guideline on Validation of Analytical Procedures is most readily applied to assay and impurities methods because it specifically covers those methods (notably chromatographic methods). Therefore, this section contains a brief discussion of the provisions of the guideline. These, together with other approaches are also addressed in Chapters 3 and 4.

The ICH validation guideline covers:

- Identification tests
- Quantitative tests for impurities content
- Limit tests for control of impurities
- Quantitative tests for the active moiety in samples of the drug substance and the drug product assay or other selected component(s) in the drug product (e.g. antioxidants and preservatives)

Although not the specific focus of the guideline, methods for quantification of the active drug substance in dissolution media are also covered by this guideline since the assay methods are often used for the analysis of the dissolution samples (see Chapter 12 for more details). Although the ICH guideline was designed primarily for assay, impurities and identification methods, the general principles have been adopted for many other tests, and are discussed throughout the present publication. It is convenient to discuss the validation of assay and impurity methods at the same time because they frequently use the same chromatographic conditions, or often are combined into a single method. If the assay and impurities determinations are combined into a single method, the validation of both can also be combined into a single exercise. If the methods are different, then the two are validated separately. Characteristics to consider in a method validation exercise include:

- Accuracy
- Precision
  - Repeatability
  - Intermediate Precision
- Specificity
- Detection Limit (DL)

- Quantitation Limit (QL)
- Linearity
- Range

The tests to be validated for assay and impurities (both limit tests and quantitative tests) are provided in Table 6.1 and recommended ranges for validation of methods are shown in Table 6.2. Not included in the above list or in Table 6.1 are robustness and reproducibility, which, for completeness, are also discussed in this chapter as well as in more detail in Chapters 3 and 4.

6.2.4.1 Accuracy

Both the accuracy and the precision of an analytical method are described by relative errors. Accuracy describes the deviation from the expected result or true value. The accuracy of a drug substance assay may be determined in one of the two ways. The first involves comparison with a reference standard of known purity. The second involves comparison of the results obtained by the method with the results obtained by a well-characterized method of known accuracy. The accuracy of a drug product assay may also be determined in one of the two ways similar to those used for the drug substance. The first method involves application of the method to synthetic mixtures (e.g. a placebo) to which known quantities of the drug substance have been added. If it is impossible to obtain all the components of the formulation, it may be acceptable to either add known quantities of the drug substance to the drug product or compare the results obtained by the method with the results obtained by a well-characterized method of known accuracy. The accuracy of the assay methods may also be inferred from the precision, linearity and specificity.

The accuracy of quantitative impurity methods is determined by analyzing samples of drug substance or drug product spiked with known amounts of impurities. If authentic samples of the impurity

Table 6.1 Validation Elements for Assay and Impurities

		Impurities	
Analytical Procedure	**Assay***	**Quantitative**	**Limit Test**
Accuracy	+	+	−
Precision			
Repeatability	+	+	−
Intermediate precision	+†	+†	−
Specificity‡	+	+	+
Detection limit (DL)	−	−§	+
Quantitation limit	−	+	−
Linearity	+	+	−
Range	+	+	−

*Includes measurement of active ingredient in a dissolution method.
†In case where reproducibility has been performed, intermediate precision is not needed.
‡Lack of specificity of one analytical method could be compensated for by other supporting analytical procedures.
§May be needed in some cases.

Table 6.2 Recommended Maximum Ranges for Assay, Impurity Determination and Related Methods

	Range*	
Analytical Procedure	**Lower Limit**	**Upper Limit**
Assay	80%	120%
Content uniformity	70%	130%
Dissolution testing	−20% of lower value	+20% of upper value
Impurities	Reporting limit	120% of specification limit

**Relative to test concentration.*

are unavailable, then it is acceptable to compare the results obtained with the method to those obtained by another well-characterized method. In the latter case, it is acceptable to assume the same response factor of the drug substance in both methods.

For the determination of the accuracy of assay and impurities, ICH recommends a minimum of nine determinations covering the range of the procedure—typically three concentrations with three replicates at each concentration.

6.2.4.2 Precision

Four types of precision determinations should be considered depending on the phase of development. In the early development, repeatability and to a lesser extent intermediate precision should be considered. In the later development, robustness and reproducibility should also be considered. For each type of precision, the data should be represented by the use of the relative error term, the relative standard deviation (RSD) also known as the coefficient of variation (CV):

$$\mathrm{RSD} = \frac{s}{\bar{x}} \times 100\%, \tag{6.3}$$

Where s is the standard deviation and $\bar{x}$ is the mean of the determination.

6.2.4.2.1 Repeatability

Repeatability should be determined for at least nine determinations covering the specified range of the procedure—typically three determinations at each of three different concentrations. Alternatively, at least six determinations should be made at the 100% (or target) concentration. Note: this experiment can be combined with the accuracy experiment described above.

Although not explicitly required for method validation, the precision of the analytical system and the precision of the method (repeatability) can be calculated separately. This is because the former provides information on errors associated with the instrumentation and the latter provides information on the complete method. The difference between the two generally arises from sample preparation. For example, in HPLC, the system precision may be determined from repetitive injections of the sample solution. The system RSD may be assumed to be a function of the random errors arising from the column, the injector, the detector and the integrator. Reasonable estimates of the RSD attributable to these components might be:

$$\mathrm{RSD}_{\mathrm{column}} = 0.1\%$$
$$\mathrm{RSD}_{\mathrm{injector}} = 0.3\%$$

$RSD_{detector} = 0.3\%$
$RSD_{integrator} = 0.1\%$

In which case the system RSD is given by:

$$RSD_{system} = \sqrt{0.1^2 + 0.3^2 + 0.3^2 + 0.1^2} = 0.45. \tag{6.4}$$

The error attributable to sample preparation will vary considerably depending upon the number of steps, the complexity of each step and the concentration of the analyte. For a simple HPLC assay to determine the potency of the drug substance, the sample preparation might be relatively straightforward involving weighing the drug (or the reference standard), dissolving in a suitable solvent and adjusting to volume. In this case, the RSD for the sample preparation (RSD_{prep}) might be in the range of 1.0%. The method RSD, or the repeatability is then given by:

$$RSD_{method} = \text{Repeatability} = \sqrt{1.0^2 + 0.1^2 + 0.3^2 + 0.3^2 + 0.1^2} = 1.13\%. \tag{6.5}$$

The trace analysis of degradants in complex drug formulations may be more complicated and involve multiple extraction steps. The precision is further eroded by the fact the concentration of the impurities will be much lower than the concentration of the drug substance itself. Therefore, the RSD for the sample preparation might be greater than 5%. In this case, the repeatability is given by:

$$RSD_{method} = \text{Repeatability} = \sqrt{5.0^2 + 0.1^2 + 0.3^2 + 0.3^2 + 0.1^2} = 5.02\%. \tag{6.6}$$

Equations (6.4)–(6.6) illustrate a very important point that overall random error associated with an analytical method is dominated by the least precise step or component. Therefore, measures designed to improve the repeatability should always be directed toward the step or component having the highest degree of random errors, which in the examples shown here is a sample preparation.

6.2.4.2.2 Intermediate precision

Intermediate precision is established by demonstrating the effects of random events on the precision of the method. Such random events include days, analysts and instrumentation. The extent to which the intermediate precision of a method should be established depends upon the phase of development. For example, limited or no assessment of intermediate precision may be required for methods to support Phase 1 clinical trials. In contrast, complete assessment of intermediate precision should be conducted and the results presented in a marketing application (NDA, MAA). It may not be necessary to change all the parameters separately, and a matrixed design of experiments (DOE) approach is recommended (see Chapter 3 for more details).

6.2.4.2.3 Robustness

Similar in concept to intermediate precision, robustness is the reliability of the analysis in response to deliberate changes in method parameters. As with intermediate precision, it may not be necessary to change all the parameters separately, and a matrixed design of experiments (DOE) approach is recommended (see Chapter 3 for more details). Examples of deliberate variations include:

- Age of analytical solutions and reagents
- Extraction Time
- Chromatographic parameters (pH, mobile phase composition, flow rate, columns, temperature)

6.2.4.2.4 Reproducibility

The reproducibility of the method is generally not discussed in marketing applications because it is usually assessed by means of interlaboratory trials conducted with the aim of standardization of the method—for example, for inclusion in a pharmacopeia (see Chapter 14). Reproducibility may also be assessed as part of the method transfer activity from a development laboratory to a routine quality control laboratory. In the latter case, the results should be included in the Method Transfer Report.

6.2.4.3 Specificity (selectivity)

The specificity of chromatographic methods is normally demonstrated by spiking experiments in which the drug substance and all known or suspected potential interferences are added to the sample matrix at appropriate concentrations. The spiked sample is then compared with representative blanks (e.g. a placebo and a solvent blank). The elution times of all significant compounds are recorded and the resolution of the critical pair or pair of peaks is determined for inclusion in the system suitability check. One of the main challenges in the determination of specificity arises from the presence of unknown interferences that are not present in the blank or placebo sample. The assay for both drug substance and drug product should be "stability indicating"; that is, they should be capable of detecting decreases in the concentration of the active ingredient as well as the appearance of the degradants—ideally in the same chromatographic run. Thus, peak homogeneity plays a key role in the assessment of chromatographic specificity. Peak homogeneity may be demonstrated by the use of in-line detectors such as a UV–visible diode array detector (DAD) or a mass spectrometer (See also Section 6.3.5). An alternative approach is the use of an "orthogonal" technique in which the results obtained by the method are compared with the results obtained either by a well-characterized method or a method that relies on a completely different mode of separation (e.g. HPLC vs. capillary electrophoresis).

Another challenge arises in the validation of assay and impurity methods when authentic standards are not available or the potential impurities (especially degradants) are unknown. In this case, the validation experiments may include comparison with a well-characterized method. If an alternative method is not available, stress testing of the drug substance itself, as well as the drug product (if feasible) under a variety of conditions (e.g. heat, light, acid/base hydrolysis and oxidation), should be conducted. Figure 6.1 shows a complex separation of esomeprazole magnesium from seven known impurities and several forced degradation peaks produced by acid hydrolysis or peroxide oxidation.

The question often arises as to how much stressed degradation is sufficient to demonstrate specificity. It is recommended here that the drug be degraded where possible to 90% of its original concentration. This is the best way to produce a degradation profile that is consistent with 10% degradation of the active ingredient. Stressing the samples to a greater extent increases the chances of producing secondary degradants that are not likely to be produced in real time stability studies.

6.2.4.4 Detection limit (DL)

Although frequently included in validation exercises for assay or quantitative impurity methods, the detection limit is applicable only to limit tests for impurities.[13] It should also be noted that if a limit test is employed, the result should be reported as pass/fail and not reported quantitatively. The DL of a chromatographic method is generally calculated from the signal-to-noise (S/N) ratio. A S/N ratio of 3:1 is generally accepted for estimating the DL. Other methods of calculating the DL are based on the

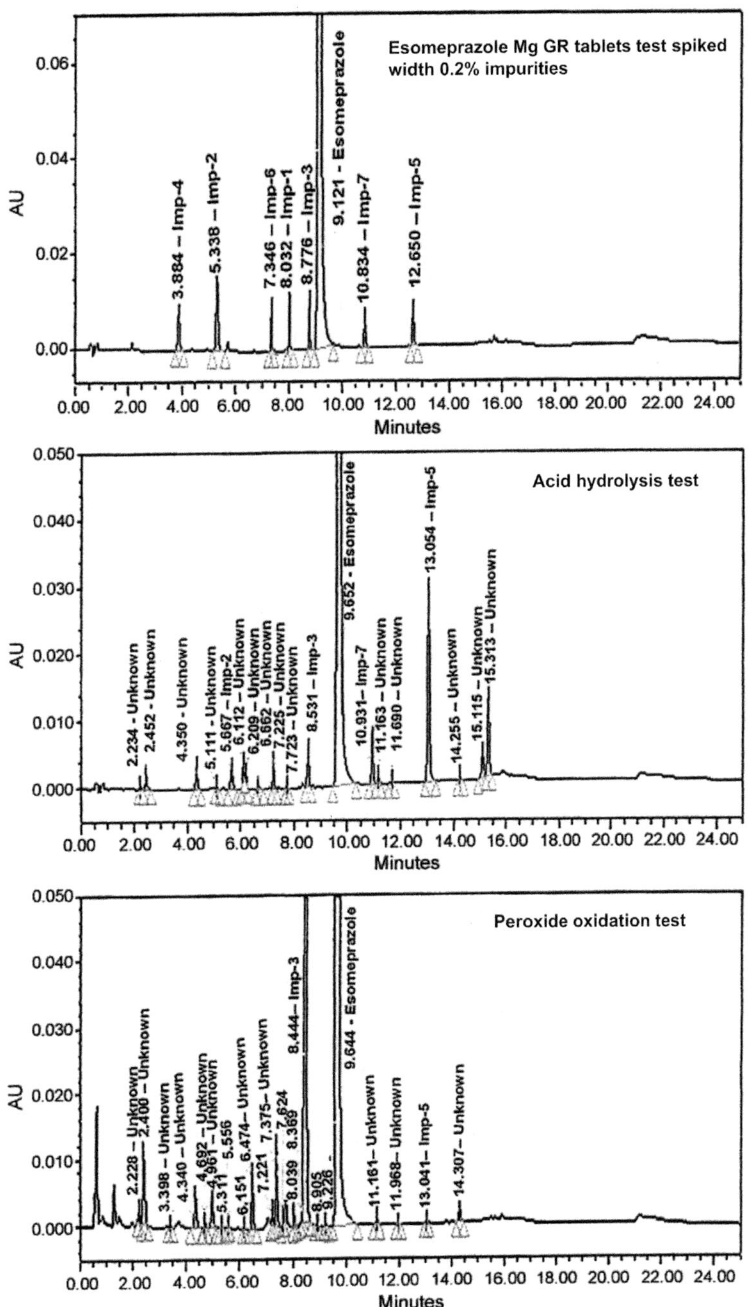

FIGURE 6.1

Separation of esomeprazole magnesium from the seven known impurities and forced degradation samples (acid hydrolysis and peroxide oxidation).

Source: Reprinted with permission from Nalwade, S. U; Reddy, V. R; Rao, D. D; Morisetti, N. K. J. Pharm. Biomed. Anal. 2012, 57, 109–114. Copyright Elsevier 2011.

standard deviation of the response and the slope. Readers interested in learning more about those methods of calculating DL are referred to the Q2 (R1) guideline.[13]

6.2.4.5 Quantitation limit (QL)

The quantitation limit is the lowest value that may be reliably reported with a numerical value and is only applicable to quantitative methods for impurities. For the reportable results to be reliable, the QL must be equal to or less than the Reporting Threshold (RT) (see Section 6.3.2.1). The QL of a chromatographic method is generally calculated from the signal-to-noise (S/N) ratio. A S/N ratio of 10:1 is generally accepted for estimating the QL. Other methods of calculating the QL are based on the standard deviation of the response and the slope. Readers interested in learning more about these methods of calculating QL are referred to the Q2 (R1) guideline.[13]

6.2.4.6 Range and linearity

The range and linearity of the method are generally established in the same validation experiment. The range of the method depends upon the application, but must span the expected concentrations in the sample (i.e. extrapolation outside the linear range is to be avoided). The maximum specified ranges for assay, impurity and related methods are summarized in Table 6.2. When a method is to be used for both assay and impurities and the drug substance itself is used as a reference, two concentration ranges should be established: one for the drug substance and one for the impurities.

Ordinarily, in the absence of chromatographic artifacts, such as solvent incompatibility or column overload, the response (peak height or peak area) of a chromatographic method is linearly related to the amount of analyte injected. Thus the concentration of the sample and the standard injected is adjusted during sample preparation so that it falls within the linear range. Generally, five or six points are used to establish linearity of the method (e.g. for assay—80, 90, 100, 110 and 120% of the nominal test solution concentration). Linearity for impurities ($n = 5$ or 6) should be established from the reporting limit to 120% of the limit in the specification. The assay method is frequently used for content uniformity and dissolution testing. In these cases, linearity should be established for content uniformity over 70–130% of the test concentration and for dissolution ±20% over the specified range.

Linearity should be confirmed by visual inspection of a plot of response versus concentration and by statistical analysis. In addition to the use of the correlation coefficient (r) to establish linearity, it is useful to analyze the residuals (the relative difference between calculated versus actual concentration) that should be normally distributed. This is because the correlation coefficient measures both curvature and random errors about the line (see Chapter 3). Linearity should always be established even if the method is to be used for a single-point calibration because a nonlinear response relationship may introduce bias in the determination. Another source of bias in the use of single-point calibrations is a nonzero intercept that can be indicative of interference in the peak of interest. Some analytical methods, such bioassays, are inherently nonlinear, in which case a calibration curve is essential. For ease of analysis, the calibration data may either be fitted to a nonlinear equation or transposed to a linear relationship.

6.2.5 Specification setting for assay

By convention, the acceptance criteria for assay (potency) in the release specification are typically 95–105% and the typical acceptance criteria in the stability specification are 90–110%. These conventional limits are typically used throughout development; however, wider limits may be justified, for example, in the assay of biologicals. Tightening the release specification decreases the likelihood that the product will fail the stability specification during storage. However, a release specification that is too tight will result in more batches failing at release. Conversely, too loose a release specification can result in a greater number of batches failing the stability specification during storage.

The use of release and stability specifications is one area in ICH Q6A that is not harmonized. In the US, a product must meet a single specification (typically assay = 90.0–110.0%) at release and throughout the shelf life. By contrast, in the EU, separate specifications are required at release and on stability. However, in the absence of a regulatory release specification, tighter "in-house" release specifications are generally used to ensure that the product will meet the regulatory specification throughout the shelf life. If this is done, it should be noted that the regulators may treat the "in-house" release specifications just like any specification when determining if a reported value is out-of-specification (OOS) and requires a formal investigation.

6.3 IMPURITIES

6.3.1 Drug substance

According to ICH Q3A (R2), impurities in the drug substances include organic impurities, inorganic impurities and residual solvents. Residual solvents are covered by Q3C (R5)[11] and discussed in Chapter 7, and inorganic impurities (elemental impurities) are covered by ICH Q3D[12] and discussed in Chapter 8. Organic impurities in the drug substance include starting materials, by-products, intermediates, degradation products, reagents, ligands and catalysts. Previously registered drugs are specifically excluded from the guidelines; however, the guidelines are generally applicable to any new application for or amendments to registrations of previously registered drugs. Although, the ICH guidelines do not apply to clinical-trial materials, they do provide a general framework for the control of impurities in development drugs—this aspect is discussed in more detail in Section 6.3.7.1. Extraneous contaminants are not covered by the guidelines and are considered to be controlled though the application of current Good Manufacturing Practices (cGMPs). Also excluded from Q3A (R2) (and Q3B (R2)) are polymorphs and enantiomers, which are both covered in Q6A and discussed in Chapters 9 and 10, respectively. Another large group of drug substances, including biologics (macromolecules) are not included in the impurities guidelines: these types of compounds are covered by Q5A and Q6B. So-called "unusually toxic compounds" are also considered outside the scope of the guidelines and, as a result of their exclusion, have received special attention over the past 10 years. Included in this class of compounds are neurotoxins and genotoxic (mutagenic) impurities, which will be discussed in more detail in Section 6.3.6.2. A final group of drug substances not specifically covered in ICH Q3A (R2) (and Q3B (R2)) and for which there is little ICH guidance include oligonucleotides, radiopharmaceuticals, herbal products and crude products of animal and plant origin. This last group of compounds was considered outside the scope of this publication.

6.3.2 Drug product

The ICH guideline Q3B (R2) defines organic impurities in the drug product as degradation products; that is degradants of the drug substance itself and reaction products of the drug substance and excipients and/or the immediate container-closure system. As with the drug substance, excluded from the guideline are clinical trial materials, previously registered drugs, biologics and other natural products, extraneous contaminants, polymorphs and enantiomers. Impurities in the excipients and leachables from the container-closure system are also excluded (see Chapter 13). Presumably, impurities in excipients included in the FDA GRAS or IIG databases are considered qualified. However, novel excipients are to be treated in the same fashion as new drug substances and are subject to the same ICH guidelines prior to approval for use in a formulation for human or animal use. The most important types of impurities excluded from the drug product guideline are impurities present in the drug substance, which generally need not be monitored or specified in the drug product unless they are also degradation products. This exclusion assumes that impurities in the drug substance are adequately addressed by the provisions in Q3A (R2). However, this does not mean that impurities originating in the drug substance can be ignored in the drug product. For example, a stability-indicating HPLC method for the drug product must be capable of separating the drug-substance impurities such that they do not interfere with the drug substance itself or with the degradants. Early in development, it is particularly important to keep track of the drug substance impurities seen in the drug product if it is not clear whether they are just process impurities or also degradants. In this case, they need to be monitored and specified in the drug substance and in the drug product. In this respect, the use of the same HPLC conditions for both the drug substance and drug product may be considered (where practical) to facilitate tracking of impurities.

6.3.2.1 Organic impurities and ICH thresholds

ICH Q3A (R2) and Q3B (R2) introduce the concepts of identification, qualification and reporting thresholds for impurities in the drug substance and the drug product, respectively (Tables 6.3 and 6.4). It is important to note that a threshold is a limit above which some action must be taken. For example, if the reporting threshold is 0.05%, then the first value to be reported (to the regulatory authorities) is 0.06%. A notable feature of these thresholds is that for a given daily dose of drug, the threshold values are higher for the drug product than for the drug substance. Presumably, the rationale for this difference is that impurities in the drug substance are considered generally more reactive and thus, potentially, more toxic. However, degradation of drugs can give rise to exceptionally toxic impurities, such as aldehydes, α,β-unsaturated carbonyls, epoxides, and polyaromatic hydrocarbons,[1] which are potentially genotoxic and require special consideration.

Table 6.3 ICH Thresholds for Impurities in Drug Substance (ICH Q3A (R2))

Maximum Daily Dose*	Reporting Threshold†,‡	Identification Threshold‡	Qualification Threshold‡
≤2 g/day	0.05%	0.10% or 1 mg per day intake (whichever is lower)	0.10% or 1 mg per day intake (whichever is lower)
>2 g/day	0.03%	0.05%	0.05%

*The amount of drug substance administered per day.
†Higher reporting thresholds should be scientifically justified.
‡Lower thresholds can be appropriate if the impurity is unusually toxic.

Table 6.4 ICH Thresholds for Impurities in Drug Product (ICH Q3B (R2))

Maximum Daily Dose*	Threshold†,‡
Reporting thresholds	
≤1 g/day	0.1%
>1 g	0.05%
Identification thresholds	
<1 mg	1.0% or 5 μg TDI, whichever is lower
1 mg–10 mg	0.5% or 20 μg TDI, whichever is lower
>10 mg–2 g	0.2% or 2 mg TDI, whichever is lower
>2 g	0.10%
Qualification thresholds	
<10 mg	1.0% or 50 μg TDI, whichever is lower
10 mg–100 mg	0.5% or 200 μg TDI, whichever is lower
>100 mg–2 g	0.2% or 3 mg TDI, whichever is lower
>2 g	0.15%

**The amount of drug substance administered per day.*
†Thresholds for degradation product are expressed either a percentage of the drug substance or as a total daily intake (TDI) of the degradation product. Lower thresholds can be appropriate if the degradation product is unusually toxic.
‡Higher reporting thresholds should be scientifically justified.

The thresholds for drug substance are divided into two levels depending on the maximum daily dose (MDD) of the drug substance administered (Table 6.3): (1) MDD ≤2 g/day and (2) MDD >2 g/day. All three thresholds are reduced as the MDD increases, reflecting the greater exposure of the impurity as the amount of drug administered is increased. For example, the identification threshold is reduced from 0.10% when the MDD is ≤2 g/day to 0.05% when the MDD is >2 g/day. As discussed in more detail in Section 6.3.7, it is important to note that the limit for unspecified impurities is always equal to not more (≤) than the identification threshold and that the reporting thresholds are always less than the identification thresholds. This allows the reporting of unspecified impurities in the range greater than the applicable reporting threshold but less than the identification threshold. The identification threshold (0.10%) is less than the qualification threshold when MDD ≤2 g/day, and the same when the MMD >2 g/day. The hierarchy for the reporting, identification and qualification thresholds for impurities in the drug product is more complex (Table 6.4); however, the same basic principles are followed as for drug substance. That is, for a given MDD, the reporting thresholds are less than the identification thresholds that are less than the qualification thresholds.

The common techniques used for the isolation and identification of impurities in the drug substance and drug product are discussed in detail in Section 6.3.5 and the general approaches to the (toxicologic) qualification of impurities are discussed in Section 6.3.6. Both the identification and qualification thresholds for the drug substance and the drug product contain the important caveat: "lower thresholds may be appropriate if the impurity (degradant) is unusually toxic". The qualification of "unusually toxic" impurities is described in subsequent sections.

6.3.3 "Unusually toxic" impurities

The term, "unusually toxic" impurities is generally reserved for neurotoxins and mutagens (genotoxic impurities). The absence of specific ICH guidelines for this class of compounds in the second revision of the ICH Guideline on Impurities in the early 2000s led to a series of draft regulatory guidelines and publications dealing with genotoxic and potentially genotoxic impurities. The first draft guideline (Guideline on Limits of Genotoxic Impurities) was published by the European Medicines Agency (EMA) in 2006.[14] This guideline classified genotoxic impurities as those where there are positive findings established in in vitro and in vivo genotoxicity tests that have the potential for DNA damage. The guideline distinguishes between genotoxic impurities with sufficient evidence for a threshold-related mechanism and those without sufficient evidence for a thresholded mechanism. For impurities where there is evidence of a threshold-related mechanism, a permitted daily exposure (PDE) is established using the principles outlined in Q3C (Residual Solvents) (see also Chapter 7).

The EMA draft guideline was followed by an industry paper by Müller et al.[15] in 2006, a draft FDA Guidance[16] in 2008 and a Questions and Answers document by the EMA[17] in 2010. These regulatory guidelines and the industry proposal are discussed in the following sections.

6.3.3.1 Threshold of toxicological concern (TTC)

The principle of "as low as reasonably possible" (ALARP) was introduced for non-thresholded genotoxic impurities in situations where presence of the impurity cannot be avoided.[14] For this class of compounds, a threshold of toxicological concern (TTC) of 1.5 μg/day was also introduced as defining a common exposure level for an unstudied chemical that will not pose an additional risk of significant carcinogenicity.[18–21] This TTC value is based on the threshold of regulatory concern, which was first introduced by the FDA for food-contact materials and established by analysis of 343 carcinogens and then confirmed by an expanded database to more than 700 carcinogens.[21] The original analysis of those carcinogens gave rise to a daily limit of 1.5 μg/day, which, assuming a lifetime exposure, would result in an increased incidence of cancer of 1 in 10^6 (i.e. a virtually safe dose). Subsequent analysis of more potent carcinogens led to a ten-fold reduction in the threshold of regulatory concern (0.15 μg/day). However, the EMA maintained that an exposure of 1.5 μg/day, which corresponds to a lifetime risk of cancer of 1 in 10^5, was acceptable for pharmaceuticals where some benefit to the patient can be expected.

6.3.3.2 Staged TTC (sTTC)

Unlike ordinary impurities, the permitted concentrations of genotoxic impurities are determined by the daily dose of the drug substance on a sliding scale (Eqn (6.7), which can itself present a logistical challenge for setting the limits for genotoxic impurities, especially in early development when the dose may not be known.

$$\text{Limit(ppm)} = \frac{1.5\ \mu\text{g}}{\text{daily dose (g)}}. \tag{6.7}$$

In addition to the logistical challenge of setting limits, a TTC of 1.5 μg/day corresponds to limits of less than 10 ppm range when the daily dose exceeds 150 mg, which can present significant challenges to analysts and process chemists concerned with controlling the levels of impurities. The EMA guideline was intended to apply to marketing authorizations; however, a concern arose within the pharmaceutical industry that the guideline might be applied by the regulators to drug development when exposures to potentially genotoxic impurities would be for short periods of time and thus higher TTC values might be acceptable.

Although the EMA guideline did acknowledge that higher TTC values might be acceptable for shorter exposure or life-threatening conditions, no further information was provided. This gap in the regulations led to an effort by a consortium of representatives from pharmaceutical companies in North America and Europe to propose alternative strategies for controlling genotoxic or potentially genotoxic impurities for short-term administration (either acute therapy or clinical trials). In the publication[15] that resulted from the group's efforts, Müller et al. proposed a three-step process to establish limits for genotoxic or potentially genotoxic impurities, as follows:

Step 1

Identification of structural alerts (Fig. 6.2) in the parent compound and impurities (both structurally identified and predicted) and classification into one of the five classes:

- Class 1: Known to be both genotoxic (mutagenic) and carcinogenic;
- Class 2: Known to be genotoxic (mutagenic), but with unknown carcinogenic potential;
- Class 3: Alerting structure (functional group) (e.g. Fig. 6.2) not present in the parent drug substance and unknown genotoxic (mutagenic) potential;
- Class 4: Alerting structure (functional group) present in the parent drug substance; and
- Class 5: No alerting structure (functional group) or indication of genotoxic potential.

Figure 6.2 gives examples of the most common structural alerts that might be found in impurities in the drug substance or the drug product. In practice, computer programs such as Toxtree, Multicase or Derek are used to identify structural alerts in impurities.

Step 2

Establishment of a qualification strategy based on the classification (see decision tree in Fig. 6.3)

Step3

Establishment of acceptable limits of the impurity in the drug substance based daily allowable intake (Table 6.5).

Table 6.5 shows that the staged threshold of toxicological concern (sTTC) values proposed by the industry ranged from 120 μg/day for exposures of 14 days to one month, to 1.5 μg/day for exposures of greater than 12 months. It is important to note that the sTTC approach is more conservative than the earlier EMA approach, since it assumes a life time cancer risk of 1 in 10^6 (compared with 1 in 10^5) for exposures of less than 12 months, reflecting the fact that these impurities may be administered to healthy volunteers for whom no therapeutic benefit would be realized. The sTTC concept and the values produced by Müller et al.[15] were subsequently accepted by the EMA (Table 6.5) in a Question and Answer Document.[17] The FDA guidance[16] generally accepts the principles laid out previously by the industry and EMA with some notable additions. FDA acknowledges that the sTTC values are acceptable for oral and inhaled products; however, further consideration may be necessary for ophthalmic and dermal products. The guidance also stated that the sTTC values should be reduced in the pediatric population by a factor 10 for children less than two years of age and a factor of 3 for children between the ages of two and 16.

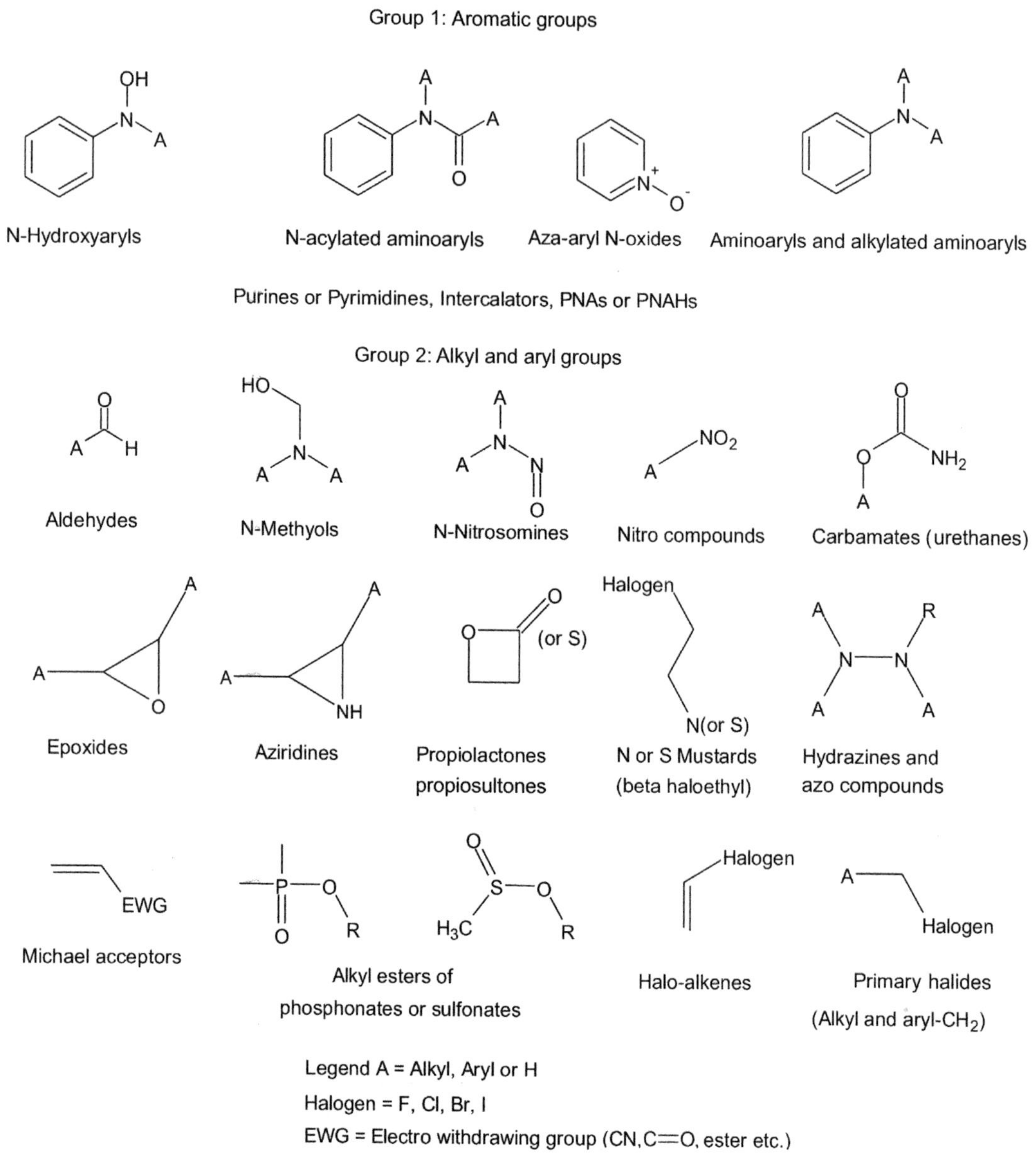

Group 1: Aromatic groups

N-Hydroxyaryls | N-acylated aminoaryls | Aza-aryl N-oxides | Aminoaryls and alkylated aminoaryls

Purines or Pyrimidines, Intercalators, PNAs or PNAHs

Group 2: Alkyl and aryl groups

Aldehydes | N-Methyols | N-Nitrosomines | Nitro compounds | Carbamates (urethanes)

Epoxides | Aziridines | Propiolactones propiosultones | N or S Mustards (beta haloethyl) | Hydrazines and azo compounds

Michael acceptors | Alkyl esters of phosphonates or sulfonates | Halo-alkenes | Primary halides (Alkyl and aryl-CH_2)

Legend A = Alkyl, Aryl or H

Halogen = F, Cl, Br, I

EWG = Electro withdrawing group (CN, C=O, ester etc.)

FIGURE 6.2

Examples of structural alerts for mutagenicity.

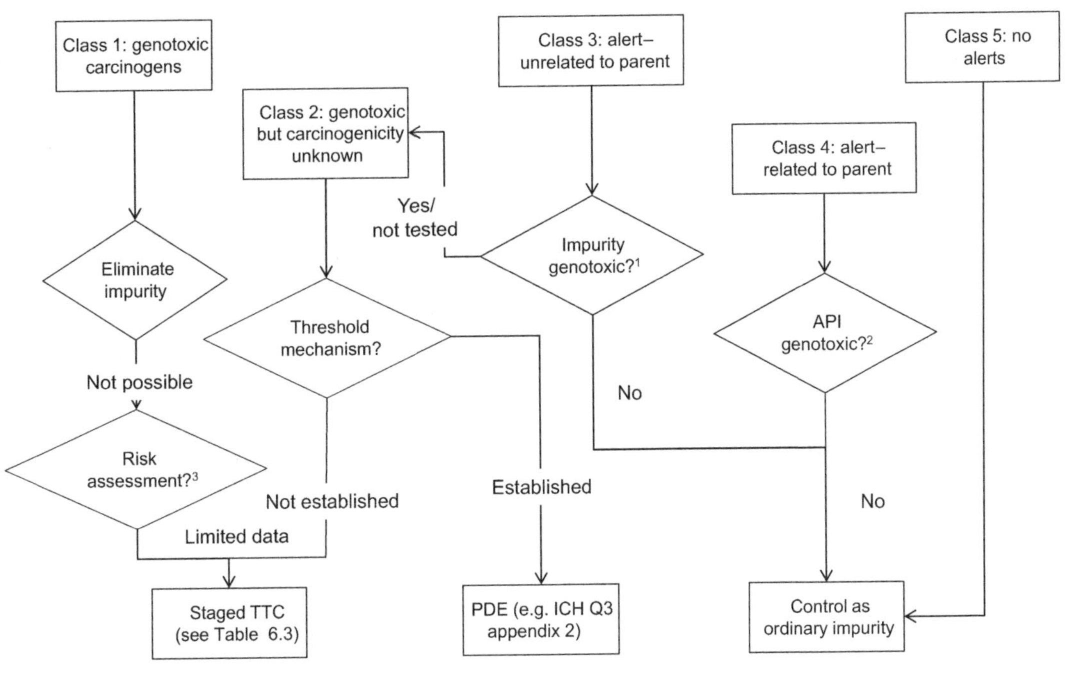

FIGURE 6.3

Decision tree for categorization, qualification and risk assessment of impurities (Note: Impurity levels may always be controlled to the staged TTC levels in Table 6.5). [1]Either tested neat or spiked into Active Pharmaceutical Ingredient (API) and tested ≤250 μg/plate. [2]If API is positive, then a risk-benefit assessment is required. [3]Quantitative risk assessment to determine acceptable daily intake.

Source: Reprinted with permission from Müller, L.; Mauthe, R. J.; Riley, C. M.; Andino, M. A.; De Antonis, D.; Beels,C.; DeGeorge, J.; DeKnaep, A. G. M.; Ellison, D.; Fagerland, J. A.; Frank, R.; Fritschel, B.; Galloway, S.; Harpur, E.; Humphrey, C. D. N.; Jacks, A. S.; Jagota, N.; Mackinnon, J.; Mohan, G.; Ness, D. K.; O'Donovan, M. R.; Smith, M. D.; Vudathala, G.; Yotti, L. Regul. Toxicol. Pharmacol. 2006, 44, 198–211. Copyright Elsevier 2005.

Table 6.5 Evolution of the Values of Staged Thresholds of Toxicological Concern

	Maximum Daily Intake and Duration of Exposure (μg/day)						
Source	**Single Dose**	**<14 Days**	**14 days–1 month**	**1–3 months**	**3–6 months**	**6–12 months**	**>12 months**
Industry	–	–	120	40	20	10	1.5
EMA	120	–	60	20	10	5	–
FDA (draft)	–	120	60	20	10	5	1.5

6.3.3.3 Comparison of ICH thresholds and sTTC values

One of the biggest challenges facing analytical chemists, process chemists and formulators is controlling the levels of genotoxic impurities in drug substances and products when the maximum daily dose is high, because this leads to very low limits (Table 6.6) that are much less than the ICH identification thresholds. The analytical challenge is best exemplified by the work of Horwitz et al. in the 1970s[22] which showed that for the analysis of aflatoxins in food, the between-laboratory relative standard deviation (RSD) is approximately related to the fraction (f_a) of the analyte in the sample by:

$$\mathrm{RSD}(\%) = 1.5\mathrm{e}^{(1-\log f_a)}. \tag{6.8}$$

Equation (6.8) predicts an RSD of 81% for the analysis of aflatoxins in food at the 1 ppm level. Clearly, this level of between-laboratory precision is unacceptable for the analysis of genotoxic impurities in drug substance or drug product and much improved analytical methodology with greater sensitivity and precision is required than that described by Horwitz et al.[22] Nevertheless, the basic principle still holds that analytical precision decreases as the analyte concentration decreases. Thus, HPLC with UV detection is frequently inadequate for the trace determination of genotoxic impurities at the ppm level and more sensitive techniques such as LC-MS, LC-EC, headspace GC, or GC-MS are frequently necessary. For example, Fig. 6.4[23] shows the complex separation of two genotoxic impurities by HPLC with UV detection and a very clean chromatogram of volatile genotoxic impurities by headspace GC-FID.

Because the limits of genotoxic impurities are dictated by daily intakes, there reaches a point when the daily dose of the drug is sufficiently low that the sTTC value exceeds the ICH identification threshold of impurity as shown in Table 6.6. Another difference between the sTTC approach for limiting impurities and the ICH guidelines occurs when the genotoxic impurity is an impurity in the drug substance and a degradant in the drug product. Generally, the ICH guidelines allow higher limits for impurities in the drug product than in the drug substance. In the case of genotoxic impurities, the limits in the drug substance and the drug product are the same. Therefore, if the genotoxic impurity is also a degradant, a lower limit than that calculated from the sTTC value may be necessary in the drug substance so that the sTTC-derived limit is not exceeded on storage of the drug product.

6.3.4 Analytical methods and calculations

Profiling of impurities in the drug substance and the drug product involves the isolation, identification and quantification in the sample of interest. Given the need to separate impurities from each other and from the active drug, HPLC with UV detection is by far the technique of choice for the quantitative determination of impurities in the drug substance and drug product because of its versatility and suitability for wide ranges of compounds. Other chromatographic and related techniques for the analysis of impurities have been reviewed extensively elsewhere and include high-performance thin layer chromatography, capillary electrophoresis, gas chromatography and electrokinetic chromatography.[4]

Two methods may be employed for the calculation of the concentration of impurities in a sample of drug substance or drug product, depending on what is known about the impurity itself.

Table 6.6 Comparison of ICH Identification Thresholds with the Concentration of Impurities Calculated from the sTTC Values proposed by FDA

	ICH Identification Threshold		Maximum Concentration of Genotoxic Impurities Calculated from the Staged TTC Values (%)					
Daily Dose (mg)	**Drug Substance**	**Drug Product**	**<14 days MDI: 120 μg**	**14 days–1 month MDI: 60 μg**	**1–3 months MDI: 20 μg**	**3–6 months MDI: 10 μg**	**6–12 months MDI: 5 μg**	**>12 months MDI: 1.5 μg**
3000	0.05	0.10	0.0040	0.0020	0.00067	0.00033	0.00017	0.00005
1000	0.10	0.2	0.012	0.0060	0.0020	0.0010	0.0005	0.00015
300	0.10	0.2	0.040	0.020	0.0067	0.0033	0.0017	0.00050
100	0.10	0.2	**0.12**	0.060	0.020	0.010	0.005	0.0015
30	0.10	0.2	**0.40**	**0.20**	0.067	0.033	0.017	0.0050
10	0.10	0.2	**1.2**	**0.60**	**0.20**	**0.10**	0.050	0.015
3	0.10	0.5	**4.0**	**2.0**	**0.67**	**0.33**	**0.17**	0.050
1	0.10	0.5	**12**	**6.0**	**2.0**	**1.0**	**0.50**	**0.15**
0.3	0.10	1.0	**40**	**20**	**6.7**	**3.3**	**1.7**	**0.50**

Values in bold are equal to or greater than the ICH Identification Threshold for the Drug Substance. Values underlined are equal to or greater than the ICH Identification Threshold for the Drug Product.

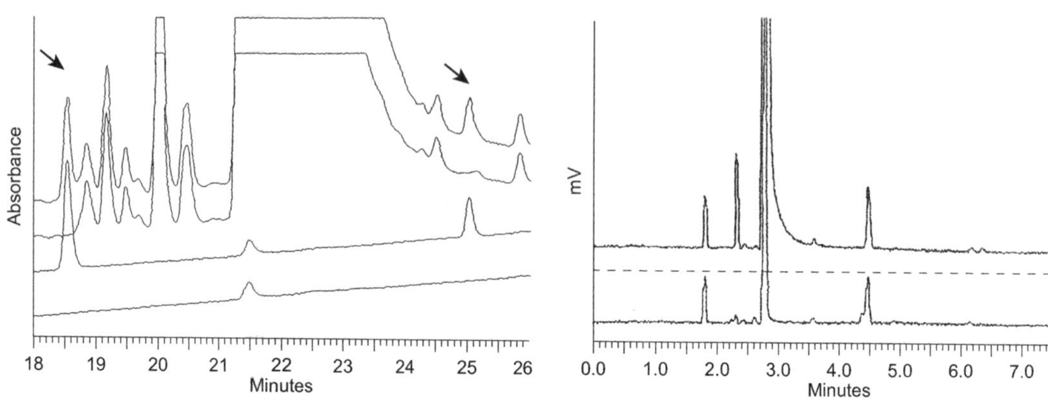

FIGURE 6.4

Examples of trace analysis of genotoxic impurities by HPLC-UV (left) and headspace GC-FID.

Source: Reprinted with permission from Pierson, D. A.; Olsen, B. A.; Robbins, D.K.; DeVries, K.M.; Vane, D.L. Org. Process Res. Dev. 2009, 13, 285–291. Copyright 2008 American Chemical Society.

6.3.4.1 Area percent

If the impurity is unidentified (or reference material is unavailable) the "area percent" ($A\%$) method can be used (for chromatographic techniques) in which the peak area of the impurity is compared with the total area of the peaks in the chromatogram, i.e.

$$A\% = \frac{A_i}{A_{tot}} \times 100\%, \tag{6.9}$$

where A_i and A_{tot} are the peak area of the impurity and the total peak area, respectively. An interesting question to ponder is exactly which peaks should be included in the total peak area. To be consistent with ICH Q3A (R2) and Q3B (R2), only the peak area of the active ingredient plus the peak areas of the reportable impurities should be included in the totals. Therefore, in the drug substance, all peaks that exceed the reporting threshold should be included in the total. In the drug product, only the peaks that are degradants in the drug product that also exceed the reporting threshold should be included in the totals. Excluded from the total peak area of the drug product are peaks equal to or less than the reporting threshold and peaks arising from impurities in the drug substance that are carried over into the drug product.

6.3.4.2 Weight percent

The impurity content can be calculated on a weight percent basis if either authentic standards are available or the response factor of the impurity relative to the drug substance is known. Calibration may be conducted either by comparison with a calibration curve or a single standard. The calibration curve should span from the reporting limit to at least 125% of the specification limit.[15] If an authentic standard of the impurity is unavailable or in short supply, the calibration curve may be prepared using the drug substance itself, provided the relative response factor is known. If a

Table 6.7 Maximum Allowable Repeatabilities (RSDs) for Single or Duplicate Determinations using a Single Calibration Solution

Acceptance Range (% of Label)	Single Determinations		Duplicate Determinations	
	95% CI	99% CI	95% CI	99% CI
98.5–101.5	0.77	0.58	1.12	0.82
97–103	1.53	1.16	2.23	1.64
95–105	2.55	1.94	3.72	2.74
90–110	5.10	3.88	7.44	5.48
85–115	7.65	5.81	11.1	8.22
75–125	12.8	9.69	18.6	13.7
50–150	252.5	19.4	37.2	27.4

single-point standard is used, the calculation is similar to that used for the assay of the drug substance, i.e.

$$\text{Impurity} = \frac{R_i \times C_s \times \text{DF} \times \text{RF}}{R_s \times C_i \times P} \times 100\%, \tag{6.10}$$

where R_i is the response of the impurity in the sample (peak height or area), C_s is the concentration of the impurity or the drug (depending on which is used as the reference) in the standard solution, DF is the dilution factor, R_s is the response of the reference standard, C_i is the nominal concentration drug in the sample, P is the purity of the reference standard (if used) and RF is the relative response factor. Early in development, samples of the impurity may be unavailable; however, it is still possible to report the results on a weight percent basis. In this case, a relative response factor of unity is assumed (this assumption should be footnoted when reporting results).

Debesis et al.[24] have calculated the maximum method repeatability and system precision (RSD_{max}) allowable, using the single-point calibration approach for HPLC assays, as a function of the acceptable assay range (LSL–USL), where LSL and USL are the lower and upper specification limits, respectively. This was accomplished by assuming that the absolute difference between the true mean (μ) and the sample mean ($\bar{x}$) is no more than 50% of the specified acceptable range.[ii] The maximum allowable repeatability (Table 6.7) is given by:

$$\text{RSD}_{\text{max}} = \frac{|\bar{x} - \mu|\sqrt{n}}{z}, \tag{6.3}$$

where n is the number of sample measurements and z is taken from tabulated values for the normal distribution, i.e. 1.96 or 2.58 for 95 or 99% confidence limits, respectively.

6.3.5 Isolation and identification of impurities and degradants

Organic impurities originating from the drug substance or drug product manufacture are typically detected during analysis of drug substance intermediates, or during method development for the drug

[ii] This approach is similar to the capability analysis described in Chapter 2.

substance and the drug product. Whereas impurities stemming from the degradation of the drug may be detected during method development, they may also be discovered during other activities such as excipient compatibility, forced degradation or formal stability studies in which stability-indicating HPLC or other chromatographic methods are utilized. Once an impurity is detected, the question arises as to whether the impurity should be isolated and the chemical structure identified. Certainly impurities that exceed the threshold levels for identification of impurities as specified in the ICH guidelines must be identified prior to submission of the marketing application. However, in early development (Phase 1), higher identification thresholds (3× ICH) have recently been proposed by the IQ Consortium.[25] In contrast to higher identification thresholds, there may be other situations that warrant the identification of impurities that may fall below the 0.1% (dose of <2 g/day) or the 0.05% threshold level (dose of >2 g/day). For example, impurities that are highly conjugated and thus exhibit strong chromophoric properties can cause discoloration of the drug substance/product at extremely low levels and may require identification to determine the origin of the contaminant and strategies to eliminate its presence. The same holds true for unusually toxic impurities or for impurities that may potentially elicit a pharmacological response, and thus pose a safety and/or efficacy concern to regulatory agencies (see Section 6.3.3). Any knowledge that may be gained regarding the source of the impurity, as well as the chemical structure, is critical to those involved in the drug substance and drug product manufacture, as it allows for optimization of processing conditions to either eliminate or minimize the levels of the impurities. Furthermore, structural information on degradants can provide mechanistic information regarding the route(s) of drug substance degradation that will be essential to formulators and processing chemists/engineers allowing for the rational design of the formulation and processing conditions to help stabilize the drug substance.

The process of identification of an impurity may be a relatively simple exercise in the case where an authentic standard of the impurity exists. Positive identification may be made by matching retention times of peaks within the chromatogram along with supporting UV-DAD (diode array detection) and LC-MS (mass spectrometry) data. Conversely, in many cases, impurity identification, particularly for trace levels of an impurity, is a very resource and labor-intensive process requiring a multidisciplinary approach that may require up to several months to complete. For this reason, each situation must be evaluated on a case-by-case basis with input from those on the development team to determine potential sources of the impurities, and whether the team should proceed with the identification process. Once a decision has been made to identify an impurity or degradation product, there are strategies and iterative processes that outline the workflow that may be undertaken to collect data and establish the structure of the impurity.[3,4] The process in many cases will be driven by the availability of analytical techniques to address the issue.

6.3.5.1 *Initial assessment of the impurity*

In those cases when there is not an authentic standard for an impurity, the first step in the identification process is to utilize any HPLC-DAD information that is already in hand as well as implementation of other hyphenated methods that can reveal spectroscopic data (LC-MS, LC-NMR) without having to first isolate the impurity. Advances in LC-MS technology have allowed this analytical technique to become more routinely used in laboratories and it is heavily relied upon in the initial assessment of impurities. Often, the original HPLC method used to initially detect and quantitate an impurity may be utilized for LC-MS evaluation, provided the mobile phase contains only volatile buffers or modifiers. Given that the spectroscopic properties of the drug substance have typically been well characterized, assessment and comparison of the spectroscopic properties of the impurity can reveal useful information to determine if the impurity is structurally related to the active drug substance. Based on this information, one may surmise whether the impurity is a degradation product, or an unrelated compound that has been introduced at some point during the processing of drug substance or product. It may be possible to propose the chemical structure of the impurity based on this information without actually isolating the impurity.

6.3.5.2 Preparative isolation of impurities

In cases where the unambiguous identification of an impurity cannot be made with hyphenated analytical methods, preparative isolation of the impurity must be performed to provide sufficient quantities of the impurity for use in additional spectroscopic experiments (mass spectrometry, NMR, and vibrational spectroscopy). A number of different chromatographic techniques may be used and scaled up to render microgram to milligram quantities of the impurity. An in-depth discussion reviewing the approaches used in the preparative isolation (e.g. HPLC, Supercritical Fluid Chromatography), experimental constraints such as solubility and stability of the impurity and method development required for scale up is presented elsewhere.[26] Once the preparative separation method has been established, fraction collection of the impurity may be prompted based on mass, UV response, elution time or other detector responses. Typically, milligram quantities would be required for spectroscopic characterization, whereas gram quantities may be desired if the isolated impurity will ultimately be used as a reference standard for future analyses.

6.3.5.3 Spectroscopic characterization of the impurity isolate

Once the impurity has been isolated, mass spectral data are obtained to ensure that the mass of the isolated substance matches that observed in the initial LC-MS analysis and that the correct fraction has indeed been collected. There are a number of different mass analyzers that may be utilized for spectral characterization,[3] all of which require very little sample for analysis.

One of the more powerful tools for structural characterization of impurities is tandem mass spectrometry (MS–MS), which consists of a series of three quadrupole mass analyzers. The first quadrupole acts as a mass filter that allows the user to select the ion of interest for further fragmentation. The selected ion is then sent to a collision cell in which additional fragmentation of the ion occurs, and finally those fragments are separated in the third quadrupole and subsequently detected. Information regarding the mass of the impurity itself along with fragmentation data can provide additional insight into the structure of the impurity, and can be compared to the fragmentation pattern of the active drug substance to determine which portions of the molecule may have been altered. Although MS–MS is amenable to use in hyphenated methods (LC-MS-MS), the analysis of larger quantities of the isolated impurity may facilitate this analysis compared to attempts to analyze trace quantities of the impurity in drug substance/product samples. In addition, time of flight (TOF) mass analyzers allow high-resolution mass spectral analysis to give the exact mass of the impurity from which an empirical formula may be calculated. The determination of the empirical formula will be particularly important in dealing with impurities that are not structurally related to the active drug substance.

Provided the availability of milligram quantities of an isolated impurity, acquisition of NMR data will ultimately prove to be the most effective analytical method for a structural determination. Simple one-dimensional NMR experiments (^{1}H and ^{13}C) may be sufficient to confirm minor structural modifications of the drug substance. However, for more complex structural alterations due to degradation processes, or for chemical structures that are unrelated to the active drug substance, additional two-dimensional homonuclear (^{1}H) or heteronuclear (^{13}C) NMR experiments may be required to gain information not only on the chemical environment and the number of protons, but perhaps more importantly the short- and long-range connectivities of protons within the chemical structure.

While reliance of structure determination is almost always based on MS and NMR data, vibrational spectroscopic data (e.g. FTIR) can also prove useful in the identification of functional groups within a molecule. The structural complexity of any impurity will dictate the type and extent of spectroscopic characterization experiments required for structure elucidation.

6.3.5.4 Case study: isolation and identification of an unusually toxic impurity

In the early stages of development, the synthetic process to manufacture drug substance is evolving and may include changes in the synthetic route as well as scale-up of the process. Consequently, the impurity profile of the drug substance may differ from lot to lot. During the development of the investigational anticancer drug XP315 (Fig. 6.5), a highly purified lot (XP315-00) was dosed in initial genotoxicity and dose-ranging studies with no observations of any immediate adverse toxicity.[27] Subsequent toxicity studies used a different lot of drug substance (XP315-01) that elicited a severe toxic reaction (anaphylaxis) in dogs immediately after IV administration. A comparison of the HPLC chromatograms of the two different lots of drug substance revealed significant differences in the impurity profile (Fig. 6.6). Given the number of new impurities in the XP315-01 lot of drug substance relative to lot XP315-00, assignment of the toxic impurity based on chromatographic data alone was not feasible. Determination of the impurity responsible for the severe toxicity, along with elucidation of its chemical structure, would however, be required to move forward in development.

The first step in addressing this issue was to determine which impurity was causing the toxic response. A logical approach to minimize the number of dosing studies was employed using preparative HPLC to isolate fractions, followed by IV dosing of the isolates. Collection of each impurity for dosing from a single series of HPLC runs was considered impractical. Therefore, the chromatogram was first divided up into three main fractions each containing several impurities, and each fraction was then dosed in dogs (Fig. 6.7). It was determined that fraction 3 resulted in a toxic response, and thus further fractionation of this sample was performed to give three new

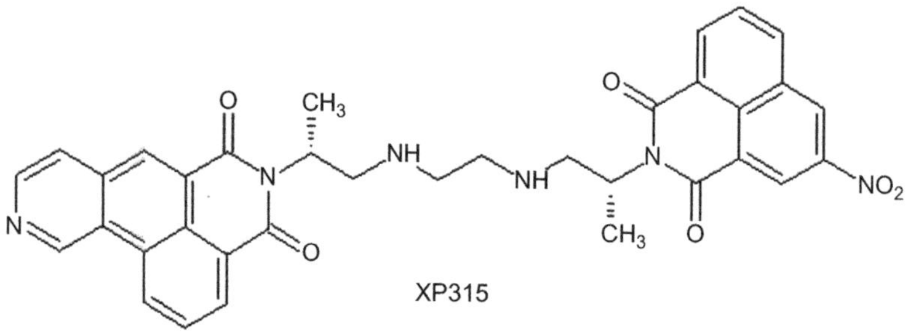

FIGURE 6.5

The chemical structure of XP315.

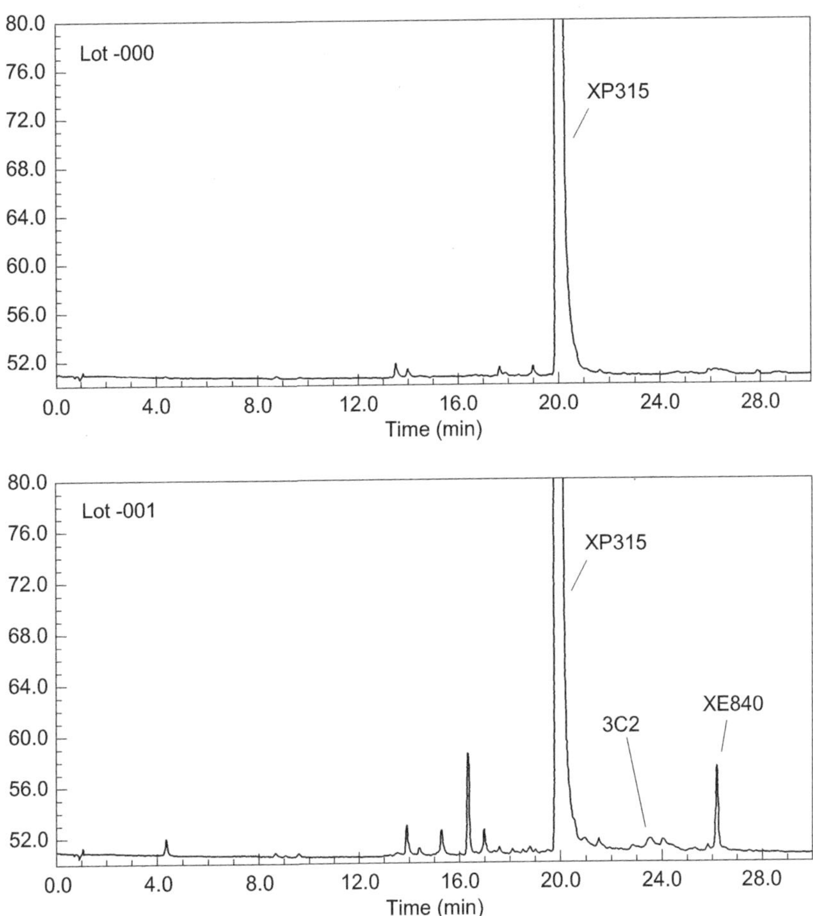

FIGURE 6.6

HPLC chromatograms showing the differences in the impurity profiles between two different lots of XP315.

Source: Reprinted with permission from Sigvardson, K. W.; Adams, S. P.; Barnes, T. B.; Blom, K. F.; Fortunak, J. M.; Haas, M. J.; Reilly, K. L.; Repta, A. J.; Nemeth, G. A. J. Pharm. Biomed. Anal. 2002, 27, 327–334. Copyright Elsevier 2002.

isolates (3A, 3B, 3C), again with each fraction containing multiple impurities. The new fractions were dosed in dogs, where fraction 3C elicited the toxic response. The impurity profile had been narrowed down to two impurities contained in fraction 3C (3C1 and 3C2). The fractions containing these two impurities were collected and again dosed to reveal that fraction 3C2 produced the toxic response soon after the IV administration was initiated. Sufficient quantities of the highly toxic impurity were collected by preparative HPLC for further characterization and structural determination. It is worth noting that the order of elution of peaks in an analytical HPLC method may very well differ from that in the preparative HPLC chromatogram; therefore, caution should be

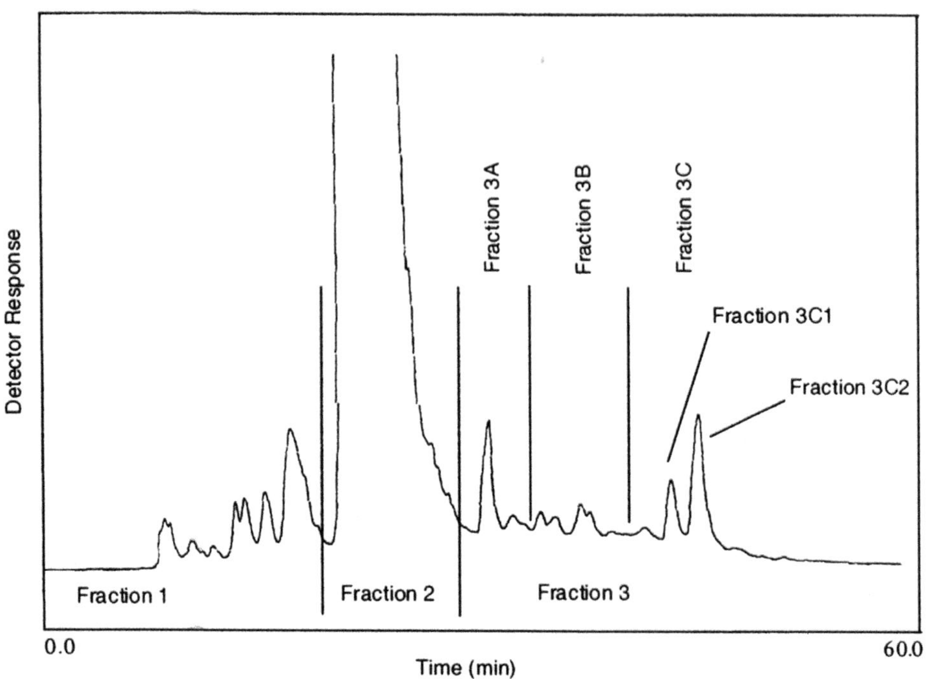

FIGURE 6.7

HPLC chromatograms showing the various fractions taken during the preparative isolation of the dimeric impurity of XP315.

Source: Reprinted with permission from Sigvardson, K.W.; Adams, S.P.; Barnes, T.B.; Blom, K.F.; Fortunak, J. M.; Haas, M. J.; Reilly, K.L.; Repta, A.J.; Nemeth, G.A. J. Pharm. Biomed. Anal. 2002, 27, 327–334 (2002). Copyright Elsevier 2002.

exercised in cross-referencing peaks between the two methods when isolating the fraction of interest. In the case of this toxic impurity, the peak was the last to elute in the preparative HPLC, whereas it eluted earlier in the analytical HPLC chromatogram based on experiments where the isolated fraction was spiked into the XP315 drug substance and subjected to the analytical HPLC separation method (Fig. 6.6).

The next step was to determine the structure of the isolated impurity, which was determined to be toxic at levels as low as 0.01% in the active drug substance. The isolated impurity was further characterized by mass spectrometry and NMR spectroscopy. Although the base peak in the positive-ion electrospray ionization (ES^+) mass spectrum was present at an m/z of 598, closer inspection of the spectrum in the 400–1400 Da range uncovered low-intensity peaks in the m/z range of 1195–1199. Based on these data, the mass spectrum results were consistent with a large molecule with a molecular weight of 1194 Da. High-resolution mass spectral data revealed an exact mass of 1196.2382 ± 0.0026 Da and a molecular formula of $C_{70}H_{58}N_{12}O_8$ for the impurity. NMR analysis utilizing gradient enhanced homonuclear chemical shift-correlated spectrum (gCOSY) as well as single- and multiple-bond gradient-enhanced heteronuclear chemical shift-correlated spectra (gHSQC

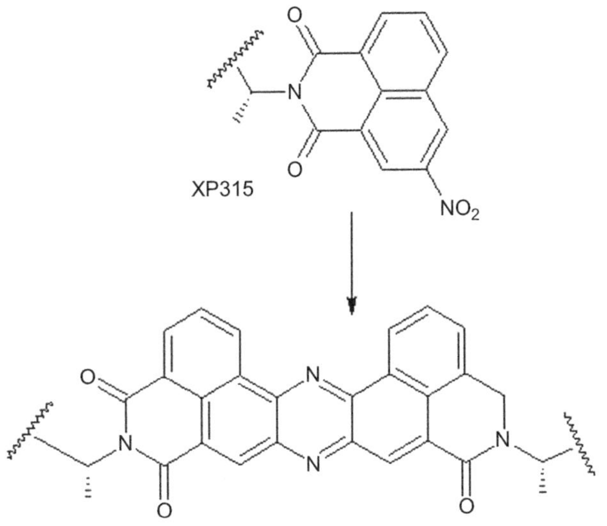

FIGURE 6.8

Polyaromatic rings in XP315 and in the dimer impurity isolated by preparative HPLC.

and gHMBC) allowed assignments of protons and carbons within the impurity, which could be compared to the assignments in the XP315 drug substance. Based on the simplicity of the NMR spectra and the large molecular weight of 1194 Da, it was concluded that the impurity must be a symmetrical dimer of XP315. NMR data along with the molecular formula suggested dimer formation via condensation of the nitro-naphthylimide groups to form the heterocylic structure linking the two XP315 molecules as shown in Fig. 6.8. Once the chemical structure of the impurity had been identified, future work was focused on elimination of the highly toxic impurity from the synthetic process.

6.3.6 Qualification of impurities and degradants

The toxicologic qualification of drug substance impurities and drug product degradants involves the acquisition and evaluation of in vitro *and* in vivo data, which collectively establish the safety of an individual impurity upon exposure to man. As discussed in the sections above, impurities in the drug substance or organic impurities (degradants) in the drug product are initially evaluated for safety in accordance with ICH Q3A (R2) and ICH Q3B (R2) guidelines. These guidelines enable categorization of impurities, e.g. reporting, identification or qualification (toxicologic evaluation), based on threshold amounts in relation to the anticipated exposure to man during therapeutic administration. While the drug substance and drug product impurity threshold limits differ (Tables 6.3 and 6.4), the approach to toxicologic safety assessment is similar and follows a common logical decision tree.

6.3.6.1 Ordinary impurities

Profiling and qualification of drug substance impurities and drug product degradants are typically focused on batches produced for registration and marketing. However, it is important to note that quantification and identification of impurities in nonclinical and clinical-trial drug batches is invaluable to the development process since these exposures can assist in the eventual overall toxicologic qualification strategy. Indeed, a discussion of the impurity profiles observed in both the nonclinical (safety) and clinical development batches is needed to support the rationale for inclusion or exclusion of impurities, and for establishing specification limits in drug substance and drug product produced by commercial processes.[9,10,28] Although impurities observed in early drug batches may be reduced or removed through subsequent improvements and refinements in manufacturing, careful documentation of batch–impurity correlations and the analytical methods utilized for detection provides for retrospective calculation of exposure margins with respect to the intended clinical dose regimen. If any impurity in a commercial drug substance can be shown to have been present at the same or higher levels in previously conducted safety and/or clinical studies, it would be considered to be adequately tested and hence qualified. Additionally, impurities that are also significant metabolites present in animal and/or human studies are generally considered to be qualified. Notably, a level of a qualified impurity higher than that present in a commercial drug substance or drug product batch can also be justified based on the actual amount of impurity administered in previous relevant safety studies.

For newly identified drug impurities or impurities for which nonclinical or clinical data are unavailable to qualify the proposed acceptance criterion (specification limit), a tiered approach to toxicologic evaluation is used. Specified impurities can be identified or unidentified in the batches produced by the commercial process. Unspecified impurities are generally limited to levels of not more than ($\leq$) the identification thresholds (drug substance and/or drug product) (Tables 6.3 and 6.4), thus only reporting is required and no further toxicologic evaluation is generally undertaken. For an identified impurity, if the level at which it is present in the commercial batch does not exceed the identification threshold, no further action is undertaken unless there is scientific evidence that it is an unusually toxic compound.

Qualification studies of specified impurities are conducted to support the proposed acceptance criteria when the qualification threshold is exceeded and previous safety data are unavailable. Additional safety testing is conducted in a step-wise fashion for both drug substance and drug product. The studies considered appropriate to qualify an impurity depend on the patient population, daily dose, administration route and duration of clinical exposure. In general, drugs intended for limited dose administration (low doses, short durations) may be considered to be qualified by genotoxicity assessments alone while therapeutic indications requiring extended or high drug exposures, or ones for which drug class effects indicate potential toxicity, may require conduct of general toxicity studies. In either case, studies are normally performed by comparing the new drug substance (or drug product produced from the new drug substance) containing the impurities to be controlled to previously qualified material levels. However, studies of isolated impurities may be warranted in some instances, particularly in in vitro evaluations of genotoxic potential.

Whether an impurity is present at levels below the ICH qualification thresholds or higher specification limits are desired, the initial approach to assessment of genotoxic potential is the same.[28,29] The initial assessment is focused on evaluation in a comprehensive structure-based assessment using structure-activity relationships (SARs), and in an in silico computational toxicology assessment (e.g. MultiCase MC4PC, Derek, Toxtree).[30,31] These assessments are designed to identify impurities that may be DNA-reactive mutagens with carcinogenic potential (SAR) or that possess structural features associated with known mutagenic mechanisms of action (in silico models). Upon completion of the initial structure-based assessment, compounds that are shown not to contain a structural alert are then generally subjected to a second review for confirmation of the in silico findings. This review can consist of either a second in silico study using a different methodology to ensure that no alerts are found using an alternative approach, or of a literature review for similar compounds for which mutagenicity data might imply that the impurity possesses or does not possess mutagenic potential. If structure-based concerns are not identified during the secondary review, it is considered sufficient to conclude that the impurity is of no concern and conduct of in vitro or in vivo genotoxicity studies are not normally suggested. In contrast, if positive findings are identified in the initial in silico assessment, the recommended follow-up actions differ as the compound would be considered to possess genotoxic potential. The recommended course of action for potentially genotoxic impurities are discussed separately in Section 6.3.6.2 below.

For impurities that were not demonstrated to be genotoxic in the initial in silico screenings, the progression through additional toxicologic assessment becomes dependent upon the desired limits of specification in commercial drug substance and/or drug product. For specification limits that do not exceed qualification thresholds per the ICH Q3A (R2) and ICH Q3B (R2) guidelines, the impurity is considered to be qualified and no additional toxicity assessments are normally performed. However, for specification levels that exceed the qualification thresholds in the relevant ICH guidelines, further qualification studies are required. These studies may include in vitro (e.g. Ames, Chromosomal Aberration assays) and in vivo (e.g. Micronucleus assay) to detect, at a minimum, potential point mutations and chromosomal effects in biologic systems.[32] Additionally, general in vivo toxicity studies may be considered to assess non-genotoxic effects.

In vitro and in vivo genotoxicity studies are most appropriately conducted with the impurity in isolation. However, when synthesis of sufficient amounts is infeasible, drug substance containing or spiked with the impurity can be utilized. For general in vivo toxicity studies, the design should allow for direct comparison of unqualified to qualified (previously evaluated) material. While these studies are most often conducted in a single species (normally rodents), they should be performed in the species most likely to detect the toxicity of the impurity (e.g. toxicities based on class effects or other literature information). The study should be designed for dosing by the intended clinical route of exposure with repeated administration from two weeks to three months in duration, depending on the intended therapeutic regimen. In some cases, for example, a single therapeutic dose drug, a single-dose toxicity study may be considered to be appropriate.

Following review of the results of the genotoxicity and toxicity studies, if adverse effects are identified, then more specific toxicity evaluations may be undertaken to further assess target systems. In practice, if impurity-related and clinically relevant adverse effects are observed at the tested level in any study, the impurity is not considered to be qualified at that level, and the simpler course of action is to reduce the impurity to a safe level through changes in the manufacturing process or modification of the formulation.

6.3.6.2 Genotoxic impurities

The existence of the two draft regulatory documents and the industry proposal containing different proposals led, logically, to the establishment of an ICH (M7) Expert Working Group (EWG) on Assessment and Control of DNA Reactive (Mutagenic) Impurities in Pharmaceuticals to Limit

Potential Carcinogenic Risk.[29] At the time of the writing of this chapter, ICH M7 was at Step 2, precluding a detailed discussion. Although still at Step 2, the toxicologic qualification of potentially genotoxic impurities follows the principles of the draft guidance. The recommendations for manufacturing control of mutagenic impurities were discussed in Section 6.3.6.2; thus, only the toxicologic risk assessment is considered here.

As introduced in the discussion of ordinary impurities, identification of structural alerts (positive finding) in the in silico SAR and computational assessments, or identification of genotoxic or carcinogenic potential in the available literature, is followed up by conduct of an in vitro mutation assay (i.e. bacterial reverse mutation [Ames] assay). This assay is generally acceptable as a screen for impurities with an identified alert since positive signals in computational toxicity evaluations are often derived from the results of bacterial mutation assays and mutagenic carcinogens are considered to operate through mechanisms that are not threshold-related. Depending on the opinion of the in silico expert toxicologist, a mammalian cell assay may also be recommended for impurities that contain specific structural groups that have not been well characterized in bacterial assays or for impurities that are cytotoxic to bacteria.

It is important to consider the study designs for impurity assessment, particularly if the assays are conducted with drug substance containing, or spiked with, the impurity due to limitations in impurity synthesis. In this case, the impurity should be evaluated at a level that is commensurate with that observed in clinical, stability and/or production batches (and in consideration of batch variability). The acceptance criterion would then not exceed the level present in the batch tested in the genotoxicity assay and would be supported by the relevant qualification thresholds and supporting general toxicity information. If the bacterial reverse mutation assay (and/or mammalian cell assay if warranted) demonstrates that the impurity is negative for genotoxic potential, no further genetic toxicity studies are recommended and the material is considered to be qualified with respect to genotoxicity.

For impurities that are demonstrated to show positive results via SAR and computational assessments and are positive in one or more genotoxicity assay, or for impurities for which other information (i.e. literature, carcinogenic study results) indicating carcinogenic potential is identified, a toxicologic weight-of-evidence (WOE) approach is used to establish a risk.[32] Additionally, the impurity is considered with respect to adequate evidence for a threshold mechanism associated with genotoxic activity. If a threshold mechanism can be established based on the WOE information, the specification limit is set based on calculated PDE thresholds as discussed in Section 6.3.3.2 In contrast, without evidence of a threshold mechanism, the impurity would be considered in the context of the clinical therapeutic risk-benefit and may be restricted or rejected in the drug substance/drug product. In either case, control of the impurity becomes focused on elimination or limitation through manufacturing processes rather than further toxicologic assessment.

These recommendations for toxicologic assessment of potentially genotoxic or carcinogenic impurities are somewhat flexible in practice because each drug development program and clinical indication is different. When applying the recommended approaches, the proposed therapeutic indication (e.g. life-threatening disease versus less serious illness), the patient population (e.g. adult, geriatric, pediatric) and route and duration of exposure are all considered. Further, the feasibility of controlling impurity levels and the manufacturing process capabilities are taken into account and will influence the overall determination of acceptance criteria.

6.3.7 Specification of impurities

6.3.7.1 Early development

As discussed elsewhere in this publication, the ICH Quality Guidelines were never intended to apply to the clinical phase of development. Nevertheless, the ICH guidelines form a useful framework for the development of specifications for the drug substance and the drug product. Very recently, the IQ Consortium[iii] has proposed controlling impurities in the drug substance and the drug product at levels that are three times higher than those defined in the described in ICH Q3A and Q3B.[32] For example, if the maximum daily dose is equal to or less than 2 g (which covers most situations), the identification and qualification thresholds in the drug substance are set at 0.3% and 0.5%, respectively. The higher thresholds in the early development are justified by the limited exposure to the clinical candidate and the low number of individuals that participate in early phase clinical studies. The IQ Consortium proposal goes on to propose that higher levels of impurities may be justified, provided those levels are supported by the appropriate nonclinical studies. For individual impurities that exceed the 0.5% qualification threshold, but supported by nonclinical data, an upper limit of 1.0% is proposed. In some situations, a limit greater than 1.0% may be justified if supported by nonclinical data or if the impurity is a known metabolite. Interestingly, the IQ Consortium proposal makes no mention of the reporting thresholds and it is proposed here that the same reporting thresholds described in Q3A and B are used to ensure that impurities are appropriately tracked in the drug substance and the drug product. Of course, these recommendations for early development do not apply to unusually potent compounds that are discussed elsewhere in this chapter.

6.3.7.2 Full development and marketing applications

6.3.7.2.1 Drug substance

As development proceeds, more information is gathered about the impurity profile in the final drug substance to be produced by the commercial route of synthesis and process of manufacturing. In late development and in the marketing application, ICH Q3A and Q3B become more applicable and the thresholds described in those guidelines are appropriate.

Four types of impurity are to be specified in the drug substance specification and a rationale for the inclusion or exclusion of each impurity should be provided in the dossier. The four classifications of impurity in the drug substance are:

- Each specified identified impurity;
- Each specified unidentified impurity;
- Each unspecified impurity with an acceptance criterion of not more that ($\leq$) the identification threshold (see Table 6.3); and
- Total Impurities.

The rationale for inclusion (and exclusion) of impurities in the specification should be based on those batches used in the key nonclinical and clinical studies, together with a consideration of the impurity profile of batches produced by the proposed commercial process. Similarly, the acceptance

[iii]The IQ Consortium is the International Consortium on Innovation and Quality in Pharmaceutical Development.

criteria should be based on those batches used in pivotal nonclinical and clinical bathes, and the three batches included in the registration stability studies. Although the number of applicable batches may be limited at the time of the marketing application (NDA, MAA etc.), ICH Q6A recommends the mean plus three standard deviations of the levels of impurity found in the key drug substance batches may be used to establish the acceptance criteria, provided the acceptance criteria are justified by the nonclinical data. Higher acceptance criteria may be proposed if qualified in the nonclinical studies or if the impurity is a known metabolite.

6.3.7.2.2 Drug product

As for the drug substance, four types of impurity are to be specified for the drug product:

- Each specified identified degradation product;
- Each specified unidentified degradation product (signified by its relative retention time);
- Any unspecified degradation product with an acceptance criterion of not more than ($\leq$) the identification threshold (see Table 6.4); and
- Total degradation products.

The selection of those degradation products to be included in the specification should be based on degradation products found in those batches manufactured by the proposed manufacturing process (in the case of the marketing application this is the proposed commercial process). The rationale for inclusion of impurities in the drug product specification should include a discussion of the degradation profiles of the drug product in safety studies, clinical studies and formal stability studies. Potential degradants should also be discussed based on the literature research and in silico approaches such as Pharma D3. Provided there are no safety concerns, acceptance criteria should be established from batches produced by the proposed manufacturing process, allowing sufficient latitude for normal manufacturing and analytical variability. For the marketing application (NDA, MAA etc.), it is standard practice as prescribed in Q1E to submit stability data based on six months of accelerated data (e.g. 40 °C/75%RH) and 9 or 12 months of real time data at the intended storage condition. Provided the levels are qualified in nonclinical studies, acceptance criteria can be justified by extrapolation of the real time data to the proposed shelf life (see Chapter 2, section 2.2.3 for more details).

References

1. Raillard, S. P.; Bercu, J.; Baertschi, S. W.; Riley, C. M. *Org. Process Res. Dev.* **2010,** *14,* 1015–1020.
2. Pierson, D. A.; Olsen, B. A.; Robbins, D. K.; DeVries, K. M.; Vane, D. L. *Org. Process Res. Dev.* **2009,** *13,* 285–291.
3. Qui, F.; Norwood, D. L. *J. Liq. Chromatogr. Relat. Technol.* **2007,** *30,* 877–935.
4. Martin, G. E. In *Analysis of Drug Impurities*, Smith, R.J., Webb, M.I., Eds.; Blackwell, Oxford, pp 124–155.
5. Guy, R. C. In *Preclinical Development Handbook: Toxicology;* Gad, S. C., Ed.; Wiley: New York, 2008; pp 1015–1046.
6. Bartos, D.; Curr, S. G. *Pharm. Anal.* **2008,** *47,* 215–230.
7. Impurities in New Drugs Substances Q3A (R2). *The International Conference on Harmonization of Technical Requirements for Registration of Pharmaceuticals for Human Use*, Second Revision, 2002.

8. Impurities in New Drugs Products Q3B (R2). *The International Conference on Harmonization of Technical Requirements for Registration of Pharmaceuticals for Human Use*, Second Revision, 2002.
9. Specifications: Test Procedures and Acceptance Criteria for New Drug Substances and Drug Products: Chemical Substances (Q6A), *The International Conference on Harmonization of Technical Requirements for Registration of Pharmaceuticals for Human Use*, 1994.
10. Specifications: Test Procedures and Acceptance Criteria for Biotechnological/Biological Products (Q6B), *The International Conference on Harmonization of Technical Requirements for Registration of Pharmaceuticals for Human Use*, 1994.
11. Impurities: Guideline for Residual Solvents (Q3C) (R5), *The International Conference on Harmonization of Technical Requirements for Registration of Pharmaceuticals for Human Use*, Fifth Revision, 2010.
12. Impurities: Metals (Q3D), Concept Paper. *The International Conference on Harmonization of Technical Requirements for Registration of Pharmaceuticals for Human Use*, 2009.
13. Validation of Analytical Procedures: Text and Methodology (Q2 (R1)), *The International Conference on Harmonization of Technical Requirements for Registration of Pharmaceuticals for Human Use*, First Revision, 2005.
14. *Guideline on the Limits of Genotoxic Impurities (CHMP);* European Medicines Agency, 2006 (draft).
15. Müller, L.; Mauthe, R. J.; Riley, C. M.; Andino, M. A.; De Antonis, D.; Beels, C.; DeGeorge, J.; DeKnaep, A. G. M.; Ellison, D.; Fagerland, J. A.; Frank, R.; Fritschel, B.; Galloway, S.; Harpur, E.; Humphrey, C. D. N.; Jacks, A. S.; Jagota, N.; Mackinnon, J.; Mohan, G.; Ness, D. K.; O'Donovan, M. R.; Smith, M. D.; Vudathala, G.; Yotti, L. *Regul. Toxicol. Pharmacol.* **2006,** *44,* 198–211.
16. FDA Guidance for Industry. *Genotoxic and Carcinogenic Impurities in Drug Substances and Products.* Recommended Approaches (draft); Center for Drug Evaluation and Research, US Department of Health and Human Services, 2008.
17. *Questions and Answers on the Guideline on the Limits of Genotoxic Impurities;* European Medicines Agency, 2010 (draft).
18. Giordani, A.; Koebel, W.; Gally, H. U. *Eur. J. Pharm. Sci.* **2011,** *43,* 1–15.
19. Barlow, S. M.; Kozianowski, G.; Würtzen, G.; Schatter, J. *Food Chem. Toxicol.* **2001,** *39,* 893–905.
20. Sofuni, T.; Hayashi, M.; Nohmi, T.; Matsuka, A.; Yamada, M.; Kamata, E. *Mutat. Res.* **2000,** *464,* 97–104.
21. Kroes, R.; Renwick, A. G.; Cheesman, M.; Kleiner, J.; Mangelsdorf, I.; Piersma, A.; Schilter, B.; Schlatter, J.; von Schothorst, F.; Vos, J. G.; Würtzen, G. *Food Chem. Toxicol.* **2004,** *42,* 65–83.
22. Horwitz, W.; Kamps, L. R.; Boyer, K. W. *J. Assoc. Off. Anal. Chem.* **1980,** *63,* 1344–1355.
23. Nalwade, S. U.; Reddy, V. R.; Rao, D. D.; Morisetti, N. K. *J. Pharm. Biomed. Anal.* **2012,** *57,* 109–114.
24. Debessis, E.; Boehlert, J. P.; Givand, T. E.; Sheridan, J. C. *Pharm. Technol.* **September 1982,** 120–137.
25. Coutant, M.; Ge, Z.; McElvain, J. S.; Miller, S. A.; O'Connor, D.; Swanek, F.; Szulc, M.; Trone, M. D.; Wong-Moon, K.; Yazdanian, M.; Yehl, P.; Zhang, Sh. *Pharm. Technol.* **2012,** *36* (10), 86–94.
26. Tefloth, G. In *Analysis of Drug Impurities;* Smith, R. J., Webb, M. I., Eds.; Blackwell: Oxford, 2007; pp 215–234.
27. Sigvardson, K. W.; Adams, S. P.; Barnes, T. B.; Blom, K. F.; Fortunak, J. M.; Haas, M. J.; Reilly, K. L.; Repta, A. J.; Nemeth, G. A. *J. Pharm. Biomed. Anal.* **2002,** *27,* 327–334.
28. FDA Guidance for Industry. *M3(R2): Nonclinical Studies for the Conduct of Human Clinical Trials and Marketing Authorization for Pharmaceuticals;* Center for Drug Evaluation and Research, US Department of Health and Human Services, January, 2010.
29. ICH M7 Step 1 Document: Assessment and Control of DNA Reactive (Mutagenic) Impurities in Pharmaceuticals to Limit Potential Carcinogenic Risk. *The International Conference on Harmonization,* November 15, 2011.

30. Fellows, M. D.; Boyer, S.; O'Donovan, M. R. The Incidence of Positive Results in the Mouse Lymphoma TK Assay (MLA) in Pharmaceutical Screening and their Prediction by MultiCase MC4PC. *Mutagenesis* **2011,** *26* (4), 529–532.
31. Cariello, N. F.; Wilson, J. D.; Britt, B. H.; Wedd, D. J.; Burlinson, B.; Gombar, V. Comparison of the Computer Programs DEREK and TOPKAT to Predict Bacterial Mutagenicity. *Mutagenesis* **2002,** *17* (4), 321–329.
32. FDA Guidance for Industry. *Recommended Approaches to Integration of Genetic Toxicology Study Results;* Center for Drug Evaluation and Research, US Department of Health and Human Services, January, 2006.

CHAPTER

Residual solvents

7

James V. McArdle

McArdle & Associates, LLC, Carlsbad, CA, USA

CHAPTER OUTLINE

7.1 INTRODUCTION

By definition, residual solvents provide no therapeutic benefit.[1] They are present in drug substances, excipients, and drug products only because they cannot be removed completely by practical manufacturing operations. However, the acknowledgment that residual solvents will be present in drug product is accompanied by the expectation that manufacturers will reduce residual solvents to the lowest practical level and that solvents used will be those of lowest toxicity that will serve the intended purpose.

An example of the principle of using the solvent of lowest toxicity that will serve the intended purpose is provided by the fictitious drug S-(+) xenplifir mesylate (Exemplifi™), introduced in Chapter 2 of this book. Note that the drug substance has five known polymorphic forms, and form III exclusively is specified for use. Note also that the drug substance is recrystallized from ethanol:hexane (5:95, v/v). Ethanol is a solvent of low toxic potential, but hexane has two neurotoxic metabolites.[2] Thus, the manufacturer of Exemplifi should be prepared to defend the use of hexane rather than a

Specification of Drug Substances and Products. http://dx.doi.org/10.1016/B978-0-08-098350-9.00007-2

solvent of lower toxic potential, such as heptane, in the recrystallization step. For example, the manufacturer might present data that show that an acceptable pharmacological response is obtained only with polymorphic form III, and that form III is obtained in purest form from the stated ethanol:hexane mixture.

According to the International Conference on Harmonization of Technical Requirements for Registration of Pharmaceuticals for Human Use (ICH) Q3 definition,[3] a solvent used as an excipient in a drug product formulation and a solvent that is a component of a drug-substance solvate are not residual solvents. Control of those solvents is still expected, and validated analytical methods will be required, but the limits stated in ICH Q3C will not apply. The amount of solvent to which a patient is exposed in those cases should be justified by data accumulated during the development stages and a thorough risk-benefit assessment.

7.2 CLASSES OF RESIDUAL SOLVENTS

The ICH Q3 guidance lists a number of potential residual solvents by class according to reported toxicity. The available safety data upon which the original determinations of class were made are summarized in a supplemental edition of Pharmeuropa (a publication of the European Directorate for the Quality of Medicines & HealthCare) as already cited.[2] Of course, additional safety data on solvents already included in the ICH guideline or new safety data on solvents not already included in the guideline might be published at any time. In such an event, the stated classification or associated limit for a given solvent might be changed or a new solvent might be added to the ICH Q3 guideline. Indeed, this process has already been implemented and the current version of that guideline is in its fifth revision.

Residual solvents are one of the three main classes of impurities described by ICH guidance Q3A, Impurities in New Drug Substances (see Chapter 6).[4] This guidance defines the concept of qualifying impurities by acquiring and analyzing data that demonstrate that an established level of an impurity is safe. Qualification of a solvent not already listed in ICH guidance Q3C may be accomplished according to the principles in Q3A, the corresponding guideline for drug product (Q3B, Impurities in New Drug Products),[5] by the acquisition of toxicity data according to the process used in Q3C, or by consideration of the principles in more than one of the guidelines.

Solvent Class 1 comprises those solvents that generally should not be used in the manufacture of drug products or their components except in extraordinary circumstances. Solvents in Class 2 are safer than Class 1 solvents to use within defined limits, and those solvents in Class 3 are of low toxicity. No safety limits are stated for solvents in Class 3 since available toxicity data indicate that they are safe for consumption at 50 mg per day or more. Class 1 and Class 2 solvents are considered in more detail below.

7.2.1 Class 1 solvents

Class 1 includes solvents known to cause unacceptable toxicities (benzene, carbon tetrachloride, 1,2,-dichloroethane, and 1,1-dichloroethene) or to present environmental hazards (1,1, 1-trichloroethane). Class 1 solvents should be avoided in the synthesis of drug substances, in the production of excipients, and in the manufacture of drug products. A manufacturer that lists a Class 1 solvent as a potential residual solvent because it is used in the production of a component of the drug product or in the manufacture of the product itself should be prepared to present a very strong risk-benefit assessment to justify even a very low level of a Class 1 solvent. The risk-benefit assessment would be expected to demonstrate that a significant therapeutic advance would simply not be possible without use of the Class 1 solvent. Since there are only five solvents listed in Class 1, a manufacturer would be expected to demonstrate that none of the larger number of Class 2 or Class 3 solvents will serve the intended purpose and that a Class 1 solvent must be used. Allowance of a Class 1 residual solvent in a drug product is likely to be an extremely rare occurrence. However, if a product might contain a Class 1 solvent, the challenge to method development and validation could be acute because of the very low limits established for some of the Class 1 solvents.

7.2.2 Class 2 solvents

7.2.2.1 Calculation of the PDE

Class 2 solvents have less severe toxicity than the Class 1 solvents as demonstrated by studies published in the refereed literature and a few in-house studies that were made available to the Q3C Expert Working Group. It is important to understand how the toxicity data were treated to derive the amount of solvent to which one could be exposed safely on a daily basis.[2] Such an understanding is important in reviewing additional toxicity data that might become available and in consideration of application of the published data to specific drug products and dosing regimens that might not fit the assumptions made in deriving the amounts in the first place.

The maximum pharmaceutically acceptable daily intake of a residual solvent is termed the permitted daily exposure (PDE). The process of deriving a PDE begins with a review of available toxicity data, and taking from each study the highest dose at which no toxic effect was observed (the no-observed effect level, NOEL). If the study did not produce a NOEL, then the lowest observed effect level (LOEL) was used. The NOEL is given in units of daily exposure to mg of solvent per kg of body weight mg/kg per day. However, the NOEL is not applied directly to human exposure. A number of safety factors are needed to allow for the extrapolation of the data from the test species to humans and to compensate for perceived weaknesses in the toxicity study and for the severity of the observed toxicity. The safety factors used are the following:

1. $F1$ compensates for interspecies differences in the ratio of surface area to body weight between the test species and humans. The range of $F1$ was 2 (for studies in dogs) to 12 (for studies in mice).
2. $F2$ allows for differences among individual humans and is always 10.
3. $F3$ accounts for the duration of the study. A value of 1 is used for a study that lasts at least one-half of the expected life time of the test species, and a value of up to 10 is used for a study of shorter duration.
4. $F4$ is assigned as 1, 5, or 10 based on the nature of the observed toxicity. For example, a value of 10 is used when neurotoxicity or oncogenicity is found.
5. $F5$ allows for the variable quality of data among the different studies. A value of up to 10 might be applied, for example, in a study that did not establish the NOEL.

The final factor in converting from the NOEL or LOEL to the PDE is the assumption of the human body mass of 50 kg. The relatively low mass used in these calculations provides another margin of safety for the adult population but might not be appropriate, for example, for a drug intended for a pediatric population. In such a case, an adjustment to the assumed body mass should be considered. The conservative nature of the safety factors is also intended to allow for the likelihood that some patients take more than one medication on a daily basis.

Thus, the calculation of the PDE from the NOEL (or LOEL) is given by Eqn (7.1);

$$\mathrm{PDE} = \frac{\mathrm{NOEL} \times 50\ \mathrm{kg}}{F1 \times F2 \times F3 \times F4 \times F5} \tag{7.1}$$

where the NOEL is given in units of mg/kg per day and the calculated PDE is in units of mg/day. There are a number of additional assumptions and details presented in reference 2, and that publication should be consulted for example calculations and summaries of the toxicity data for the classified solvents.

7.2.2.2 Application of the PDE

The PDE of a given solvent can be related to the allowed concentration of that solvent in the drug product through Eqn (7.2):

$$\mathrm{Concentration} = \frac{\mathrm{PDE} \times 10^{-3} \times 10^{6}}{\mathrm{dose}} \tag{7.2}$$

In this equation, the factor 10^{-3} converts the expression of PDE from mg/day to g/day, dose is the maximum dose of the drug product in g/day, and the factor 10^{6} converts the decimal fraction to concentration in units of ppm.

This calculation can be used in one of two ways. The first option assumes that all drug products are administered daily at 10 g of total product mass. Thus, acetonitrile, which has a PDE of 4.1 mg/day, has a limit of 410 ppm in the drug product under Option 1. There are two significant advantages in using Option 1. The first is that this option can be used when the maximum daily dose is not yet fixed, as might be the case in an early development. The second advantage comes from simplified inventory control. Consider an excipient that meets the Option 1 limit for acetonitrile and any other potential residual solvent. This excipient can be used in any proportion with any other excipients and drug substances that also meet the Option 1 limits for all potential residual solvents as long as the maximum amount of the drug product total mass does not exceed 10 g. Such drug products might not even have to be tested for the residual solvents; however, see section 7.3.1 below.

The second option for applying Eqn (7.2) utilizes the known maximum daily dose of the drug product and the PDEs of each potential residual solvent in the drug product. To illustrate the use of Option 2, we will extend the example given in reference 2 to Exemplifi, the fictitious drug already considered in this chapter. The drug product is supplied in 200-mg extended release tablets, and the daily dose is 400 mg. The components of Exemplifi are presented in Table 7.1. In this example, the maximum content of each residual solvent as limited by the component's specification and the corresponding amount is stated. The manufacturing process for the drug product uses no organic solvents.

There are a number of important observations to be made by perusal of Table 7.1. Most importantly, the daily exposure to the Class 2 solvents hexane and acetonitrile is below their respective PDEs of 2.9 mg/day and 4.1 mg/day, respectively. Note that the limit established for hexane in the drug substance is well above the Option 1 limit of 290 ppm. The manufacturer however is free to choose to apply Option 2 to the justification of the limit of 2500 ppm hexane in the drug substance because a patient taking the established dose of 400 mg is still exposed to less hexane than the calculated PDE. Likewise, the supplier of excipient 3 established a limit for hexane above the Option 1 limit. Excipient

Table 7.1 Components of Exemplifi 200-mg Extended Release (ER) Tablets

		Hexane		Ethanol		Acetonitrile	
Component	**Amount (mg)**	**Limit (ppm)**	**Amount (mg)**	**Content (ppm)**	**Amount (mg)**	**Content (ppm)**	**Amount (mg)**
Xenplifir	200	2500	0.50	2000	0.40	ND*	ND
Excipient 1	325	NP†	0	NP	0	410	0.13
Excipient 2	80	NP	0	10,000	0.80	410	0.03
Excipient 3	125	5000	0.625	NP	0	410	0.05
Excipient 4	50	NP	0	NP	0	NP	0
Total amounts for two tablets	1560		2.25		2.40		0.42

*None detected.
†Not a potential residual solvent.

3 is acceptable in this formulation for the maximum stated dose, but excipient 3 provided under this limit might or might not be acceptable in a different formulation.

Ethanol is a Class 3 solvent and well below the generally accepted limit of 50 mg/day for that class. The manufacturer of xenplifir might recognize that there remains a large safety margin and be tempted to raise the limit on ethanol in drug substance. It is clear, however, that the limit for any residual solvent must be based on more than just safety. The limit should also meet expectations for modern manufacturing technology and have a solid basis in the batch history of the drug substance and in the range of level of residual solvent that has proven to yield a drug substance of acceptable manufacturability, efficacy, and safety.

7.3 ANALYTICAL METHODS AND THEIR VALIDATION

This section considers various testing strategies for residual solvents and appropriate validation parameters for each. The strategies are based on the least-safe residual solvent that might be present. A solvent used in a manufacturing process is regarded as likely to be present unless adequate data exist to demonstrate that the solvent is removed consistently by a validated process. The burden of providing adequate data to show that a solvent used somewhere in the process for an excipient, a drug substance, or a drug product is not likely to be present might increase depending on the class of the potential solvent and where it is used in the manufacturing process. Thus, in the unlikely event that a Class 1 solvent was used in the manufacture of a drug product, there is a high probability that testing of the drug product for that solvent would be required. On the other hand, if a Class 1 solvent was employed in an early step in the production of an excipient used in a small amount in a drug product, and data exist to show that the Class 1 solvent does not appear in the excipient (assuming an adequate limit of detection in the test method), then the manufacturer might be able to present a convincing case that the solvent is not likely to be present in the drug product and routine testing of the drug product for that solvent might not be required.

7.3.1 Test articles

One of the discussion points that the ICH Expert Working Group had to resolve was what the drug product manufacturer would be required to test. Industry representatives in the Expert Working Group felt that the production of excipients, unlike synthesis of a new chemical drug substance, was not usually under the direct control of the drug product manufacturer. Therefore, the first-hand knowledge of which residual solvents might be in an excipient and at what level those solvents might be present was not always held by the drug product manufacturer. Regulators expressed the concern that since the patient was exposed to the drug product comprising all of its components, the level of residual solvent in the drug product was the paramount safety concern.

The quandary was resolved in discussions with the International Pharmaceutical Excipients Council (IPEC), the industry association of excipient producers. Companies that are members of IPEC understand the responsibility that drug product manufacturers have with respect to residual solvents and other quality parameters. Indeed, IPEC has developed over recent years a number of guidances regarding quality of excipients,[6] and these guidances are recommended for both excipient manufacturers and drug product manufacturers. Among these documents are guidances that relate to the current good manufacturing practice (cGMP) for excipients, to certificates of analysis, and to auditing of excipient suppliers. All of these documents include some discussion of residual solvents. Thus, IPEC-member companies and those companies that abide by IPEC standards expect to share information with their customers about aspects of quality of their products, including discussion of potential residual solvents.

Likewise, manufacturers of generic drugs might not always have first-hand knowledge of all the potential residual solvents that might be present in the drug substances that they purchase. Certainly, consumers of the generic drugs are entitled to the same level of protection from residual solvents as are consumers of the brand-name drugs.

Of course, cGMP requires producers of the drug products to audit suppliers of excipients and drug substances, and to test excipients and drug substances on some defined schedule. Such testing should include tests for the residual solvents. An excipient or drug substance from an established supplier that has provided data that has been shown to be reliable might be placed on a reduced testing schedule. In those cases, the drug product manufacturer's own data on residual solvents in those substances or data on the certificate of analysis provided by an accredited supplier could be used in calculating the total exposure to solvents that a patient might receive.

Test articles thus can be drug substances, excipients, or the drug products themselves. If the sponsor chooses to test the components of the drug product, then Option 2 may be applied to ensure that patients are not exposed to more than a safe level of any residual solvent. In some cases, the sponsor may choose to test the drug product directly. A single test on the drug product may be simpler, less time-consuming, and less expensive than testing the components individually. However, discovering that an excipient has more than the expected amount of a residual solvent at the drug product stage is far more problematic than making that discovery on the excipient before it is used. At times, it may be to the sponsor's advantage to test the drug product. For example, the total of a residual solvent from all components of a drug product might add up to an amount that exceeds the Option 2 limit, but a step in the manufacturing process such as a drying step or a lyophilization step might reduce the actual total in the drug product to below the Option 2 limit.

7.3.2 Test methods

The European Pharmacopoeia (EP),[7] the Japanese Pharmacopoeia (JP),[8] and the United States Pharmacopeia (USP)[9] should be the primary references for analytical methods for residual solvents in drug products and their components. The USP provides additional guidance in the form of frequently asked questions.[10] Gas chromatography (GC) is the standard test method for Class 1 and Class 2 residual solvents,[11] although other types of methods might be allowed in some circumstances. Sufficient harmonization exists among the pharmacopeias that a single method might suffice in all the three regions. In all cases, the pharmacopeial method should be shown to be suitable for the particular test article, or a modified pharmacopeial method or an alternative method should be validated.

A method based on GC was presented in 1991 in a paper in Pharmacopeial Forum,[12] and a year later in Pharmeuropa.[13] This method became the basis for the now obsolete USP Method IV, and the current USP methods A and C share the same column and other characteristics as the method presented in the 1991 paper. Related methods were shown to separate 40 of the solvents listed in the ICH guidance plus a few additional solvents in only a few minutes.[14,15] Such methods are well suited in performing routine tests or periodic checks on drug substances or excipients, especially if one is screening for a wide range of unknown solvents. A chromatogram and accompanying table from reference 14 are shown in Fig. 7.1 and Table 7.2, respectively.

The testing strategy in the pharmcopeias is built on the assumption that the drug product manufacturer does not have first-hand knowledge of all likely residual solvents in purchased components. For example, Procedure A in the USP has the ability to screen for Class 1 and Class 2 solvents. Procedure A is run as a limit test against a standard mix of Class 1 and Class 2 solvents. If a residual solvent is detected by Procedure A in the test article above its limit, then Procedure B, which is also a limit test but uses a different column, is run to confirm the presence of the residual solvent detected by

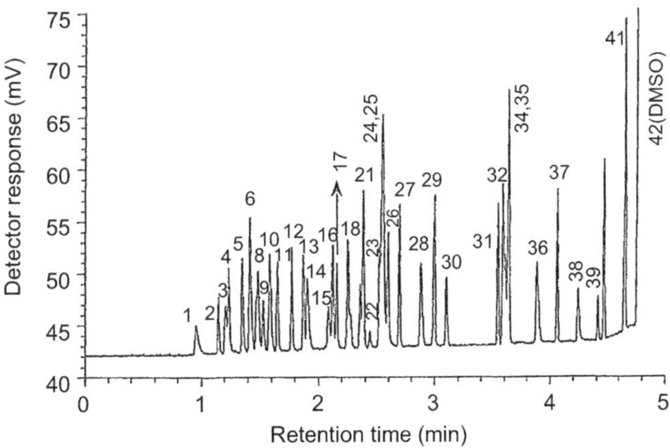

FIGURE 7.1

Gas chromatogram demonstrating resolution of 41 potential residual solvents in less than 5 min.

Source: Reprinted with permission from Ted K Chen, Joseph G Phillips, William Durr. Analysis of Residual Solvents by Fast Gas Chromatography. J. Chromatogr., A. *1998, 811, 145–150.*

Table 7.2 Identification of Solvents by Peak Number in the Chromatogram in Figure 7.1

Separation of Commonly Used ICH Class 2 and Class 3 Organic Solvents			
Peak No	**Solvent**	**ICH Solvent Class**	**Retention Time (min)**
1	Methanol*	2	0.96
2	Pentane	3	1.16
3	Ethanol*	3	1.22
4	Diethyl ether*	3	1.25
5	Acetone*	3	1.36
6	2-Propanol*	3	1.43
7	Methyl acetate	3	1.49
8	Acetonitrile*	2	1.50
9	Methylene chloride*	2	1.55
10	*tert.*-Butanol*	†	1.60
11	Methyl *tert.*-butyl ether (MTBE)	3	1.67
12	*n*-Hexane*	2	1.79
13	Isopropyl ether*	†	1.89
14	1-propanol*	3	1.92
15	Nitromethane	2	2.10
16	Methyl ethyl ketone (MEK)*	3	2.14
17	Ethyl acetate*	3	2.18
18	Tetrahydrofuran (THF)*	3	2.27
19	Chloroform*	2	2.30
20	1,1,1-Trichloroethene	1	2.38
21	Cyclohexane	2	2.41
22	Carbon tetrachloride	1	2.47
23	2-Methyl-1-propanol	3	2.55
24	Benzene	1	2.57
25	1,2-Dichloroethene	1	2.57
26	2-Methoxyethanol	2	2.57
27	Isopropyl acetate	3	2.62
28	*n*-Heptane*	3	2.72
29	*n*-Butanol*	3	2.91
30	Methylyclohexane	2	3.03
31	1,4-Dioxane*	2	3.13
32	Methyl isobutyl ketone*	3	3.58
33	Pyridine	2	3.62
34	3-Methyl-1-butanol*	3	3.64
35	Toluene*	2	3.67
36	1-Pentanol (n-amyl alcohol)	3	3.92
37	*n*-Butyl acetate*	3	4.11

Table 7.2 Identification of Solvents by Peak Number in the Chromatogram in Figure 7.1 *(continued)*

Separation of Commonly Used ICH Class 2 and Class 3 Organic Solvents			
Peak No	**Solvent**	**ICH Solvent Class**	**Retention Time (min)**
38	Dimethylformaride (DMF)*	2	4.28
39	*m*-Xylene*	2	4.45
40	*p*-Xylene*	2	4.51
41	*o*-Xylene*	2	4.70
42	Dimethyl sulfoxide (DMSO)	3	4.91

*Analyte solvents included in the mixture used in precision study.
†No classification from ICH (International Committee on Harmonization).
Source: Reprinted with permission from Ted K Chen, Joseph G Phillips, William Durr. Analysis of Residual Solvents by Fast Gas Chromatography. J. Chromatogr. A. *1998, 811, 145–150.*

Procedure A. If the identity of the solvent is confirmed and the solvent is above its limit by Procedure B also, then the amount of that residual solvent is quantitated by Procedure C.

Water and dimethylformamide are common solvents for extraction of residual solvents from pharmaceutical test articles but dimethyl sulfoxide (DMSO) and other suitable solvents can also be used.[16] The USP states a preference for dissolving the test article, but complete dissolution is not a requirement as long as the method can be validated. Mixtures of water and organic solvents have been used and shown to have better sensitivity than organic solvents alone.[17] Ionic liquids such as 1-butyl-3-methylimidazolium tetrafluoroborate and 1-n-butyl-3-methylimidazolium dimethyl phosphate also have been used as solvents for GC analysis because of their excellent dissolution characteristics, including their ability to dissolve excipients derived from carbohydrates, their extremely low vapor pressures, and their thermal stability.[18,19] Thermal desorption coupled with GC may be used to measure residual solvents without the need for an extraction step.[20] Methods based on GC with multiple headspace single-drop microextractions[21,22] and dispersive liquid–liquid microextraction[23] are advanced techniques that offer the potential to meet the low levels of quantitation required for even the Class 1 solvents, to require only short analysis times, and to analyze solid dosage forms with little sample manipulation.

7.3.3 Class 1 solvents

If a Class 1 solvent is likely to be present in the test article, perhaps among Class 2 and Class 3 solvents, a quantitative test for that Class 1 solvent is required. If Class 2 solvents are likely to be present in the same drug product, there also should be a test for those solvents. The test for the Class 2 solvents could be a limit test but a single method to quantitate both the Class 1 and Class 2 solvents would often be most economical. If Class 3 solvents are likely to be present in addition to the Class 1 solvent with or without the likely presence of Class 2 solvents, the Class 3 solvents probably would be most economically tested by the same GC method. The sponsor may choose to test the Class 3 solvents by loss on drying, with proper compensation for the lack of specificity in the loss on drying test by other data such as Karl Fisher titration for moisture and GC data for other solvents.

7.3.4 Class 2 solvents

If a Class 2 solvent is likely to be present in the test article, perhaps among Class 3 solvents but in the absence of Class 1 solvents, a test for that Class 2 solvent is required. That test could be a limit test or a quantitative test, except that Class 2 solvents that are present above their Option 1 limits must be quantitated. Limit tests have fewer parameters to evaluate as part of validation, but yield less information. A quantitative test requires a more extensive validation; however, the quantitative data may prove useful, for example, for statistical process control (see Chapters 2 and 3). Direct application of the pharmacopeial methods in this case presents the opportunity for minimal method qualification and favorable regulatory acceptance.

7.3.5 Class 3 solvents

If only Class 3 solvents are likely to be present in the test article, the loss on drying test may be used if the sum of those solvents is less than 0.5%, the Option 1 limit for any single Class 3 residual solvent. If any single residual Class 3 solvent or the sum of the Class 3 solvents might be present above the PDE of 50 mg/day for any single Class 3 solvent, then a test that is specific for each solvent should be run. The accepted limit for the loss on drying test is 0.5%, which is, of course, itself a limit test. This test is a viable option if the sum of Class 3 residual solvents and water is less than 0.5%. The loss-on-drying test result in some instances can be a bit less straightforward to interpret due to its nonspecific nature (see Chapter 11). Nevertheless, loss on drying could still be the method of choice, even if, for example, data from a separate test for moisture were needed to interpret correctly the loss on drying test.

7.3.6 Validation

As with any other official pharmacopeial test method, the method must be shown to be appropriate for the particular test article. Although a full validation is not generally expected for a pharmacopeial test method, basic performance characteristics always should be confirmed.

Table 7.3 Characteristics for Validation of GC and Loss on Drying Methods

Characteristic	Quantitative GC	GC Limit Test and Loss on Drying
Accuracy	√	–
Repeatability	√	–
Intermediate precision or reproducibility	√	–
Specificity	√	√
Detection limit	–	√
Quantitation limit	√	–
Linearity	√	–
Range	√	–

√, include this characteristic in validation of the method; –, not necessary to include this characteristic in validation of the method.

Validation parameters for modified pharmacopeial methods or novel methods may be taken directly from ICH guideline Q2(R1), Validation of Analytical Procedures: Text and Methodology,[24] which states that either quantitative tests or limit tests can be applied to impurities (such as residual solvents) in drug substances and drug products, and that flexibility was considered in the discussion above. The objective of the method should be clearly and explicitly stated in the method description, and that stated objective should be used to determine the characteristics that are included in the validation. However, guideline Q2(R1) provides typical characteristics for the type of method (see also Chapters 4 and 6), and those characteristics are reflected in Table 7.3 above for both GC methods and the loss on drying test.

Note that ICH Q2(R1) expects that specificity for a limit test such as loss on drying is demonstrated, but also allows for another test to compensate for lack of specificity. As discussed, the loss on drying test is not specific for residual solvents. But if water is shown to be the only other significant component that is lost on drying, then data from a specific test for moisture such as Karl Fisher titration (see Chapter 11) could be used to compensate for that lack of specificity.

References

1. http://www.ich.org/fileadmin/Public_Web_Site/ICH_Products/Guidelines/Quality/Q3C/Step4/Q3C_R5_Step4.pdf, page 1.
2. Pharmeuropa, Vol. 9. No. 1. Supplement, April, 1997 and references cited therein pp. S37-S38. A link to reprints of this paper is available here: http://www.edqm.eu/en/pharmeuropa-bio-and-scientific-notes-584.html.
3. http://www.ich.org/fileadmin/Public_Web_Site/ICH_Products/Guidelines/Quality/Q3C/Step4/Q3C_R5_Step4.pdf.
4. http://www.ich.org/fileadmin/Public_Web_Site/ICH_Products/Guidelines/Quality/Q3A_R2/Step4/Q3A_R2__Guideline.pdf.
5. http://www.ich.org/fileadmin/Public_Web_Site/ICH_Products/Guidelines/Quality/Q3B_R2/Step4/Q3B_R2__Guideline.pdf.
6. https://ipecamericas.org/ipec-store.
7. *European Pharmacopoeia*, 2.4.24 Identification and Control of Residual Solvents.
8. *Japanese Pharmacopeia* XV, Guideline for Residual Solvents, p. 1682. http://jpdb.nihs.go.jp/jp15e/JP15.pdf
9. *United States Pharmacopeia*, Chapter <467> Residual Solvents.
10. http://www.usp.org/support-home/frequently-asked-questions/general-chapter-residual-solvents.
11. For an introduction to the principles of gas chromatography a number of excellent texts are available. See for example David Sparkman, O.; Penton, Zelda E.; Kitson, Fulton G. *Gas Chromatography and Mass Spectrometry: A Practical Guide;* Academic Press: Burlington, MA, 2011.
12. Chen, T. K.; Moeckel, W.; Surprenant, H. L.; Ho, M. Y. K. Proposed Changes to Method I for Organic Volatile Impurities. *Pharm. Forum* **1991,** *17,* 1475–1479.
13. Chen, T. K.; Moeckel, W.; Surprenant, H. L.; Ho, M. Y. K. Modifications to the European Pharmacopoeia's Proposal on Residual Solvent Analysis. *Pharmeuropa* **1992,** *4* (1), 62–67.
14. Chen, Ted K.; Phillips, Joseph G.; Durr, William Analysis of Residual Solvents by Fast Gas Chromatography. *J. Chromatogr., A* **1998,** *811,* 145–150.
15. Rocheleau, Marie-Josée; Titley, Mélanie; Bolduc, Julie. Measuring Residual Solvents in Pharmaceutical Samples using Fast Gas Chromatography Techniques. *J. Chromatogr., B* **2004,** *805,* 77–86.

16. Smith, Ian D.; Waters, David G. Determination of Residual Solvent Levels in Bulk Pharmaceuticals by Capillary Gas Chromatography. *Analyst* **1991,** *116,* 1327–1331.
17. D'Autry, Ward; Zheng, Chao; Wolfs, Kris; Yarramraju, Sitaramaraju; Hoogmartens, Jos; Van Schepdael, Ann; Adams, Erwin Mixed Aqueous Solutions as Dilution Media, in the Determination of Residual Solvents by Static Headspace Gas Chromatography. *J. Sep. Sci.* **2011,** *34,* 1299–1308.
18. Liu, Feng-hua; Jiang, Ye Room Temperature Ionic Liquid as Matrix Medium for the, Determination of Residual Solvents in Pharmaceuticals by Static Headspace Gas Chromatography. *J. Chromatogr. A* **2007,** *1167,* 116–119.
19. Laus, Gerhard; Andre, Max; Bentivoglio, Gino; Schottenberger, Herwig Ionic Liquids as Superior Solvents for Headspace Gas Chromatography of Residual Solvents with Very Low Vapor Pressure, Relevant for Pharmaceutical Final Dosage Forms. *J. Chromatogr. A* **2009,** *1216,* 6020–6023.
20. Hashimoto, Keiji; Urakami, Koji; Fujiwara, Yasuhiro; Terada, Syunji; Watanabe, Chuichi Determination of Residual Solvents in Pharmaceuticals by Thermal Desorption-GC/MS. *Anal. Sci.* **2001,** *17,* 645.
21. Yu, Yingjia; Chen, Bin; Shen, Cidan; Cai, Yi; Xie, Meifen; Zhou, Wei; Chen, Yile; Li, Yan; Duan, Gengli Multiple Headspace Single-Drop Microextraction Coupled with Gas Chromatography for Direct Determination of Residual Solvents in Solid Drug Product. *J. Chromatogr. A* **2010,** *1217,* 5158.
22. Saraji, Mohammad; Khayamian, Taghi; Hassanzadeh Siahpoosh, Zahra; Farajmand, Bahman Determination of Volatile Residual Solvents in Pharmaceutical Products by Static and Dynamic Headspace Liquid-Phase Microextraction Combined with Gas Chromatography–Flame Ionization Detection. *Anal. Methods* **2012,** *4,* 1552.
23. Farajzadeh, Mir Ali; Goushjuii, Leila; Djozan, Djavanshir; Mohammadi, Javad Kompani Dispersive Liquid–Liquid Microextraction Combined with Gas Chromatography for Extraction and Determination of Class 1 Residual Solvents in Pharmaceuticals. *J. Sep. Sci.* **2012,** *35,* 1027.
24. http://www.ich.org/fileadmin/Public_Web_Site/ICH_Products/Guidelines/Quality/Q2_R1/Step4/Q2_R1__Guideline.pdf.

CHAPTER

Validation of procedures for elemental impurities

Todd L. Cecil

United States Pharmacopeial Convention, Inc., Rockville, Maryland, USA

CHAPTER OUTLINE

8.1 INTRODUCTION

Validation is the demonstration that a procedure is capable of delivering a result that meets the intended purpose of an analytical test. This statement is often used as the definition of validation. It is actually a very good description of the practice when the special terms and phrases have been defined.

Specification of Drug Substances and Products. http://dx.doi.org/10.1016/B978-0-08-098350-9.00008-4

For example, the term "demonstration" indicates that a validation cannot occur on paper or in theory alone—physical measurements are required. The term "procedure" is shorthand for all of the details of a physical measurement. It further indicates the expectation of a set of criteria that must be met in order to adequately demonstrate that a procedure is valid. These details may include sample and standard preparations, calibration curves, daily system suitability considerations, and instrumental installation, maintenance, settings, and configuration. The phrase "delivering a result" means that the output of the process is not simply data, but is an actionable result. This result must be decisional and that decision must be relevant and appropriate for the intended purpose of the test. The overused phrase "intended purpose of a test" is a convenient way of stating that every analytical technique can be used for a number of different needs. The needs of the measurement and decision must be factored into determining the success or failure of a result, and, by extension, the validation. In order to really describe the complexity of validation, it is helpful to apply the principle to an analytical technique. The goal of this chapter is to examine the principles of validation when applied to a variety of procedures that can be used to measure elemental impurities. Hopefully, this will provide greater clarity and a set of rules and conditions to aid the reader in the application of these principles.

8.2 ELEMENTAL IMPURITIES

Elemental Impurities is a term defined in the United States Pharmacopeia (USP) General Chapters <232> and <233>[1] and describes a group of potentially toxic elements that are commonly considered in the evaluation of pharmaceutical products. The "big four" toxic elements are lead, cadmium, arsenic, and mercury. Other elements with potential toxicity include iridium, osmium, palladium, platinum, rhodium, ruthenium, chromium, molybdenum, nickel, vanadium, and copper. These elemental impurities can be introduced through starting materials, from the environment, through process operations, or through synthetic processes (catalysis). The decisions concerning the inclusion of specific metals and an understanding of the toxicity parameters that led to the general chapters are included in a Pharmacopeial Forum (PF) Stimulus Article.[2] Due to the potential toxicity of these impurities, limits range from sub-ppb to 1000 ppm, depending on the element and the application. There are several methods that can be considered for measurement of these elements in a pharmaceutical product matrix. These methods range from basic colorimetric procedures to advanced atomic spectroscopic and mass spectrometric procedures. Each has strengths and weaknesses, and very different validation requirements using the International Conference on Harmonization (ICH) Q2 (R1)[3] and USP General Chapter <1225>[4] approaches.

For example, a colorimetric procedure, such as the sulfide precipitation procedure, will produce a colored solution that is compared to a similarly prepared solution with a known concentration of the standard measurand (metal to be measured). If the color of the sample is darker than the standard, the sample will fail. This procedure is considered as a category II—*Limit Test* requiring only specificity and detection limit measurements using the traditional validation approaches. An instrumental procedure could be substituted for the wet chemistry procedure, and using very limited validation requirements be said to be equivalent and interchangeable with the colorimetric procedure. The instrumental procedure could also be validated as a category II—*Quantitative Procedure* requiring a full range of measurements to complete the validation. So how does an analyst choose the correct validation approach? In this case, it would depend upon the reportable value in the specification. If it is a pass/fail specification, then the Limit Test is sufficient. If it is a numerical limit, the Quantitative Procedure is to be used.

To address the weakness of direction in USP General Chapter <1225> for inorganic impurities, the authors of General Chapter <233> included a series of validation requirements that must be satisfied in order to allow a procedure to demonstrate compliance to the General Chapter. These validation principles meet the definition of validation in the introduction, but differ in substance and scope from the validation standards of ICH.

8.3 VALIDATION PRINCIPLES (TRADITIONAL AND STANDARDIZED)

When the definition in the introduction is applied in the laboratory, one of the first requirements is for the analyst to determine what measurements will be made and when those measurements will provide sufficient assurance of the validity of the data obtained from the procedure when testing an unknown sample. This is a step that is often skipped in favor of the "checking-the-box" approach to validation. The checking-the-box approach entails reviewing ICH Q2 (R1), and completing the minimum amount of testing to meet the requirements of the guidance. With no standards of acceptability in the guidance, the act of completing the testing can be described as meeting the validation. Even the most well-intentioned analysts can choose to forego replicates and statistical rigor for time and material savings. This section will provide best practices for each validation variable from the Traditional and the Standardized (<233>) approaches.

8.3.1 Analytical performance characteristics

USP General Chapter <1225> provides a table to indicate the data elements required for validation. These are also called the Analytical Performance Characteristics. The portion of the table that applies to the testing of Elemental Impurities is included in Table 8.1.

Table 8.1 is included as a reference for the discussion of each of the Analytical Performance Characteristics. From the table, it is clear that a quantitative procedure is a more demanding validation challenge, but is the most useful for the majority of the demands of elemental impurity control. The remainder of this section will be dedicated to a description of the individual performance characteristics, similarities or differences between the two conflicting approaches, and the best practice for the readers to consider for each situation. For the sake of clarity, it is beneficial to describe the differences between a limit test and a quantitative test.

Table 8.1 Analytical Performance Characteristics for a Traditional Validation Protocol

Analytical Performance Characteristic	Category II	
	Quantitative	**Limit Tests**
Accuracy	Yes	Maybe
Precision	Yes	No
Specificity	Yes	Yes
Detection limit	No	Yes
Quantitation limit	Yes	No
Linearity	Yes	No
Range	Yes	Maybe

8.3.1.1 Limit test

A limit test typically compares the signal obtained from a standard solution having the measurand (elemental impurity) present at the maximum acceptable concentration to a sample solution of the analyte (drug substance or excipient). Where the signal obtained from the sample solution is less than that of the standard solution, the sample is said to pass the test. The results of this type of test are usually indicated in a certificate of analysis with "passes" or "meets the requirement".

8.3.1.2 Quantitative test

A quantitative test compares the standard and sample solution in such a manner that an estimate of the true value of the measurand can be calculated. An elemental impurity evaluation will often use a calibration curve prepared from a series of standard solutions of a high-quality reference standard to convert the signal obtained from the instrument into a concentration value of the measurand. These calibration curves are typically measured each day and in many cases are repeated during the analysis to ensure that any instrumental drift (bias) has been accounted for. The results of this type of test may be reported with a value, a "passes" statement, or a "<x ppm" notation in a certificate.

8.3.2 Accuracy

The accuracy of a procedure is a simple measure of the difference between a measured value and the true value. This difference has in the past been referred to as bias. The bias of a measurement can be determined several different ways, including: (1) comparison of a result of a certified reference material to calibration curve of the measurand; (2) comparison of common preparations to previously validated orthogonal analytical procedures; and (3) a standard addition method or spike recovery study using a reference standard of the measurand.

Measurement Approach (1) would require a certified reference material (CRM) of the pharmaceutical matrix with known contamination with the impurities of interest. This type of CRM is available from the National Institute of Standards and Technology (NIST) for soil samples and other environmental applications, but is rare for pharmaceutical ingredients.

Measurement Approach (2) requires a second, fully validated procedure having a similar level of uncertainty as the procedure being validated. This is not generally a cost-effective approach to validation, and ascribes all sources of variability across both instrumental approaches to bias.

Measurement Approach (3)—spike recovery—is the most commonly used approach for the validation of elemental impurity procedures. High-quality standard materials are freely available from NIST and other National Measurement Organizations (NMOs). However, spikes of multiple elements can be problematic due to dilution effects of the combination of the standards.

Once the Measurement Approach is selected, the next question centers on the concentration of the spike and the acceptable level of the variability that can be expected from the measurement. The two validation approaches require a similar number of concentrations and independent solutions to be prepared and measured. The Traditional approach suggests that the results of the spike recovery should be plotted and the slope of the linear regression should be close to 1.

The Standardized approach includes some additional information and requirements. In this approach, it is specified that a range of spiked samples should be centered on the analytical target value (limit for the element), and include a range from 50% to 150% of the target value. Samples must be spiked before any sample manipulation (e.g. digestion) occurs. The average spike recovery value is

Table 8.2 Accuracy Comparison

Validation Approach	No. and Type of Samples to be Examined	Measurement Approach	Acceptable Measure
Traditional	Three concentrations with three replicates	Spike recovery	Slope of spike recovery over the three concentrations is close to 1.0
Standardized	Minimum of three spiked concentrations ranging from 50% to 150% of limit Three Independent solutions at each concentration	Average spike recovery at each concentration	70–150% recovery at each concentration
Best practice	See standardized approach	See standardized approach	See standardized approach

determined and must be between 70% and 150% at each concentration value. The wide range is provided because of the typical variability (precision) associated with determination of measurands in the subpart-per-million range, which is common for elemental impurity analyses. The asymmetrical range is used to allow greater freedom for a procedure that overestimates elemental concentration, thereby reducing the patient risk. The specified range for the determinations may be much smaller than the linear dynamic range of many of the instruments used for this analysis, but it assures the analyst that a critical measurement near the limit will provide an accurate result with an acceptable level of bias.

The types and number of samples, measurement approaches, and acceptable measures for accuracy determination of each approach have been compiled in Table 8.2. The third row in the table contains a best practice recommendation when applying the validation approaches to procedures used to evaluate elemental impurities.

8.3.2.1 Best practice

The additional details in the Standardized approach are critical to obtaining a meaningful validation result. However, an even greater level of detail would benefit the independent analyst. The improved descriptions included below are method specific, but can be generalized to a certain degree.

1. Use at least 5 concentrations in the accuracy determination (50%, 75%, 100%, 125%, and 150%).
2. Consider increasing the range by including spike concentrations at about the 10% level to allow reportability of the low-level metals as necessary.
3. Either purchase mixtures of concentrated elements of interest or prepare a spiked solution from concentrated single elements, thereby reducing the dilution effects. Extra care must be exercised, however, in transferring the concentrated solutions because small errors will cause exaggerated bias values.
4. After preparing 3 independent spiked stock solutions, use aliquots to prepare the independent spiked solutions.

8.3.3 Precision

The precision of a measurement is typically considered to be dominated by a random error associated with making a scientific measurement. This error is used to capture nonsystematic sources of errors such as environmental, instrument–to-instrument, and analyst-to-analyst. The error is usually calculated by determining the standard deviation of a series of measurements and then determining the percentage of that error relative to the measurement, or the relative standard deviation (RSD). These values are also called the coefficient of variance or CV. Due to the ease of the measurement, this term can be used to filter out errors from different sources and provide a better understanding of the ability of the procedure to produce a precise result. In the traditional approach, three different measurements are described. These include:

Repeatability: Multiple determinations within a short period of time using the same equipment and personnel (e.g. Standard 1, rinse, Standard 1, rinse, Standard 1, could be considered a repeatability experiment).
Reproducibility: Multiple measurements across laboratories.
Intermediate precision: Multiple determinations within a lab, using different instruments, analysts, or days.

The Standardized approach only specifies the evaluation of repeatability (with additional details) and introduces the term ruggedness.

Repeatability: Six independent samples (from the same lot) spiked at the analytical target level (limit value for the metal).
Ruggedness: Conduct the repeatability experiment over three independent events; different days, different instruments, or different analysts.

The requirement of six independent samples is specified to provide enough degrees of freedom (5) to allow a meaningful statistical evaluation of the data. The ruggedness requirement is intended to be a compromise to help the industry move validation through more quickly. Although USP General Chapter <1225> does not indicate it, statisticians usually require at least six independent events to statistically consider intermediate precision. The ruggedness requirement takes advantage of the peculiar nature of a pharmacopeia to reduce the testing needs. The pharmacopeia is a pass/fail requirement (e.g. the unknown is tested using a validation procedure and compared to a standard material, and it either passes the requirement or it doesn't) and therefore, multiple laboratory studies are not necessarily needed for a pharmacopeial standard, because these methods will not be used in that manner. Further, it recognizes that many manufacturers do not have multiple instruments and even multiple trained analysts. So the reduced degrees of freedom are justified and still provide an adequate estimate of the amount of error likely to be found over the course of a typical pharmacopeial testing regime.

In addition to repeatability, intermediate precision, and ruggedness, there has been one more term that has been used at various times in the past. The missing term is robustness, which is defined as the evaluation of the repeatability of the measurement after deliberate changes have been made to the analytical instrument setup. This was excluded from ICH Q2 (R1) based upon the determination that this was a component of analytical development and it did not need to be demonstrated during the validation of the procedure. This change was adopted by the USP and therefore robustness is not included in either the Traditional or Standardized approaches. However, this is a source of some debate in the industry, especially during technology transfer across sites. The robustness study can clearly define conditions that will perturb the results and others that will not. This information can be critical to tracking down the source of a reproducibility problem. Therefore, the author recommends including robustness in the design of experiments where there is a possibility that a procedure will be transferred.

Table 8.3 Precision Comparison

Validation Approach		No. and Type of Samples to be Examined	Measurement Approach	Acceptable Measure
Traditional	Repeatability	Three concentrations with three replicates or six determinations at 100%	RSD	None
	Reproducibility	–	–	–
	Intermediate precision	–	–	–
Standardized	Repeatability	Six independent solutions at measurand limit value	RSD	NMT* 20%
	Ruggedness	*Repeatability* at three independent events	RSD	NMT 25%
Best practice	Repeatability	Six independent solutions at measurand limit value	RSD	NMT 20%
	Reproducibility/ Robustness	*Ruggedness* at NLT† two independent labs/*repeatability* across typical instrumental condition variations	RSD	NMT 25%/ NMT 20% at each condition
	Intermediate precision	*Repeatability* at six independent events	RSD	NMT 25%

**Not more than.*
†Not less than.

The types and numbers of samples, measurement approaches, and acceptable measures for precision determination of each approach have been compiled in Table 8.3. The third row in the table contains a best practice recommendation when applying the validation approaches to procedures used to evaluate elemental impurities.

8.3.3.1 Best practices

The Standardized approach is preferred for a compendial validation, but it misses a couple of important points when the validation is not targeted at compliance to the USP standard. Those missing pieces (reproducibility and intermediate precision) are therefore recommended for a basic validation protocol. The reproducibility study should include the reduced testing set described by the ruggedness requirement at each site that may be asked to implement the procedure. The data from all the sites are combined and the relative standard deviation is calculated. The RSD should not exceed 25% when examining limits in the sub-ppm range. In addition to the reproducibility study, an understanding of intermediate precision is recommended. This is a single laboratory evaluation that should examine not fewer than 6 independent events. These six events are usually collected in a 3×2 matrix (e.g. three different days and two analysts/day). Each of the independent events should include six (6) individual spiked samples. Remember that spikes should always occur before any digestion takes place. The intermediate precision study should demonstrate an RSD of not more than 25% across all of the data. These recommendations represent additional work for the laboratory, but will ensure that the procedure will pass both the Traditional and Standardized approaches for precision, while providing greater assurance of long-term applicability of the analytical procedure. In addition to reproducibility and intermediate precision studies, the majority of multinational manufacturers would benefit from an extensive robustness study to ease the transition of the procedures to external sites or third-party vendors. The robustness study is of greater importance when there is lack of homogeneity in the instrument type and age, as is often the case for Atomic Absorption (AA) spectrometry and Inductively-coupled Plasma (ICP) spectrometry instrumentation.

8.3.4 Specificity

The specificity determination is, at its root, another measure of accuracy. This point is indicated in the discussion in USP General Chapter <1225>. However, even this discussion leaves a number of unresolved details for measurement and acceptance. The Standardized approach is equally vague and perhaps even more uncertain. Both of the approaches fail to provide a means to adequately address specificity requirements. The main reason for the lack of definition in the standards is that the application of the principle is instrument- and category-specific. The specificity requirements needed for an identification procedure are very different from an assay. In the case of elemental impurities, the specificity is linked directly to the ability to determine the presence and content of individual trace impurities in the presence of other trace metal impurities and complex matrix components. The majority of the analytical methods used for the actual determination of the metals will eliminate or significantly reduce the interference from organic impurities. This is accomplished through the digestion process, where applied, and the excitation/emission processes of the instrumental approaches. However, even modern techniques may be susceptible to interferences from other metals in the matrix. For example, the measurement of cadmium in a dosage form that contains a large quantity of aluminum as an excipient in an Inductively-Coupled Plasma Optical Emission Spectrometry (ICP-OES) procedure presents significant difficulties. The principal analytical emission line for aluminum and cadmium are the same, meaning that an analyst must move to an alternative emission line to detect the cadmium content. This is neither a surprise nor a difficult action for a trained analyst, but it does illustrate the need to know the materials that will undergo testing and to ensure specificity for common and potential interferences.

8.3.4.1 Measuring specificity

The level of specificity can usually be determined by measuring the amount of bias caused by the addition of interferences relative to the true value. This should be measured using a spike recovery study. It is important that the validation process includes a step that demonstrates the specificity of each impurity relative to the sample matrix and a further study of the interaction between the elemental impurities that may be present in a sample, as determined by the material's control strategy. In other words, it is important to demonstrate that the matrix will not affect the ability to quantify the element being measured, but it is equally important to demonstrate that other elements that could be present will not also interfere with the quantification. A series of spiked samples can be used to challenge the procedure, but how is it possible to determine when a potential interference becomes a significant problem? For impurity procedures, an interferent becomes problematic once it changes the ability to accurately quantify the measurand.

8.3.4.2 Limit tests

In the case of a limit test, a false positive is the likely outcome of lack of specificity. This is best indicated using a mixture of the all of the likely elemental impurities targeted for a given analyte.

In this solution, each of the impurities should be spiked at the maximum allowable level for each element. If the signals associated with each element in the spiked sample are found to be similar to the signal from the standard solution, then the procedure can be considered to be acceptable. Because of the inherent variability of the analyses at the trace levels usually used in these measurements, the level of acceptable similarity needs to link to the variability. Therefore, an acceptable procedure should be able to quantify a spiked solution to within 20% of the target limit. For instance, a spiked solution at the maximum target levels of three metals might contain 5 ppm lead, 15 ppm arsenic, and 5 ppm cadmium and the analyte (e.g. aspirin). When measured, the lead, arsenic, and cadmium signals should be less than the signals from separate standards at 6 ppm, 18 ppm, and 6 ppm, respectively.

8.3.4.3 Quantitative tests

The requirements and rationale for the quantitative procedures and the limit procedures are very similar. The variability of the procedure must be considered when examining specificity. Unlike limit tests (such as sulfide precipitation), a lack of specificity for quantitative tests is rarely an additive function across all elements in a test, but is usually caused by specific interactions, overlaps, or sampling issues. For instance, when measuring arsenic by inductively-coupled plasma mass spectrometry (ICP-MS), care must be taken to ensure that argon chloride is not mistaken for arsenic (each has the same m/z). However, the most effective way to measure for a lack of specificity is comparison of the spiked solutions to the calibration curves for each target element. If the measured content of the spiked solution differs from the known value of the spike by more than 20%, the procedure lacks sufficient specificity.

The types and numbers of samples, measurement approaches, and acceptable measures for accuracy determination of each approach have been compiled in Table 8.4. The third row in the table contains a best practice recommendation when applying the validation approaches to procedures used to evaluate elemental impurities.

8.3.5 Quantification limit

The limit of quantification (quantitation) is defined using two primary means in the Traditional approach. These include the signal-to-noise (S/N) ratio and analysis of the calibration curve.

Table 8.4 Specificity Comparison

Validation Approach	No. and Type of Samples to be Examined	Measurement Approach	Acceptable Measure
Traditional	–	–	–
Standardized	–	–	–
Best practice	N + 1 independent spiked solutions (N = No. of target elements)	(Limit test): comparison of spiked solutions to standard solutions (Quantitative test): comparison of spiked solutions to calibration curve	(Limit test): signal from spiked solutions at 100% target values are less than signals from standards at 120% (Quantitative test): calculated value of each element in the spiked solutions differs from the spiked value by <20%

Table 8.5 LOQ Comparison

Validation Approach	No. and Type of Samples to be Examined	Measurement Approach	Acceptable Measure
Traditional	Suitable no. near quantitation limit	Signal-to-noise ratio or Standard deviation of blank and slope	S/N of >10 —
Standardized	See accuracy for the standardized approach	See accuracy for the standardized approach	See accuracy for the standardized approach
Best practice	See accuracy for the standardized approach	See accuracy for the standardized approach	See accuracy for the standardized approach

The S/N ratio approach compares the signal obtained from an appropriate portion of the baseline and the signal of samples at a low concentration. Alternatively, the standard deviation of a blank and slope of the calibration curve can be used to calculate the quantification limit. The Standardized approach indicates that the limit of quantitation (LOQ) need not be measured. The Standardized approach takes advantage of the unique application of the standards in the pharmacopeia to indicate that the LOQ has no bearing upon the ability of a procedure to provide a meaningful response. In the pharmacopeia, the limit is known, which means that the target concentration is always defined before the validation efforts begin. These limits are often well removed from the limits of the instrumentation used to measure the elements. Therefore, the accuracy studies are sufficient to ensure that the ability of a procedure validated as described will be capable of measuring the target elements.

The needs of an analytical lab are not always linked solely to the analysis of samples to comply with a pharmacopeial standard. Often an analytical system (ICP or AA) is set up to measure the content of elemental impurities at much lower and even higher levels than those described by the pharmacopeia. Therefore, it is beneficial to attempt to identify the limit of quantification for elements that are often measured in a laboratory. Once a limit of detection (LOD) is identified and a calibration curve leading to the LOD is prepared, the analytical system becomes much more flexible to the individual tasks asked of the analyst. The use of an LOD therefore serves a practical need even if knowledge of the LOD does not specifically affect the evaluation of an acceptable system as defined by the USP. So an extension toward the absolute limit of the ability of an instrumental procedure should be considered on a case-by-case basis, but a systematic and extensive study to determine the LOD may be wasteful and largely unnecessary.

The types and numbers of samples, measurement approaches, and acceptable measures for limit of quantification determination of each approach have been compiled in Table 8.5. The third row in the table contains a best practice recommendation when applying the validation approaches to procedures used to evaluate elemental impurities.

8.3.6 Linearity

The linearity of a procedure is a critical piece of information in most traditional validations. Once again, the outcome of a linearity study is used as a surrogate estimation of the accuracy of the procedure.

The linearity is typically measured with a series of standard solutions that span the entire range of expected values of the measurand. These data are then subjected to linear regression and the slopes, y-intercepts and correlation coefficients are considered. A linear regression fit is an approximation of concentration of a solution with a given signal obtained. A large number of factors play into the adequacy of the curve, and a small bias in the slope or intercept will lead to a systematic bias in the results that is difficult to predict. The slope and intercept are not directly correlated to the amount of random error present in a system. However, there is a loose correlation between the correlation coefficient (R^2) and systematic bias. This loose correlation can be used as a rule of thumb, but none of the measures of linearity provide any greater insight into, or information about the performance of the procedure than that already obtained in the accuracy studies presented in the Standardized procedures described previously. In general terms, the analytical procedures used to measure elemental impurities provide extremely large linear dynamic ranges across several orders of magnitude. The accuracy approach will ensure that a procedure will have adequate ability to quantify an impurity at 50% of the target limit. When a user wants to extend the range and application of their procedure across several analytes (and therefore differing target limits), extending beyond the 50–150% range is justified, but the acceptance criteria described in the Accuracy section should be maintained across the entire range.

8.3.6.1 Best practice

In the evaluation of elemental impurities, a majority of the instrumental procedures have very large linear dynamic ranges, suggesting that an extensive study of the linearity is often unnecessary. In addition, all of these procedures will require calibration using standards across the range of interest. Therefore, the linearity is evaluated on a regular basis even before a validation is initiated. Conducting a linearity study will, however, provide an indication of a matrix interaction through the difference between the values obtained from the standards and spiked samples. However, this information is also provided by the accuracy studies described in the standardized approach. The extension of the linearity study will provide greater flexibility than the standardized procedure for other application of the procedures, so an extension of the range from 50% to 10% is a value-added proposition (See Section 8.3.5). Otherwise, the Standardized approach reduces the amount of work without sacrificing capability or confidence in the procedure.

The types and numbers of samples, measurement approaches, and acceptable measures for linearity determination of each approach have been compiled in Table 8.6. The third row in the table

Table 8.6 Linearity Comparison

Validation Approach	No. and Type of Samples to be Examined	Measurement Approach	Acceptable Measure
Traditional	Five Concentrations	Slope, intercept and correlation coefficient calculation	–
Standardized	See accuracy for the standardized approach	See accuracy for the standardized approach	See accuracy for the standardized approach
Best practice	Five or more concentrations from 10% to 150% of the target value	See accuracy for the standardized approach	See accuracy for the standardized approach

contains a best practice recommendation when applying the validation approaches to procedures used to evaluate elemental impurities.

8.3.7 Range

The range study is often considered in conjunction with the linearity study. In the Traditional approach, the range is evaluated in order to demonstrate that the procedure is capable of providing sufficient accuracy and precision across the interval of expected target limits. There is no specific guidance on the acceptable range of a procedure, but in the case of elemental impurity procedures, the range will typically exceed any acceptance criteria by several orders of magnitude. The Standardized approach recognizes the typical range of the procedures and indicates that a range study is not necessary because the accuracy study has provided all of the data that is necessary.

8.3.7.1 Best practice

As stated in the Linearity Section 8.3.6, there is no reason to complete a separate range study; instead the accuracy study as indicated in the Standardized approach is sufficient to ensure that the procedure being validated is capable of consistently providing acceptable results.

The types and numbers of samples, measurement approaches, and acceptable measures for range determination of each approach have been compiled in Table 8.7. The third row in the table contains a best practice recommendation when applying the validation approaches to procedures used to evaluate elemental impurities.

Table 8.7 Range Comparison

Validation Approach	No. and Type of Samples to be Examined	Measurement Approach	Acceptable Measure
Traditional	–	–	Adequate accuracy, precision, and linearity
Standardized	See accuracy for the standardized approach	See accuracy for the standardized approach	See accuracy for the standardized approach
Best practice	See accuracy for the standardized approach	See accuracy for the standardized approach	See accuracy for the standardized approach

8.3.8 Detection limit

The LOD is a mainstay of the Traditional approach and is required only for a limit test. The LOD is the concentration at which a measurand can be detected but not necessarily quantitated. The Traditional approach as described in USP General Chapter <1225> indicates that the LOD does not actually need to be measured—instead, the ability to show that a signal is obtained at the target limit of the procedure is sufficient. Alternatively, the LOD can be estimated to be twice to three times the level of the noise. However, this would be demonstrated by an extrapolation from the linearity (which is not a requirement for a limit procedure).

The Standardized approach eschews the concept of LOD in favor of a new term, Detectability (see Section 8.3.9). The rationale for the movement away from the LOD procedures is particularly important in the evaluation of elemental impurities. In a limit procedure, the critical measurement must demonstrate the smallest gradation between a passing value and a failing value at the target limit of the procedure. This value is not reflected by the LOD, which measures the point at which the analytical signal diminishes to the point that it can no longer be distinguished from the noise. Instead, it may be determined by measuring a series of standards that approach the pass/fail limit. These solutions would differ from the limit by smaller and smaller values, and the results would be evaluated using a statistical analysis to determine when the procedure can no longer detect a difference between the passing and failing value. To effectively obtain these results, the number of replicates would need to be increased to better represent the true value, thus making this approach a time-consuming and expensive proposition. *For example*:

Analyte is acetaminophen.
Measurand is lead.
Target Limit is 10 ppm.
Analytical Procedure is Inductively Coupled Plasma—Optical Emission Spectroscopy.

The LOD is found to be 0.01 ppm. This LOD would theoretically support a specification of 10 ± 0.01 ppm. However, the precision of this procedure has been demonstrated to have an RSD of about 10%. The precision would limit the specification to 10 ± 0.1 ppm.
To determine which of these is correct, a series of 6 independent solutions each having 11 ppm, 10.1 ppm, 10.01 ppm, and 10.001 ppm would be prepared and measured. To reduce the error in the experiment, it would be recommended to increase the number of independent solutions until the confidence interval was sufficient to statistically differentiate the samples. This would likely result in upward of 20 independent solutions.

The added expense of this approach is not justified for a limit test that is intentionally intended to be a less-capable and less-costly approach than a quantitative procedure. The detectability procedure provides a less-costly and equally appropriate approach to replace LOD determinations.

8.3.8.1 Best practice

The LOD is a value that is of interest to the analyst in terms of the range of the analytical procedure, but it provides no value for a validation study. The Standardized approach provides much greater assurance that a procedure is capable of providing a valid result by the requirement of precision and detectability. The best practice is to determine LOD as a component of the procedure development efforts and use detectability in the validation of the procedure.

The types and numbers of samples, measurement approaches, and acceptable measures for LOD determination of each approach have been compiled in Table 8.8. The third row in the table contains a best practice recommendation when applying the validation approaches to procedures used to evaluate elemental impurities.

8.3.9 Detectability

The detectability measure is a new term and approach that was introduced in the USP General Chapter <233>. Detectability is measured by evaluating two solutions: (1) with an analyte spiked with the measurand at the target limit, and (2) with an analyte spiked with the measurand at a concentration

Table 8.8 LOD Comparison

Validation Approach	No. and Type of Samples to be Examined	Measurement Approach	Acceptable Measure
Traditional	–	Visual detection	Detection at a concentration lower than the limit
Standardized	Proceed as directed in detectability (Section 8.3.9)	Proceed as directed in detectability (Section 8.3.9)	Proceed as directed in detectability (Section 8.3.9)
Best practice	Proceed as directed in detectability (Section 8.3.9)	Proceed as directed in detectability (Section 8.3.9)	Proceed as directed in detectability (Section 8.3.9)

below the limit. The results of these spiked solutions are compared with the results of a standard solution at the target limit. The first solution represents a failing value. The second solution represents a passing value. The concentration difference between these solutions is defined in the chapter as 20%. The second solution should provide a passing result. If it cannot, then the procedure is incapable of adequately determining the limit value. The limit of 20% is set by typical precision of instrumental techniques near 1 ppm. Although the concepts are not difficult, the application of the concept can be tricky. Therefore, an example is included herein.

Analyte is acetaminophen.
Measurand is lead.
Target Limit is 10 ppm.
Analytical Procedure is Inductively Coupled Plasma—Optical Emission Spectroscopy.
Standard solution: 10 ppm Lead.
Sample solution 1 (Limit value): 10 ng/mL lead spiked in acetaminophen at 1 mg/mL.
Sample solution 2 (Passing value): 8 ng/mL lead spiked in acetaminophen at 1 mg/mL.

In this example, the solutions should be prepared in triplicate and the average values from each compared to the standard solution value. The emission value obtained from *Sample solution 1* should be not significantly different than that of the *Standard solution*, and the results obtained from *Sample solution 2* must be significantly less than the *Standard solution.*

It should be clear from this example that the preparation of the solution must be very carefully completed and a procedure capable of completing this validation step should be considered to be acceptable.

8.4 CONCLUSIONS

The validation of elemental impurities procedures is an area of growing importance, as evidenced by the inclusion of validation principles in the new USP general chapters. As the understanding of validation has grown over the past 20 years, it has become increasingly clear that the validation guidances and chapters are often inadequate. The new general chapters challenge the traditional approaches to validation through the introduction of a new standardized approach. The new approach provides specific procedures and acceptance criteria to clearly demonstrate that a procedure is capable of providing valid results. This approach focuses on the importance of precision and accuracy evaluations and recognizes that many of the traditional approaches were used as surrogates for the well-defined accuracy and precision measurements. However, these standardized procedures have been designed to a very specific pharmacopeial application. In the laboratories of the analysts that must validate their procedures, internal uses that extend beyond the limited (but very important) pharmacopeial uses, a broader application of the principles of validation often hold value. For these applications, this chapter has provided a series of best practice recommendations that can be used to extend the understanding of a specific procedure and its capabilities. These best practices tend to follow the recommendations of the standardized approaches with specific procedures and acceptance criteria. This is an evolving field and there are a number of workers in the area of validation with new and exciting approaches to validation. Over the next several years, there are likely to be a number of important changes in our understanding of validation, and this chapter attempts to change present perceptions and approaches to analytical validation in general and the validation of methods for elemental impurities in particular.

References

1. USP 36-NF 31, USPC, Rockville, pg 151.
2. DeStefano, A. J.; Zaidi, K.; Cecil, T. L.; Giancaspro, G. I.; the USP Elemental Impurities Advisory Panel. Elemental Impurities – Information. Jan–Feb. *Pharmacopeial Forum* **2010,** *36* (1), 298.
3. ICH Q2A(R1). *Validation of Analytical Procedures: Text and Methodology, International Conference on Harmonization.* http://www.ich.org/products/guidelines/quality/article/quality-guidelines.html.
4. USP 36-NF 31, USPC, Rockville, pg 893.

PART

Specific Tests: Drug Substance

3

CHAPTER

Solid-state characterization

Patrick A. Tishmack, Ping Chen, Pamela A. Smith

Analytical Chemistry, SSCI, A Division of Aptuit, West Lafayette, IN, USA

CHAPTER OUTLINE

9.1 INTRODUCTION

The International Conference on Harmonization (ICH) and pharmacopeial guidelines for method development and validation are primarily written to address chromatographic analyses. However, drug substances or drug products may need to be tested in the solid form to ensure that they have been consistently manufactured, stored, and handled with respect to maintaining the desired solid form for delivery. Although the various elements of method development and validation are generally very similar for solid-state techniques, there are several issues that must be addressed differently. Typical chromatographic analyses use homogeneous solutions of the reference and test analytes. Solid-state techniques also require homogeneous mixtures to get accurate and reproducible results, but homogeneous mixtures of solid materials are usually much more difficult to generate and maintain. Several intrinsic properties of solids contribute to inhomogeneity of mixtures. Various

Specification of Drug Substances and Products. http://dx.doi.org/10.1016/B978-0-08-098350-9.00009-6

combinations of particle size, shape, surface area, electrostatics, hygroscopicity, compressibility, and crystallinity (or lack thereof) may affect the degree of homogeneity for mixtures of two or more solids. The solids may be physically different (e.g. polymorphs) or chemically different (e.g. active pharmaceutical ingredient (API) and excipients). It is difficult or impossible to predict how the above-stated properties will affect any particular analytical technique, which means that some trial and error is necessary in the feasibility stage of solid-state method development. It is also important to consider how the sample is prepared, such as mixing technique and exposure to water vapor in the air. Solid-state reactivity is occasionally an issue, but chemical degradation is usually much more rapid in solutions and would have been addressed at an earlier stage in the drug development process.

Although there are numerous solid-state analytical techniques available to characterize materials, relatively few are practical for pharmaceutical method development. The ideal solid-state analytical technique preferably will have good specificity, high sensitivity, minimal time requirements, consistent response, and be nondestructive. Bulk techniques such as thermal analyses and vapor sorption have minimal specificity and are destructive. Certain techniques, such as particle size determination (PSD), are commonly used in method development and validation, but there are several ICH validation elements that cannot be easily or practically evaluated [e.g. limits of detection and quantitation (LOD and LOQ) for PSD] because no standards are available to adequately determine these validation elements. The most common solid-state analytical techniques used for method development and validation for pharmaceuticals are X-ray powder diffraction (XRPD), Raman spectroscopy, infrared (IR) spectroscopy, and solid-state nuclear magnetic resonance (SSNMR) spectroscopy. Each of these techniques has a unique specificity and sensitivity for solid materials, and these characteristics are what must be adequately tested and exploited in deciding the appropriate technique for method development.

According to the ICH validation guideline (Q2 (R1)), there are four different types of methods that may be developed[1]: (1) identification test, (2) limit test for the control of impurities, (3) quantitative test for impurities, and (4) quantitative assay (content/potency) for major component. The United States Pharmacopeia (USP) uses four categories for methods[2]: (I) analytical procedures for quantitation of major components of bulk drug substances or active ingredients (including preservatives) in finished pharmaceutical products; (II) analytical procedures (quantitative or limit tests) for the determination of impurities in bulk drug substances or degradation compounds in finished pharmaceutical products; (III) analytical procedures for determining performance characteristics (e.g. analytical procedures such as PSD, surface area, bulk and tap density, and dissolution); and (IV) identification tests. The choice of an appropriate method will depend on the stage of drug development and an assessment of the level of monitoring required of the solid-state properties of the material. Table 9.1 is a summary that consolidates the ICH and USP tables for method validation; it has been modified to be applicable to solid-state method development.

An identification test is the simplest method to develop because only one validation element (specificity) must be evaluated (see also Chapter 5). For solids, this is usually a qualitative physical identification test for the sample. This assessment is typically achieved by comparison of the sample to a reference sample tested in an identical manner. An identification test would be used to compare drug substance batches to a reference batch in the early development stage. Later in development a qualified reference standard would be used.

Table 9.1 Validation Characteristics Recommended for Various Types of Analytical Testing

	Type of Analytical Procedure				
	USP Category IV	USP Category III	USP Category II (Impurity)		USP Category I (Major Component)
Analytical Performance Characteristics	ICH Identification	ICH Assay*	ICH Quantitative Test	ICH Limit Test	ICH Assay*
Specificity	++	++	++	++	++
Linearity	–	†	++	–	++
Range	–	†	++	–	++
Accuracy	–	†	++	–	++
Precision (repeatability)	–	†	++	–	++
Intermediate precision (ruggedness)	–	†	++	+	++
Detection limit (DL)	–	†	–†	++	–
Quantitation limit (QL)	–	†	++	–	–
Robustness	–	†	+	+	+
System suitability	–	–	+	+	+
Goodness-of-fit‡	–	–	+	+	+

– Indicates that this characteristic is not normally evaluated.
+ Indicates that this characteristic is recommended for solid-state methods.
++ Indicates that this characteristic is required by the Food and Drug Administration or ICH Q2 (R1).
**USP has two categories where the ICH has only one for these tests.*
†This characteristic may be needed in some cases depending on the assay type.
‡This characteristic is only required for methods involving chemometrics or modeling.
Sources: ref 1 (ICH) and 2 (USP)

A limit test is typically used to assess the amount of an impurity (e.g. minor solid form or forms) present in a sample. It is usually important that impurity levels should be below a given value, and an exact amount is not always necessary to determine. A limit test would be used when a particular minor form must be monitored and cannot be above a certain value, for example, at lot release of the drug substance. A quantitative method is the most involved procedure to develop, and the required characteristics to evaluate during development and validation are designed to accurately determine the amount (e.g. concentration) of a component in a sample and the associated error range for the measurement. An example of where a quantitative test would be necessary is when one must accurately assess the amount of a contaminating solid form present in a sample. This type of test would be useful in the case of patent infringement. Another type of quantitative test is an assay for the amount of the major component (e.g. solid form) present in a sample. This type of quantitative assay would be used if the identity of the minor component(s) is not important, such as when monitoring processing changes and their effect on form purity.

9.2 VALIDATION

Validation is the process to establish the performance characteristics of analytical procedures to meet the requirements of the analytical application.[2] In general, all recommended validation characteristics shown in Table 9.1 must be evaluated according to the intended purpose for a solid-state method to be called "validated". More validation characteristics for solid-state methods are recommended than are defined by the various guidelines due to the unique challenges inherent in analyzing solids. If methods do not contain all of the recommended validation characteristics, they are usually considered "non-validated". Although "partial validation" is a relatively common term used when one or more specific validation elements (characteristics) are not evaluated, this category has not been defined by the ICH or compendial guidelines. Therefore, methods are either validated or non-validated. Individual validation elements are evaluated in the process of validating a method. Verification and qualification apply to compendial methods that do not require revalidation. The verification or qualification of a compendial method demonstrates that it works properly in the testing laboratory.

Analytical methods should be validated for the following instances: (1) Phase II development and beyond, (2) acceptance or release of raw materials, (3) release of drug substance and drug product, (4) stability studies, (5) setting specifications, and (6) establishing expiration dates. Non-validated methods may suffice for the following instances: (1) Phase I development (fit-for-purpose development), (2) during process validation when "qualification" is acceptable to ensure reliable, objective, and accurate results, (3) comparability studies, and (4) characterization studies. Note the potentially confusing use of the term "process validation" as one of the instances when qualification is acceptable.

9.3 SAMPLE PREPARATION

Sample preparation is a critical issue in any method development because it must ensure that the tested material is consistent and suitable for the analysis being performed. It is very important to ensure that the reference standards and validation samples are homogeneous, and are actually representative of typical samples that ultimately will be analyzed by the method. Reference standard materials used in method development must be "well characterized" and of suitable purity.[1] Solutions are homogeneous by definition. However, for solid samples, homogeneity is much more difficult to achieve. For solid samples, it is also difficult to use internal standards because this practice adds another component requiring homogeneous mixing into the sample. There is also the possibility that the standard may interact with the other components in the solid state. For solid-state methods, it is better to normalize to the total response of all components. Typically, normalization is made to the unit peak area or unit variance. It is also important to choose an appropriate number of sample batches to adequately represent production materials whenever possible. A method is only applicable for testing the type of materials that were used to develop the method.

Several procedures are available to obtain a homogeneous solid mixture. Stirring, shaking, and geometric mixing are common for dry mixing, but slurry mixing may work in certain cases where there is very little chance that a solid form conversion will occur. Geometric mixing, the most common mixing technique, is a procedure that involves mixing equal quantities of each component together starting with the smallest amount for one component. For example, assume the target is a mixture of

Table 9.2 Geometric Mixing Example for Two Solid Forms Each at 5% by Weight in 500 mg of a Drug/Excipient Mixture

Step	Form I (mg)	Form II (mg)	Excipient Blend (mg)	Total Mixture (mg)
1	25	25	25	75
2	–	–	75	150
3	–	–	150	300
4	–	–	200	500

two solid forms (e.g. form I and form II) at the same concentration (5% each) with sufficient excipient blend (a prepared mixture of all required excipients in the correct ratios) to make 500 mg of a mixture. An example of geometric mixing to obtain this mixture is summarized in Table 9.2. Equal amounts of the three components are accurately weighed onto a weighing paper or dish (i.e., 25 mg each of forms I and II and excipient blend), and the cone-and-quarter technique[3,4] is used to mix the materials together thoroughly. The total amount of material just mixed is 75 mg to which is added 75 mg of excipient blend and mixed thoroughly as before. The total weight is now 150 mg of mixture, and 150 mg of excipient blend is added and mixed again to give 300 mg of mixture. Finally, the remaining 200 mg of excipient blend is added to the mixture using the same procedure to give 500 mg of thoroughly mixed sample at the desired concentrations of materials. This procedure can take a significant amount of time depending on the concentrations and number of components, which is one reason to use an excipient blend as one component rather than attempting to mix each excipient independently. However, to avoid subsampling issues, one must ensure that the excipient blend has been properly mixed before use. Significant additional difficulties may arise if the mixture has to be prepared in the glove box or if static electricity is an issue. Also, whenever a mixture is handled, the preferential loss of one component over another is always a risk. Finer components may settle and adhere to the weighing paper when transferring to a vial, for example.

Slurry mixing may be useful if the sample can be exposed to a solvent and does not change form. Slurry mixing may be successful in cases where the two solid forms have different crystal habits such as needles and plates where a homogeneous dry mixture would be very difficult to produce. Slurrying two insoluble solid forms together creates a random suspension of the particles that will usually settle out as a homogeneously mixed material. Evaporation of the solvent results in a dry solid that is suitable for analytical method development.

9.4 FEASIBILITY TESTING

The amount of preliminary work required to characterize a material appropriately can be quite extensive prior to selecting an appropriate analytical method for development. A primary concern with analyzing a solid is particle size, particularly when working with XRPD or IR analyses. Special care needs to be taken in these cases to minimize the potential for particle size effects or to characterize them very well for the material. The best procedure is to use appropriately representative materials for all of the forms of interest. If a particle size specification exists, the analytical test method should be developed specifically for the specified range. Any change in the particle size after the method has been developed would have to be examined to see if the method is still appropriate. A typical scenario is that no particle size specification is in place when the method is developed. In this case, control of the particle size must be performed as part of method development using several feasibility experiments to determine the particle size effect as well as the handling properties of the material. It might be necessary to gently crush or grind samples to use for method development, but it must be demonstrated that this action does not affect the solid form or create disorder or amorphous material in the sample. Sieving the samples might be considered a simple means to control the particle size, but sieving may result in preferential loss of material in the sample, which could affect the results. Unanticipated loss of a critical component of the sample is particularly important if nonauthentic material was used to develop the method.

Another characteristic that must be assessed prior to method development is sample purity. An analytical test may readily detect a low level of an unrelated contaminant that should not be considered when assessing a sample for multiple solid forms of a drug substance. Chemical contamination should not be an issue when developing a solid-state method because ideally this potential issue would have been previously addressed by a chromatographic technique to ensure the purity of the sample.

9.5 IDENTIFICATION TESTS

Identification tests (see also Chapter 5) are the simplest methods to develop and validate because specificity is the only required data element for validation[1,2]. Specificity is defined as the ability to unequivocally assess the analyte in the presence of components that may be expected to be present. This means that specificity of any analytical technique should be determined for the analyte in the presence of one or more similar materials. This ensures that the analysis can distinguish the analyte unambiguously from other compounds that may reasonably be expected to appear during processing. Even when care is taken to avoid contamination by reactants, side products, degradants, and related compounds, other solid forms may arise depending on the handling processes or storage conditions. A drug product usually has been handled and processed much more than the drug substance it contains, which introduces more opportunities for solid form changes. Identification tests also become significantly more difficult when the number of components increases, which is nearly always the case going from drug substance to drug product. It is also important to be sure that a sufficient number of test materials are analyzed during method development. As noted in Section 9.3, the test materials must be representative of the expected variation that occurs during production for the method to be reliable. An example of specificity determination for an XRPD identification test is shown in Fig. 9.1, which can be compared to the specificity determination of the same materials by ^{13}C SSNMR spectroscopy in Fig. 9.2. A comparison like this is standard practice in determining the feasibility of a particular analytical technique for solid-state method development and validation. In this case, either XRPD or SSNMR spectroscopy is suitable for an identification test method given that both techniques have sufficient specificity for each solid form. This may change if a drug product is to be analyzed by the method, in which case the excipients become part of the specificity determination.

Identification of a compound can be confirmed by obtaining positive results from samples containing the analyte, obtaining negative results from samples that do not contain the analyte, and

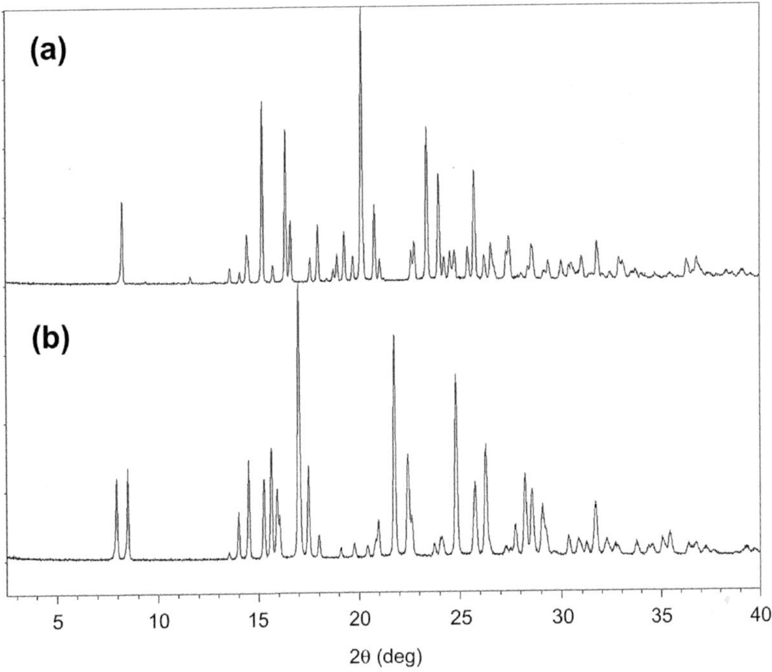

FIGURE 9.1

Specificity determination by XRPD for two polymorphs of ranitidine HCl: (a) form 2, (b) form 1.

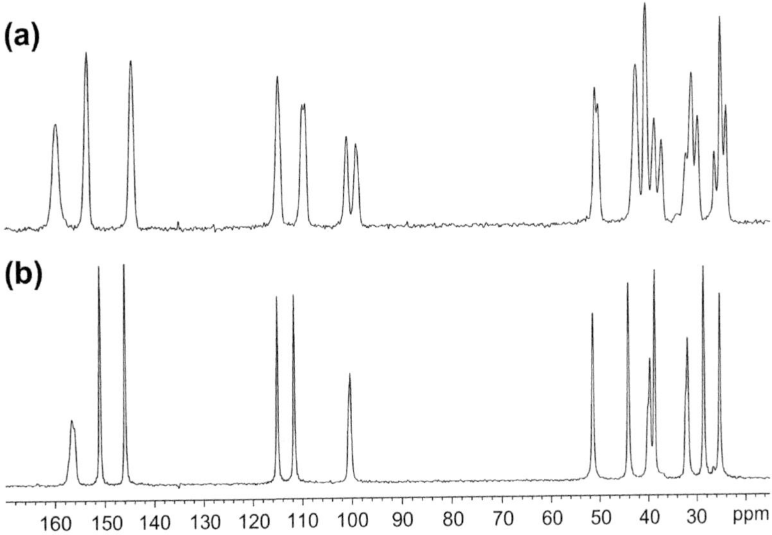

FIGURE 9.2

Specificity determination by ^{13}C SSNMR spectroscopy for two polymorphs of ranitidine HCl: (a) form 2, (b) form 1.

blinding the sample identity prior to analysis to reduce analyst bias. Some of the common analytical techniques for solid-state identity testing include XRPD, IR, Raman, SSNMR spectroscopy, differential scanning calorimetry (DSC), or dynamic vapor sorption (DVS). Nonspecific tests like DSC or DVS typically require a second test to establish specificity for the material of interest. The diffraction and spectroscopic tests are significantly better for establishing specificity in the solid-state, particularly for mixtures where bulk techniques typically fail.

There are different ways to validate identification tests. Visual examination is the most common approach. However, this approach is usually very subjective, and an objective level of disregard for low level solid phases should be incorporated into the method. For XRPD and spectroscopic methods, it is common to attempt identification by picking peaks and checking that their positions are acceptable. This approach is subject to the error of the peak picking algorithm and even to differences in the shape of peaks from sample to sample. It is usually much more efficient and reproducible to match the entire pattern or spectrum, while allowing for small differences overall. This is where the level of disregard criterion comes in to allow very minor differences in samples to be ignored as long as these are not associated with performance differences between the samples. Establishing characteristic peaks to identify a solid form is a useful exercise, but it typically requires careful assessment of the standard and multiple batches of test material to obtain a reliable set of characteristic peaks. Once the best peaks are found, one common approach is to normalize the spectrum (or pattern) and then compare these in an overlay with a reference spectrum (or pattern). Such an approach will provide some measure of objectivity and increase the reliability of the visual assessment.

An example where XRPD is used for an identification test is shown in Fig. 9.1. An important issue for identification tests is setting the acceptance criteria. One aspect that needs to be clear is the meaning of the word "conforms". The meaning may be slightly or significantly different depending on the method and the samples being tested. A decision needs to be made on how conservative (restrictive) the meaning needs to be for each particular method case. Some examples in the order of the most conservative to the least are: (1) all peaks in the reference pattern/spectrum are present in the test pattern/spectrum, and the test pattern does not contain additional peaks not observed in the reference pattern/spectrum; (2) comparison of the test and reference patterns/spectra over a defined range(s) and examination for certain "characteristic peaks"; (3) visual comparison and assessment using good scientific judgment to determine if a test pattern conforms to a reference pattern/spectrum.

9.6 LIMIT TESTS

Limit tests are somewhat more complicated than identification tests and have ICH required validation elements of specificity and DL. Although only two validation characteristics are required, there are good reasons for including robustness, ruggedness, and system suitability when developing limit tests for solid state materials. The need for specificity is obvious for limit tests, especially in the presence of other materials that may be at much greater concentrations. Limit tests are useful if the exact amount of impurity present is not required, but the minor form needs to be controlled at or below a specific level. There will usually be a level of solid-state impurity that is acceptable in a drug substance or drug product. The desired threshold level is chosen once the acceptable level and the

DL have been established. The detected impurity would be reported as "equal to or less than" or "above" this threshold value. The choice of threshold level must be set somewhere between the DL and the acceptable level, inclusive of the low and high limits. When the threshold value is equal to the DL, then the wording for the reported impurity may be changed to "detectable" or "non-detectable" according to the observed result. The threshold does not have to be set at the DL, especially if this concentration does not correspond to the determined acceptable level for the samples. The setting of the threshold value depends on the risk tolerance for obtaining false positives or false negatives from samples that may be tested.[i] Setting the threshold value closer to the DL increases the probability of reducing the number of false negatives but increases the probability of false positives. Conversely, setting the threshold level closer to the acceptable level will reduce the number of false positives but increase the risk of false negatives. Being able to set the threshold value appropriately is a good reason for including sufficient batches of materials for testing during method development.

Verifying the DL is an important part of developing a limit test. The following descriptions of the DL[ii] are taken from the ICH guidelines and are essentially identical to those in the USP.[1,2] Three different ways of estimating the DL are described below, but the last description is generally the most appropriate means to estimate the theoretical limit. The estimated limit needs to be experimentally proven because the analytical test as applied for the method may not give the expected outcome. This verification is typically done by analyzing multiple samples prepared at or very near the DL.

The DL is the lowest concentration level at which an analyte can be detected. Three procedures that may be used are:

(1) Determine the DL by visual examination of data obtained from analysis of samples with known analyte concentrations, and establish the minimum level where the analyte can be reliably detected.

(2) Determine the signal-to-noise ratio (*S*/*N*) by comparing measurements of known low concentration samples with those of blank samples, and establish the minimum concentration of analyte that can be reliably detected. A *S*/*N* of 3:1 or 2:1 is generally considered to be acceptable.

(3) Based on the standard deviation of the analyte response and the slope of the calibration curve, $DL = (3.3 \times \sigma)/S$, where σ is the standard deviation of the response and S is the slope of the calibration curve. The standard deviation (σ) may be determined from the analytical responses (noise) of an appropriate number of blank samples or from the regression line of the calibration curve that includes the analyte at the DL. It may be inappropriate to use the standard deviation from blank samples in solid-state methods because this approach mainly characterizes the variation from instruments.

---

[i] A false positive (Type I error) occurs if some undesired material is apparently detected but none is actually present. Setting the threshold higher than the mean by some multiple of the standard deviation prevents interpreting values close to the mean as having the undesired form when it is not present but was apparently detected. A false negative (Type II error) occurs if no undesired material is apparently detected but some is actually present. Setting the threshold lower than the mean by some multiple of the standard deviation prevents interpreting values close to the mean as not having the undesired component when it is present but not detected. See reference 4.

[ii] Rather than using the common terms quantify and quantification, both the ICH and USP guidelines use the terms quantitate and quantitation, which are rarely used in other scientific disciplines. These words were apparently derived from the word quantitative. We use the terminology that is consistent with the ICH and USP guidelines.

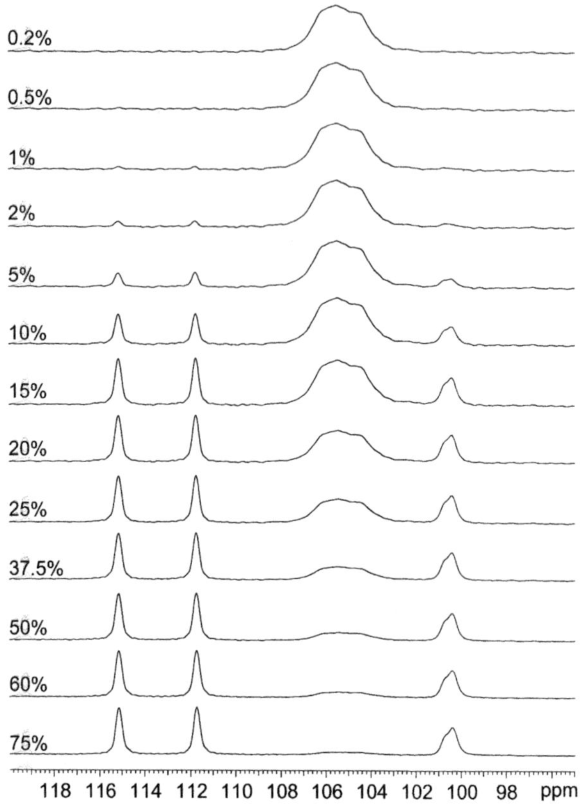

FIGURE 9.3

Estimation of the ^{13}C SSNMR DL for ranitidine HCl form 1 (peaks at 99, 101, 109.6, and 114.8 ppm) in an excipient mixture. The weight percent of ranitidine HCl form 1 in the mixtures is given above the corresponding spectrum.

However, for solid-state method development, other experimental factors such as sample and specimen preparation, unexpected peaks from an impurity, and lot-to-lot variability, may have a larger effect on the measurement.

Figures 9.3 and 9.4 are examples of how visual DLs for solid forms in a drug product may be determined using ^{13}C SSNMR spectroscopy. Form 1 in Fig. 9.3 can be detected at about 0.5% in the mixture, but form 2 in Fig. 9.4 can only be detected at about 1% in the mixture. Although the spectra for both forms were obtained using identical conditions and amounts of sample, form 1 could be detected at a lower level because it had significantly sharper peaks that resulted in a higher *S/N*.

An example of deciding on how to set a DL follows. A threshold level of 1.5% by weight of form A was acceptable. Six replicate analyses (separate samples) each of 0% form A samples (pure form B) and 1.5% form A samples were obtained. Statistical *t*-tests at the 99% confidence level were performed, which demonstrated that the 0% form A samples were distinguishable from the 1.5% form A samples (comparing 0% vs. 1.5% and 0% vs. 1.5% $-$ 2σ). The theoretical DL was estimated to be significantly

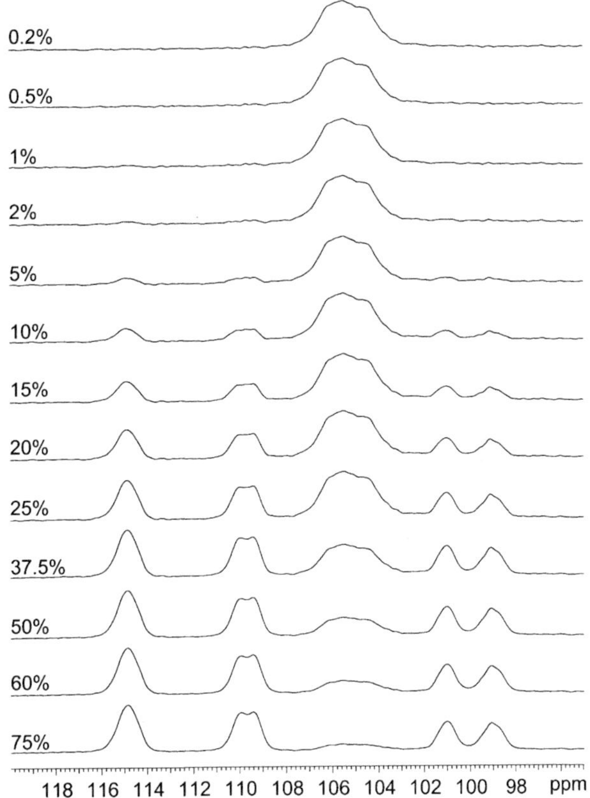

FIGURE 9.4

Estimation of the ^{13}C SSNMR DL for ranitidine HCl form 2 (peaks at 100.4, 111.8, and 115.2 ppm) in an excipient mixture. The weight percent of ranitidine HCl form 2 in the mixtures is given above the corresponding spectrum.

less than 1.5%. Some of the possible options for setting the threshold level are: (1) at the mean of 1.5% form A samples, (2) mean + σ, (3) mean − σ, (4) mean + 2σ, or (5) mean − 2σ. In this case, minimizing the risk of false negatives was necessary, so the threshold level was set at the average minus 2σ.

For solid-state method development, it is highly recommended to include the additional characteristics of robustness, intermediate precision (ruggedness), and system suitability to ensure that a reliable method results. Robustness refers to factors that are internal to the method where small changes to them would not necessarily be expected to affect the results. Robustness should be evaluated to assess whether the results from the analytical procedure remain unaffected by these slight changes. These experiments help identify areas in the method where tight control is needed. Robustness can be tested by altering data collection parameters and sample preparation procedures in a systematic way while obtaining multiple analyses. Robustness can be informally built into some methods by using different instruments, or even using different detectors on the same instrument.

Intermediate precision in the ICH guideline is called ruggedness by the USP.[1,2] These factors are all external to the method and are not expected to influence the test results. Ruggedness can be informally evaluated during method development by using multiple analysts and acquiring data on different days.

Both robustness and ruggedness can be formally addressed simultaneously by using full-factorial designs as described in Chapter 3. Full-factorial designs are generally preferred for solid-state method development because the factors tend to have unpredictable effects depending on the sample and analytical technique. Table 9.3 shows a typical example for a two-level (two options per factor), three-factor full-factorial design to evaluate ruggedness and robustness. The resulting data can be t-tested to determine if there are observed statistical differences for any of the parameters evaluated. As noted in Chapter 3, adding a factor to this design would increase the number of experiments by two times ($2^3 = 8$ vs. $2^4 = 16$) while adding a level would increase it by more than three times ($2^3 = 8$ vs. $3^3 = 27$).

The following scenario is an example of a FT-Raman method using a step-and-repeat accessory to perform a drug substance limit test. Two analysts obtained results that were statistically different from one another based on a t-test at the 95% confidence level. It is not acceptable to have a method that can be performed correctly by only a particular analyst. The two analysts in this case were interviewed to determine if there were any differences in sample preparation or analysis. The sample mass was controlled, so that was not a potential variable affecting the outcome. One difference noted was the procedure each analyst used to pack the sample cup. One analyst used the smooth side of a glass slide to pack the sample into the cup, and the other analyst used the frosted side. A second difference was the way each analyst focused the laser onto the sample. One analyst focused the laser on a stationary sample, but the other analyst focused the laser while the sample was spinning. These two differences changed the intensity of the spectra, and caused statistical differences between the results. Consistent results could be reproducibly obtained by specifying the exact sample preparation and laser focusing procedures in the method. These seemingly small differences (frosted vs. smooth glass slide, and focusing during spinning vs. static) had statistically significant impacts on the results, demonstrating the increased complexity of solid-state method development over traditional solution phase methods.

In another example, this time with an XRPD method, the sample mass was varied slightly during robustness testing. With the power level set at 35 kV/35 mA, there was a statistical difference between the

Table 9.3 Example of a Two-Level, Three-Factor Full-Factorial Design for Robustness and Ruggedness

Experiment	Analyst	Day	Instrument
1	1	1	1
2	2	1	1
3	1	2	1
4	2	2	1
5	1	1	2
6	2	1	2
7	1	2	2
8	2	2	2

results obtained for the two masses. This difference was not observed when the power level was higher (40 kV/40 mA). For this case, one could choose to control the mass carefully or specify the exact instrument power level. Controlling the mass was chosen because this approach ensured better consistency over time, given that the power settings might vary depending on the X-ray tube age. The X-ray tube output and necessary counting time could be more appropriately addressed in a system suitability study.

9.7 QUANTITATIVE METHODS

Quantitative method development and validation require the evaluation of more characteristics than any other type of method (Table 9.1). Quantitative methods are generally used to accurately determine the amount of impurity or the major component present in a sample. Note that the QL is required for validating quantitative tests for impurities, but it is not required for assays because the working range of quantitative tests for impurities usually starts at or close to the QL of the method while assays for the major component are far above these levels. Robustness, ruggedness, and system suitability characteristics are as important to evaluate for quantitative methods as for limit tests. According to the USP,[2] the QL does not necessarily need to be determined if a higher level of quantitation for an impurity is suitable for the intended purpose. Although "linearity" is one of the necessary characteristics for validation, a nonlinear response to the amount of analyte is acceptable if a consistent response to a systematically varied analyte amount is obtained during development.[1,2]

9.7.1 Calibration Curve

The results of quantitative tests may be reported as either a relative or absolute quantity of the analyte in a sample, and the choice depends on the situation. Relative quantitation may be possible if an impurity is measured and compared to the main component with the assumption that no other contaminants will be present. Because the sum of the two components must be 100% of the total material, a relative measurement can be obtained from the ratio between the analytical responses of the two components. The more common approach is an absolute quantitation by making a calibration curve of the analytical responses at various concentrations (usually weight percent) of the impurity in a synthetic mixture representing the material of interest (e.g., drug substance or drug product).

If a calibration curve is designed appropriately, it will provide information for several characteristics that must be evaluated in method development and validation. In addition to the DL and QL, the linearity and range can also be determined from the calibration curve. The linearity can be demonstrated with standard statistical curve fitting analysis. The range of the calibration curve is typically determined from the requirements of the method because it is unlikely that a range of 0–100% would be required in most cases. A smaller range will almost always improve the accuracy of the method and possibly the precision as well. For example, if one needs to detect the minor solid form at the 2% level, it would not be beneficial to develop a method with a range from 1% to 50%. A recommended starting point for accuracy is to use a range with the lowest level desired for the method approximately at the midpoint. This may not be possible if the desired level is close to the detection or quantitation limit. Calibration points should be relatively evenly spaced across the chosen range.

9.7.2 Accuracy

Accuracy is a measure of how well a method produces a prediction close to a true value for an analysis. The "trueness" of this value can be relative to a reference standard or an accepted conventional value.[1] The accuracy of a method should be established across the range over which the method will be used. For a drug substance, accuracy may be determined by performing the same analytical procedure on a reference standard of known purity or by comparison of the results with those of a reliable independent analytical procedure for which the accuracy has been established.[1,2] For a drug product, the accuracy may be determined by applying the analytical method to synthetic mixtures of materials representing the drug product formulation where the drug substance is varied within the analytical range of the method. If all of the components of the drug product are not available, it may be acceptable to use the method of standard addition or to compare the results with those of a reliable independent analytical procedure for which the accuracy has been established.[1,2] The method of standard addition is a useful approach to overcome matrix effects and establish the DL, QL, and accuracy for an analyte in a complex mixture. This technique involves adding known amounts (spiking) of the drug substance (or analyte of interest) to the drug product and measuring the response for these spiked samples as well as an unspiked drug product. The resulting calibration curve can then be used to determine the amount of analyte in the drug product. For both drug substance and drug product, the accuracy may be inferred once specificity, precision, and linearity have been established.[1]

Quantitation of impurities typically requires spiking each of them into the drug substance or drug product to determine accuracy. For quantitation of some impurities or degradation products, it is more likely to be necessary to assess accuracy by comparison to another established procedure, and the response factor of the drug substance may need to be used when this cannot be established for impurities that are extremely difficult or impossible to purify.

Accuracy is typically calculated as a percentage recovery by the assay of the known added amounts of analyte in the sample or as the difference between the mean and the accepted true value with the associated confidence interval.[2] The acceptable recoveries for solid-state methods are likely to be larger than what is typically obtainable with chromatography methods.[5] According to USP General Chapter <1225>, accuracy is assessed by evaluating the recovery of the analyte across the range of the assay or by evaluating the linearity between the estimated and actual concentrations.[2] The preferred statistical criterion is that the confidence interval for the slope encompasses 1.0 or that the slope of the calibration curve be close to 1.0. The validation protocol must provide the desired confidence interval or define closeness for the slope. For data collection to determine accuracy, the ICH guidelines recommend a minimum of nine determinations for at least three concentrations over the desired range of the method.[1] A set of three concentrations analyzed in triplicate would satisfy this recommendation, although it is sometimes difficult to obtain true triplicates when dealing with complex solid-state mixtures containing multiple components.

9.7.3 Quantitation Limit (QL)

Verifying the QL for an impurity is similar to the procedure for a limit test, and the description of the QL[ii] given below is very similar to that of the DL[1,2] described earlier. The third description is generally the most appropriate for determining the theoretical values for the limits. The estimated QL also needs to be determined experimentally to confirm whether the estimate is appropriate.

Similar to the DL, this is typically done by analyzing multiple samples prepared at or very near the QL.

The QL is the lowest level at which the analyte can be quantified with an acceptable accuracy and precision. Three procedures that may be used are:

(1) Determine the QL by visual examination of data obtained from analysis of samples with known analyte concentrations, and establish the minimum level at which the analyte can be reliably quantified.
(2) Determine the *S/N* by comparing the measurements of known low-concentration samples with those of blank samples, and establish the minimum concentration of the analyte that can be reliably quantified. A *S/N* of 10:1 is generally considered acceptable.
(3) Based on the standard deviation of the analyte response and the slope of the calibration curve, $QL = (10 \times \sigma)/S$, where the variables are defined previously for the DL.

Figures 9.5 and 9.6 show examples of calibration curves for a quantitative solid-state NMR spectroscopic method. Note that for either solid form of this particular material, the results show a relatively large scatter in the data, particularly for the blank. This leads to higher DL and QL than are desirable for this material, and it also demonstrates the real-world difficulties encountered in solid-state method development. Several issues to note were the difficulty in consistently preparing and analyzing replicates and the observation of analyst bias. Figs 9.5 and 9.6 only show a part of the calibration curves, which were actually determined up to 50–60% of each form. The linearity was significantly better ($r^2 = 0.997$) for both full-range curves due to the reduced influence from the relatively larger variations in the responses for calibrants at lower concentrations. However, the DL/QL increased to 3.7%/11.2% for form 1 and 3.0%/9.0% for form 2.

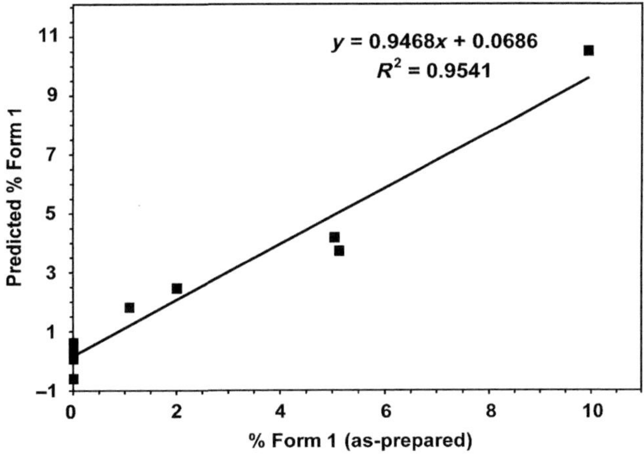

FIGURE 9.5

Calibration curve for the estimation of the DL and QL for ranitidine HCl form 1 in an excipient mixture. The estimated DL was 2.5%, and the estimated QL was 7.6%.

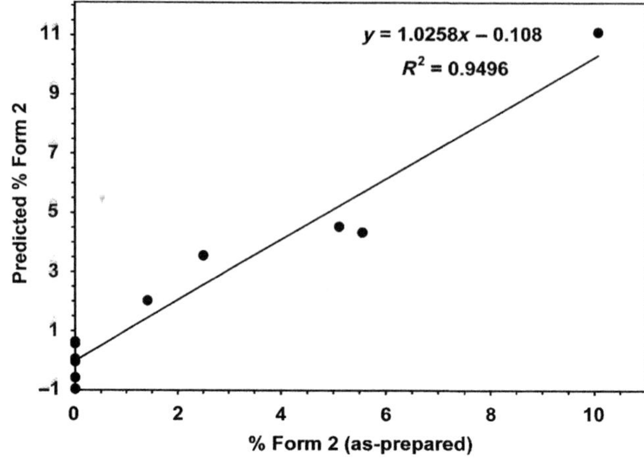

FIGURE 9.6

Calibration curve for the estimation of the DL and QL for ranitidine HCl form 2 in an excipient mixture. The estimated DL was 2.7%, and the estimated QL was 8.7%.

9.7.4 Precision

Precision is a validation characteristic required for assays of drug substances and quantitation of impurities. Precision is a measure of how close the results of multiple analyses of a homogeneous sample are to each other, and it should be assessed by determining the standard deviation and/or relative standard deviation and the confidence interval of a statistically relevant number of measurements. Precision can be viewed from several different perspectives, including reproducibility (inter-laboratory precision), repeatability (intra-laboratory precision for the same analyst and equipment), and intermediate precision or ruggedness (intra-laboratory precision where the analyst, equipment, or day of testing are varied). Other factors, such as multiple lots of reference materials or mass of the specimen, should also be considered. Table 9.3 shows an experimental design that can be used to study intermediate precision. The ICH guidelines recommend assessing repeatability using at least nine determinations in essentially the same manner as for accuracy, but an alternative approach is to perform at least six determinations at 100% of the test concentration.[1] A general recommendation is that precision should be ±2% to have an acceptable assay[6,7], but the acceptable precision would be higher for solid-state assays given the higher variability of experimental factors associated with solid-state materials. The acceptable precision for a quantitative test for impurity may be ±10–20%, depending on the range of the test.[5] Reproducibility is not required according to ICH guidelines, but it may be useful for producing a compendial monograph method.

9.7.5 Linearity and range

Linearity and range are two related validation characteristics required for assays of drug substance and quantitation of impurities. Linearity is the relationship between concentration of analyte and the assay response for it. Linearity may be obtained by performing a mathematical transformation of the response such as log, square root, or reciprocal. A nonlinear model is acceptable as long as there is a predictable relationship between the analyte concentration and the analytical response. The linearity can be visually established across the desired range using a plot of the analytical response vs. analyte concentration, and subsequently mathematically calculated by an appropriate statistical analysis (e.g. least squares linear regression). The ICH guidelines recommend establishing linearity with at least five concentrations over the range of the method.[1] Some exceptions to the requirements for evaluating certain validation characteristics are noted in the USP. For example, linearity is not appropriate for particle size analysis because a concentration range is defined such that the measured particle size distribution is not affected by changes in concentration within the defined concentration range.[2]

The range of a method describes its ability to measure analyte concentrations from the lower concentration to the upper concentration with appropriate precision, accuracy, and linearity. The ICH and USP guidelines recommend ranges for certain cases: (1) assay of drug substance or drug product—80–120% of the test concentration, (2) impurity determination—50–120% of the acceptance criterion, (3) content uniformity—70–130% of the test concentration, (4) dissolution test—±20% over the specified range (e.g. for 30–90% the range would be 10–110% of the label claim).

An example of a quantitative assay based on solid-state NMR spectroscopy of a tablet formulation follows. There are two polymorphic forms of a drug used in producing tablets by a relatively simple formulation. Forms I and II along with the excipient blend were analyzed with ^{13}C SSNMR spectroscopy. The data acquisition parameters were optimized such that good quality spectra were obtained for each material. The advantage of obtaining these preliminary spectra was to use linear combinations of them to estimate or predict some of the method development and validation characteristics. Using linear combinations of only three SSNMR spectra one can (1) determine the integral regions for the signals of interest, (2) predict whether the mixtures used for modeling have been made correctly, (3) estimate the visual DL for components of mixtures, and (4) assess whether to use multiple regression analysis or factor analysis such as partial least squares (PLS) or principal component regression.

The multiple regression model procedure was as follows: (1) obtain "raw" spectral data, (2) normalize the spectrum to the total integrated area, (3) select proper peak regions to integrate for each component, (4) sum up the characteristic regions for each component, (5) create a model based on the spectral response vs. the amount of component present, (6) perform multiple linear regression analysis on the data to obtain the fit parameters, DL, and QL, (7) analyze a validation data set and apply the model, and (8) apply the method to unknown samples. Figures 9.3–9.7 illustrate some of the results for this procedure for ranitidine HCl forms 1 and 2 in commercial tablets.

The SSNMR spectroscopic method highlights some difficulties that may be encountered in solid-state method development. Although the visually estimated DL and QL (via linear combinations) may be approximately 1% or less, predictive models must be carefully developed to attain these limits. Signal overlap in calibration spectra needs to be minimized to get statistically low DL and QL. This may require special data processing to deconvolute overlapping signals instead of standard peak integration. Prediction of the presence of the forms when none were actually in the sample was mainly due to the overlap of the various peaks of each form and the error in measuring the blank response.

9.8 SYSTEM SUITABILITY AND GOODNESS-OF-FIT

System suitability and goodness-of-fit are two related method development and validation characteristics. From the ICH guideline (Q2(R1)) and the USP, system suitability testing ensures that the

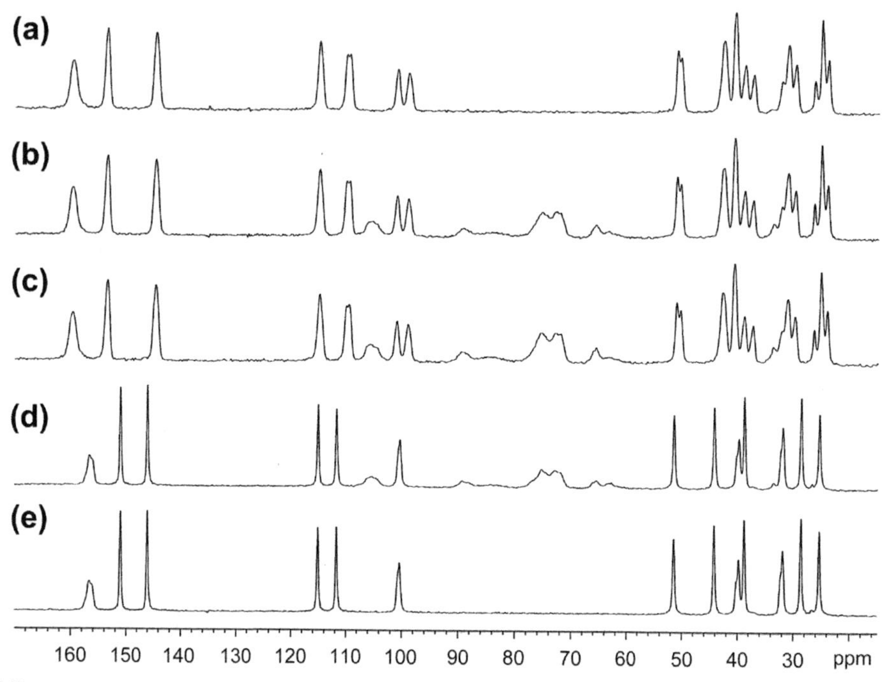

FIGURE 9.7

^{13}C SSNMR spectra of ranitidine HCl polymorphs and commercial tablets containing approximately 60% drug substance: (a) pure form 2, (b) tablet K, (c) tablet W, (d) tablet C, and (e) pure form 1. It is obvious from these spectra that tablets K and W contain form 2 and tablet C contains form 1. The additional peaks that are only in the tablets are due to excipients, which were essentially identical in each tablet.

entire system (equipment, electronics, analytical operations, and samples to be analyzed) are all acceptable for performing the method, and both ruggedness and robustness studies can help define the system suitability.[1,2] For solid-state methods, it can be a challenge to find an appropriate system suitability sample. One typically cannot use the sample of interest or a reference sample of it as is commonly done for chromatography because the solid-state characteristics of the material may change over time. It is not practical to retain standard samples for solid-state analysis, as form changes may occur, altering the expected concentration of the sample. A useful standard will be stable over long periods of time and not be related to the test samples in any way other than being solid.

The following is an example of a system suitability test for an FT-IR spectroscopy limit test where the threshold level was set at the DL. A polystyrene card was analyzed daily for the instrument performance qualification, and the *S/N* was determined for 21 days during method development and validation to give a mean value of 250 over a specified spectral range. This range did not contain any polystyrene bands, but because the sample was a thin film, there were interference fringes visible in the spectrum. The system suitability threshold was set at the mean minus 2σ (208) based on a 95% confidence level. The low threshold setting at the DL for the limit test required a system suitability test with a consistently high *S/N* to reduce the chances of an out-of-specification result. Every time the method was performed, a system suitability check of the *S/N* was conducted, and had to be ≥208 to continue with testing the sample. This requirement led to an unexpected consequence. The method was transferred internally to a separate group using the same instrument and failed immediately. A brief investigation revealed that the polystyrene card was placed into the spectrometer differently by inverting it 180° relative to the source, and the orientation of the film resulted in statistically different *S/N* values. Presumably, differences in the thickness of polystyrene or the film orientation caused a change in the spectral fringes, which in turn impacted the *S/N* and ultimately the failure of the system suitability. The system suitability section of the method was revised to specify the orientation of the polystyrene card. In subsequent versions of the method, the orientation of the polystyrene card was alternated during method development to avoid this restriction.

Goodness-of-fit in this context is a procedure to check how well the unknown samples match the calibration standards used in the method development. It is necessary when conducting chemometric analyses of data acquired during method development. This procedure is not required when routine peak area or intensity calculations are used for assessing data. The goodness-of-fit value is used to assess the validity of the predicted result. Partial least squares will always give a result, but it is necessary to determine whether this result is valid or sensible. There are several different approaches used in the industry, and some examples are given here. Statistics derived from calibration and validation samples, such as Mahalanobis distance and Hotelling T^2, could be used to for this purpose. In addition to the objective metrics, visual assessment and comparison of the pattern/spectrum of the unknown sample with that of calibration and validation standards also help to evaluate the match between unknown samples and calibration standards of the model. For example, new peaks found in the pattern/spectrum of an unknown sample but not present in those of calibration standards may indicate that the sample is different, and the model would not apply to this sample. Detailed instructions may be needed to avoid the subjectivity of this kind of visual assessment.

Another way to evaluate the goodness-of-fit of a sample is the spectral *F*-ratio (Eqn (9.1)). The spectral *F*-ratio determines the difference between the spectral residual (X_{res}) of the unknown sample and the averaged spectral residual of calibration and validation standards of the model (avg X_{res}). This ratio gives a good indication of whether the unknown sample is very different from the samples used to develop the model. Using statistics tables, one can set the threshold for the spectral *F*-ratio. If the *F*-ratio is below the threshold value, the result can be reported. If the *F*-ratio is above the threshold value, the result cannot be reported.

$$\text{Spectral } F\text{-ratio} = \frac{X_{res} \text{ in prediction object}}{\text{avg } X_{res} \text{ in validation object}} \tag{9.1}$$

Valid results will only be obtained by testing samples that are similar to those used in method development. A threshold level based on a statistical analysis of the standard and validation results can be set for an acceptable *F*-ratio value for predicted samples. If the *F*-ratio is below the threshold value, the predicted results for the tested sample can be reported. Note that given the difficulty in obtaining representative samples for a solid-state method, a threshold level should not be set based only on statistical analysis. The threshold level may be adjusted later when more data are available.

The following example is from an FT-Raman quantitative method. Partial least squares was used to determine the concentration of a contaminating solid form. The QL was set at 2.97%, and the *F*-ratio threshold was set at 2.67. Any results that were <2.97% could not be reported as quantifiable, and any results with an F-ratio >2.67 also could not be reported. This method was run for several months with no problems. There were occasional outliers, but these results were expected, given that the F-ratio was set at the 95% confidence level. After several months, the predicted results were much higher than expected, and all of them were above the F-ratio threshold. Two possibilities were that the samples had changed, or the instrument had some problem. Because there was a system suitability test in place that had not failed for any of these samples, the initial speculation was that the instrument was performing properly. However, upon examining the polystyrene spectra used for the system suitability assessment, a peak shift and a change in peak shape were observed. These changes demonstrated that the instrument and not the sample was the problem, and the *S/N* system suitability test was not sufficient for this FT-Raman method. The instrument was subsequently repaired, and the method was modified to include both *S/N* and peak position calibration as part of the system suitability test to ensure that the method would perform properly.

9.9 CONCLUSIONS

Solid-state methods are very different from chromatographic methods and require a unique approach to development and validation. Control and understanding of the samples are critical to obtain the homogeneity required for proper method development. Robustness and ruggedness are very important elements to consider even when not required by the guidelines. Many solid-state methods must adequately address system suitability and goodness-of-fit to develop dependable methods. If these factors are controlled and understood, excellent qualitative and quantitative solid-state methods can be developed and validated for drug substances and drug products.

References

1. International Conference on Harmonisation of Technical Requirements for Registration of Pharmaceuticals for Human Use, ICH Harmonised Tripartite Guideline, Validation of Analytical Procedures: Text and Methodology, Q2 (R1), Nov 2005, Current Step 4 version.
2. United States Pharmacopeia and National Formulary (USP-NF), United States Pharmacopeial Convention, Rockville, MD. USP-NF <1225>, Validation of Compendial Procedures.
3. Gerlach, R. W.; Dobb, D. E.; Raab, G. A.; Nocerino, J. M. Gy Sampling Theory in Environmental Studies 1: Assessing Soil Splitting Protocols. *J. Chemometr.* **2002,** *16,* 321–328.
4. United States Pharmacopeia and National Formulary (USP-NF), United States Pharmacopeial Convention, Rockville, MD. USP-NF <1010>, Analytical Data—Interpretation and Treatment.
5. Ermer, J.; McB. Miller, J. H. *Method Validation in Pharmaceutical Analysis. A Guide to Best Practice;* WILEY-VCH Verlag GmbH & Co. KGaA: Weinheim, 2005.
6. Nilsen, C. L. Analytical Method Validation, Design & Execution. Part IV. *Pharm. Formulation Qual.* **May 2004,** 86–88.
7. ORA Laboratory Procedure ORA-LAB.5.4.5. Methods, Method Verification and Validation. FDA Office of Regulatory Affairs. (revised Jan 25, 2012). http://www.fda.gov/ScienceResearch/FieldScience/LaboratoryManual/ucm171877.htm.

CHAPTER

Chiral methods

10

Brian He, David K. Lloyd

Analytical and Bioanalytical Development, Bristol-Myers Squibb, New Brunswick, NJ, USA

CHAPTER OUTLINE

Specification of Drug Substances and Products. http://dx.doi.org/10.1016/B978-0-08-098350-9.00010-2

10.1 INTRODUCTION

10.1.1 Chirality

A molecule having two isomeric forms that are nonsuperimposable mirror images of each other is said to be chiral, and the two individual forms are called enantiomers. Chirality may be due to one of several structural characteristics,[1,2] including:

- An asymmetric center such as a carbon atom with four different substituents (the most common origin of chirality in pharmaceuticals)
- A chiral axis within a molecule, e.g. due to hindered rotation around a single bond in atropisomers
- Molecules where twisted or helical structures are a direct consequence of the molecular framework, e.g. hexahelicine
- Higher-order structure in macromolecules, e.g. protein or polysaccharide helices

A variety of naming conventions are used to describe chiral molecules.[3] In the common case of a carbon atom with four different ligands, the pair of enantiomers are identified as (R)- or (S)-, following the Cahn–Ingold–Prelog convention[2,3] which relates the properties of the ligands to their ordering around the chiral center. Other naming schemes are also found, e.g. D- and L- designations for enantiomers using the Fischer convention, which are commonly used for sugars and amino acids.[3] Many drug molecules contain more than one chiral center, with the configuration of each being separately identified. Thus, for a species containing two chiral centers, four stereoisomers are possible: RR, RS, SR, and SS. RR and SS are one pair of enantiomers, RS and SR are another. The RR/SS and RS/SR pairs are related as diastereomers, and unlike a pair of enantiomers, have different physicochemical properties.

Although indistinguishable in a nonchiral environment, each enantiomer of a chiral molecule may interact differently with an enantiomer of another chiral molecule. Differential interactions between enantiomers form the basis of enantiomeric separations by chromatography or electrophoresis, where transient diastereomeric complexes are formed with a chiral selector in the separation column. Such separations are highly important in the analysis of chiral pharmaceuticals, and are described in Sections 10.3 and 10.4 of this chapter. Stereoisomers also behave differently in their interaction with polarized light, and various optical and other spectroscopic methods are considered briefly in Section 10.5.

10.1.2 Chiral drugs in a chiral environment

Although often exhibiting symmetry on a macroscopic scale, at a molecular level, biological systems are predominantly asymmetric, a fact recognized by Pasteur in the nineteenth century: "...many organic substances (I might say nearly all, if I were to specify only those which play an important role in plant and animal life) all of which are important substances to life, are asymmetric, and indeed have the kind of asymmetry in which the image is not superimposable with the object."[4] Pasteur also demonstrated stereoselectivity in a biological process—the ability of a yeast to transform dextro-tartarate but not levotartrate, leading him to conclude regarding chirality that "...this important characteristic is perhaps the only distinct line of demarcation which we can draw today between dead and living nature."[4] Thus, it is hardly surprising that in many cases the enantiomers of chiral compounds, including drugs, have differing biological activities in the chiral environment of the body. A simple example is that the taste of many amino acids depends on their absolute stereochemical configuration, D-asparagine being sweet while L-asparagine is tasteless, D-leucine being sweet while L-leucine is bitter.[5] In the case of chiral pharmaceuticals, desired therapeutic activity or toxicity may reside in one or the other or both enantiomers (or, if there are multiple chiral centers, in one or more of the diastereomers).[6,7] An increasing understanding of differences in the pharmacology and pharmacokinetics of chiral molecules transformed the research landscape such that, by the 1980s, it became difficult to ignore chirality as an important factor in drug development.[5,8] In contrast to the situation in the 1980s when the development of racemates was typical, today, a proposal to market a racemic mixture requires justification (e.g. if a single enantiomer underwent rapid racemization in vivo, this could be the basis for justifying the development of a racemic mixture).[9] Thus, chiral analysis methods are needed throughout drug development to achieve an appropriate level of understanding and control.

10.2 IMPLEMENTATION OF HEALTH AUTHORITY GUIDELINES TO CHIRAL DRUG DEVELOPMENT

10.2.1 Guidelines related to chiral drug development

With advances in chiral synthesis and analytical methodology enabling the development of single-enantiomer pharmaceuticals, health authorities started to introduce formal guidance related to the development of enantiomeric drugs in the early 1990s.[9] These include guidance from the US Food and Drug Administration[10] and the European Medicines Agency.[11] A variety of texts have been published that include discussion and interpretation of these and other guidelines.[9,12,13] The focus of regulatory interest is on compounds with a single chiral center, where the enantiomers can only be differentiated by chirally-selective analytical techniques.

10.2.2 Analytical testing and specifications for chiral drugs

The desire to develop and manufacture drugs that are safe and effective, with appropriate quality, is common to the regulation of all pharmaceuticals, including chiral drugs. Guidance on the required specifications for chiral drugs is given in the International Conference on Harmonization (ICH) guideline Q6A and related appendices.[14]

In comparison to nonchiral pharmaceuticals, the following additional tests are required for an enantiomeric drug substance:

- A chiral identity test. At some point in development, the determination of the absolute configuration needs to be performed, e.g. using single-crystal X-ray diffraction. However, the routine determination of chiral identity for purposes such as batch release is typically performed by a simple technique such as chiral chromatography or polarimetry, which can distinguish between the individual enantiomers and a racemic mixture.

- Chiral impurity method for the unwanted enantiomer. This may be needed to release batches, and for stability studies. This is most commonly performed using separation methods such as chiral liquid chromatography (LC). Besides release testing, a quantitative chiral method may also be used to monitor enantiomeric impurities (or diastereomeric impurities) for many other purposes, such as fate and tolerance studies, drug substance and drug product stability programs, and forced degradation studies.
- Chiral assay. This is often replaced by the combination of an achiral assay along with a chiral impurity method.

For the drug product specification, if it can be demonstrated in development that racemization does not occur during manufacture and storage, chiral assay and identity tests are not required.[9] Conversely, if significant racemization does occur, chiral analysis is required.

10.2.3 Chiral control strategies

In discovery and early development, it is not uncommon to proceed with an achiral synthesis, and then separate the desired enantiomer of the active compound by chiral preparative chromatography or by a stereoselective crystallization using a chirally-pure auxiliary compound. Analytical methods to determine identity and purity can then be put into place after the chiral resolution step. However, chiral resolution is an inefficient approach in later development or for commercial manufacture, and at these stages, a stereoselective synthesis is almost universally employed. Most small-molecule pharmaceuticals are produced via multistep synthetic processes. A chiral center (or centers) may be introduced with a starting material, or at a later step in the synthesis. The most appropriate point of control is often that at which the chiral center is introduced. If the chiral center comes from a starting material, the specification for that material would include chiral identity and purity tests. If the chiral center is introduced at a step within the synthetic process, then chiral identity and purity would be tested most typically after the next isolation step.

It is quite common to have multiple chiral centers in a drug candidate, e.g. as in the case of Brivanib shown in Fig. 10.1, which has two chiral centers, and thus four stereoisomers. Regulatory guidance speaks most clearly to drugs with single chiral centers, but logical strategies for the development of multiple chiral center drugs have also been developed.[15] Since diastereomers have differing physicochemical properties that generally allow them to be analyzed in achiral systems, they can frequently be tested as part of a single impurity method (e.g. by gradient reversed-phase (RP) LC using a C18 or similar stationary phase). However, in some cases, adequate selectivity cannot be achieved in this way and a chiral column may provide greater selectivity. The enantiomer of the desired stereoisomer may not need to be specified in the final product if it can be demonstrated that there is adequate control. For example, if the active pharmaceutical ingredient (API) is R,R and this is made via the coupling of two species which each have a composition of 99:1 R:S, the final mixture has a composition of 98.01% R,R, 0.99% each of R,S and S,R, and only 0.01% of the enantiomer S,S. Clearly, adequate control for the S,S species can be achieved by control of the input materials, and the lack of need for a test for the enantiomer in the final product can be justified in the final specification (although development data should be provided to support this). A similar analysis may obviate the need to individually monitor all

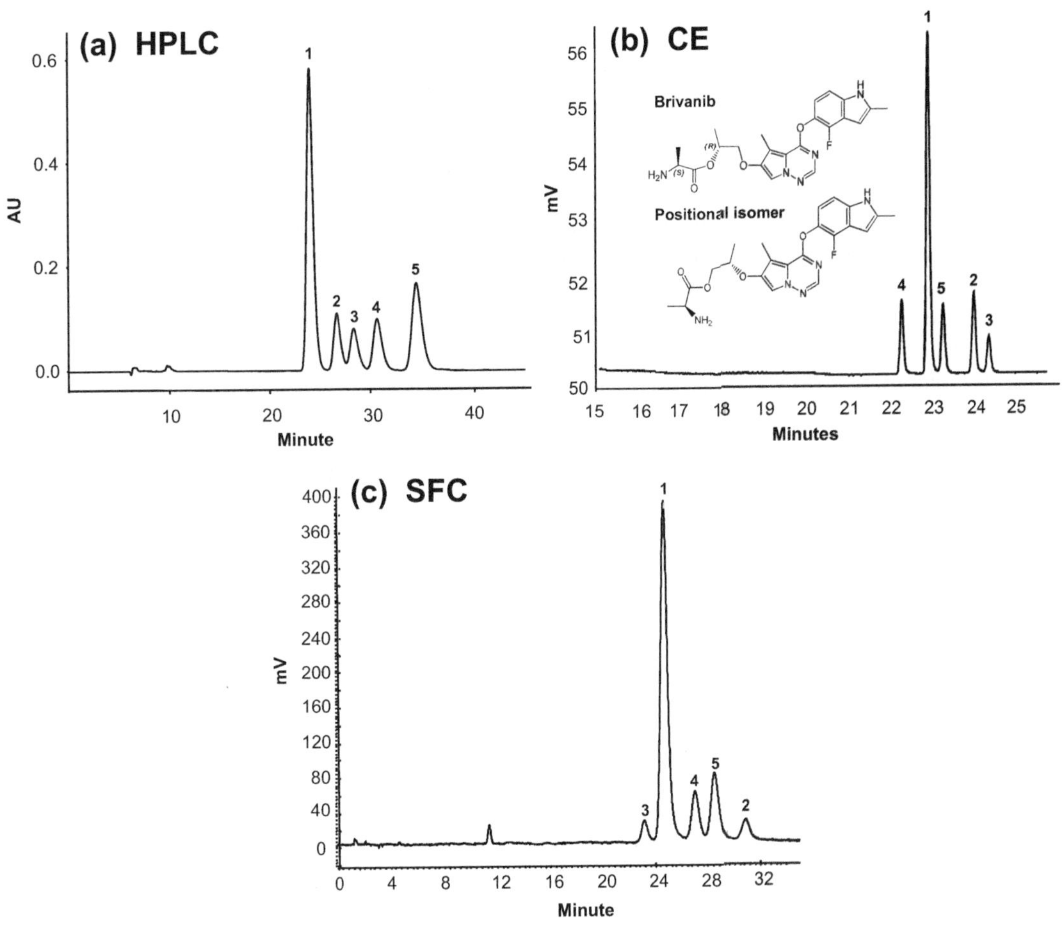

FIGURE 10.1

Simultaneous separation of Brivanib and its four stereoisomeric impurities by (a) HPLC, (b) CE, and (c) SFC. Peak identification: (1) Brivanib, (S,R) isomer; (2) Diastereomer-(R,R); (3) Positional isomer; (4) Enantiomer-(R,S); (5). Diastereomer-(S,S). Experimental conditions: (a) Chiralcel OJ-RH (150 × 4.6 mm, 5 μm), MP: methanol, 1 mL/min, 30 °C; (b) Running buffer containing 10% (w/v) highly sulfated-γ-cyclodextrin (CD), 0.3% (w/v) γ-CD, and 10% (v/v) ACN, 30 kV, 20 °C, capillary: 56 cm × 50 μm ID, fused silica capillary with extended light path; (c) Sepapak-2-HR (250 × 4.6 mm, 3 μm) MP: 79/21 (v/v) CO_2/MeOH with 0.1% DEA, 2.0 ml/min, 45 °C, outlet pressure: 150 bar.

possible diastereomeric impurities; thus, in compounds with more than two chiral centers, an individual stereoisomer is typically developed and the unwanted diastereomeric forms can normally be analyzed by achiral techniques.[15] The presence of many possible stereoisomers can likely be discounted through the knowledge gained during development, and hence they may not need to be specified in the API.

10.2.4 Determination of enantiomeric impurity

In the development of a chiral drug, a chiral impurity method is required for the specification of the drug substance (or the drug product in the case where interconversion of enantiomers or diastereomers occurs) to ensure that enantiomeric impurity is controlled to an appropriate level.

In the literature, numerous analytical techniques have been evaluated for the determination of enantiomeric impurities, such as chiroptical methods, nuclear magnetic resonance (NMR) spectroscopy, chromatography, calorimetry, or isotope dilution.[16] Today, the technologies that are routinely used in pharmaceutical laboratories are mainly separation-based techniques. High-performance liquid chromatography (HPLC) is used primarily, but separation approaches also include gas chromatography (GC), capillary electrophoresis (CE), and supercritical or subcritical fluid chromatography (SFC). Since both enantiomers possess identical physicochemical properties in achiral environments, enantiomers have to be converted into diastereomers in order to be resolved from each other. In general, there are two approaches to achieve this goal: (1) derivatization with enantiomerically pure chiral reagents to form covalently bonded diastereomers, or (2) formation of transient diastereomers with chiral mobile phase (MP) additives or chiral stationary phases (CSPs). Due to their simplicity and absence of potential interconversion risk during the derivatization process, methodologies involving the direct separation of enantiomers as transient diasteriomeric complexes through the use of MP additives or CSPs are preferred.

The advantage of separation-based technologies is that both enantiomers can be accurately determined individually. The calculation of enantiomeric impurity is straightforward, as described in Eqn (10.1):

$$\text{Enantiomeric Impurity (\%)} = \frac{A_{\text{undesired}}}{A_{\text{undesired}} + A_{\text{desired}}} \times 100 \tag{10.1}$$

where $A_{\text{undesired}}$ and A_{desired} are the area count of undesired and desired enantiomers, respectively. In some fields, enantiomeric excess (% ee) is more frequently used. It is defined as the percentage of the enantiomer in excess of its antipode, and is determined by Eqn (10.2):

$$\text{Enantiomeric Excess (\%)} = \frac{A_{\text{desired}} - A_{\text{undesired}}}{A_{\text{undesired}} + A_{\text{desired}}} \times 100 \tag{10.2}$$

Eqns (10.1) and (10.2) are valid when both enantiomers are within the linear dynamic range of the detector. Precautions must be taken when an alternative detection method other than ultraviolet–visible (UV–Vis) detection is used, such as charged aerosol detection (CAD) or evaporative light scattering detection (ELSD). The CAD typically has a quadratic response over wide concentration ranges (two or more orders of magnitude) and the ELSD response is exponentially related to sample mass. Consequently, these equations are no longer applicable. Instead, an external standard calibration curve has to be established for the quantitation of the undesired enantiomer.[17] To ensure the accuracy of a separation-based chiral method, the undesired enantiomer peak must be free of interference, i.e. separated from the desired enantiomer and from any other peaks, including system peaks, process-related impurities, and potential degradants. Hence, it is always good practice to reassess the specificity of a chiral method when any process change, which may result in impurity profile change of intermediates and API, is implemented.

10.3 CHIRAL LC

10.3.1 Selection of separation-based techniques

Abundant examples have demonstrated that chromatographic techniques, such as HPLC, GC, CE, capillary electrokinetic chromatography (CEC), and SFC, are effective for enantiomeric separation and enantiomeric impurity determination.[18] It is common that a pair of enantiomers can be resolved with more than one enantioselective technique. An example is given in Fig. 10.1 to illustrate the simultaneous separation of an anticancer prodrug, Brivanib, and its four stereoisomeric impurities, including the enantiomer, by using three different techniques, HPLC, CE, and SFC.

What type of separation technology should be chosen to develop a chiral method for specification purposes? Depending on the stage of pharmaceutical development, many factors, such as the availability of Good Manufacturing Practice (GMP) qualified instruments, chromatographic interferences, the most suitable detection method, method robustness, the technical training of personnel, and method greenness, may play a role in the decision making. A "fit-for-purpose" chiral method with limited optimization is usually adequate for a chiral drug at early phases of development. As drug development moves into the later stages, method robustness and transferability are often the primary concerns for a chiral method. Before registrational filings, chiral methods along with all other methods in the specifications of drug substance and drug product will be transferred to global manufacturing sites. Therefore, when the situation depicted in Fig. 10.1 is encountered, the inherent robustness of available enantioseparation methodologies should be carefully evaluated before the technique is chosen and method conditions are finalized. In the case of Brivanib, all four stereoisomeric impurities, i.e. the enantiomer, two diastereomers, and a major positional isomer, were required to be monitored due to moderate regioselective control in the manufacturing process. It is desirable to have an enantioselective HPLC method that directly separates all these impurities. Unfortunately, the bulky methylpyrrolo triazinyl group makes it extremely difficult to achieve such a separation with chiral HPLC. Consequently, all Brivanib analytes had to undergo precolumn derivatization through carboxybenzylation of the primary amino group prior to separation on a chiral HPLC column (Fig. 10.1(a)). From a method robustness standpoint, derivatization adds significant complexity to sample analysis, and is thus avoided by many analysts. In the meantime, the direct separation of Brivanib and its four related stereoisomeric impurities was successfully achieved without derivatization by CE and SFC, as illustrated in Fig. 10.1(b) and (c), respectively. Both separations have a similar analysis time, but chiral CE provided the preferred elution order with the enantiomeric impurity eluting before the API. This is advantageous for enantiomeric impurity determination since any tailing of the main API peak might interfere with the quantitation of trace-level enantiomeric impurity. Additionally, considering the availability of GMP-qualified CE instruments at the manufacturing sites (vs SFC instrumentation), the method sensitivity and robustness, a CE method was eventually adopted and implemented in API batch release.

Undoubtedly, HPLC has a well-established reputation as the technology of choice for achiral and chiral pharmaceutical analyses, particularly in a regulated environment, due to readily accessible, robust and GMP-compliant instruments. Therefore, the discussion of chiral method development in this chapter is mainly focused on the HPLC technique. Other analytical separation techniques, such as GC, SFC, and CE, and nonseparation techniques are utilized less frequently and briefly discussed in later sections.

10.3.2 Chiral HPLC method development

10.3.2.1 Chiral column screening

Currently, HPLC enantioseparation with CSPs is the first-choice approach for enantiomeric impurity determination in pharmaceutical development. A wide spectrum of commercially available CSPs and multiple separation modes have been developed for the HPLC separation of enantiomers.[19] Chiral HPLC method development centers on finding the right combination of chiral column and MP composition that afford adequate enantioselectivity and chemoselectivity, and separation efficiency. Since chiral separation is so sensitive to subtle variations of molecular structure and the understanding of chiral recognition at a molecular level is limited, it remains a daunting task to identify the best chiral column for the separation of a given pair of enantiomers. A practical approach widely employed in the pharmaceutical industry is to screen a small set of chiral HPLC columns followed by the optimization of the separation conditions on the most promising chiral columns. This "generic" chiral method development approach easily leads to a workable method condition for most chiral pharmaceutical compounds. It does not require extensive chiral separation experience and knowledge for an analyst to conduct a chiral column screening; however, a better understanding of how a CSP works can definitely facilitate the method optimization process and result in a more robust chiral method. The chiral recognition mechanisms on various CSPs have been summarized in a number of recent reviews and books.[20–22]

Chiral column screening serves at least two purposes. The most important is identification of proper columns with adequate enantioselectivity and separation efficiency. Unlike preparative chromatography, it may be unnecessary for trace-level chiral analysis to seek a separation with a large degree of selectivity. Ultrahigh enantioseparation often gives rise to two outcomes: either the first enantiomer is poorly retained by the column or the second enantiomer is retained too long. In the first situation, interference from the sample matrix may be encountered for the first eluted enantiomer; it is recommended to have retention factor $(k) \geq 1$, where k is defined as $(t_R - t_0)/t_0$, where t_R is the retention time of the analyte and t_0 is the void volume. The second situation may result in a low peak efficiency and difficulties in peak integration. Thus, it is a huge disadvantage if the "wrong" enantiomer is the late eluting peak, which emphasizes the importance of having the unwanted enantiomer eluting before the API. This leads to the second purpose of doing column screening, which is comparison of enantiomer elution order (EEO) on different columns. Control of EEO is an important issue in enantiomeric impurity analysis. CSPs with opposite configurations (such as (+) and (−)—(18-Crown-6)-tetracarboxylic acid) and "pseudoopposite" (such as quinine and quinidine) configurations guarantee the reversal of the elution order of two enantiomers (not diastereomers). This can be achieved with small synthetic or semisynthetic CSPs. For CSPs derived from natural sources, such as polysaccharides, proteins, and macrocyclic glycopeptides, it is impossible to control EEO since there are no antipodes of these complex molecules available, but the desired EEO may be achieved using CSPs with different chiral recognition mechanisms. An example is illustrated in Fig. 10.2. Opposite EEOs for Brivanib as 6-aminoquinolyl-*N*-hydroxysuccinimidyl carbamate (AQC) derivatives were observed on an amylose tris[(S)-1-phenylethylcarbamate] column (Chiralpak AS-RH) and a cellulose tris(3,5-dimethylphenylcarbamate) column (Chiralcel OD-RH) under RP conditions. The Chiralcel OD-RH column gave the preferred EEO.

Besides CSPs, several other factors, such as column temperature, the type and content of polar organic modifier in the MP, and the type and concentration of acidic MP additives, have been reported

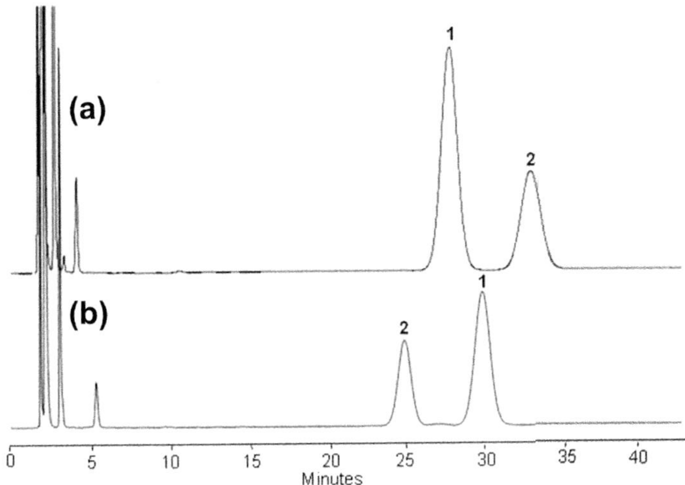

FIGURE 10.2

Enantioseparation of Brivanib as AQC derivatives on (a) AS-RH (150 mm × 4.6 mm, 5 μm) and (b) OD-RH (150 × 4.6 mm, 5 μm). Peak identification: (1) Desired enantiomer; (2) wrong enantiomer. MP: (a) 40/60 ACN/20 mM ammonium acetate (NH_4OAc) in water; (b) 45/55 ACN/20 mM NH_4OAc in water. Other conditions: flow rate 1.0 mL/min; temperature: 40 °C; UV detection: 245 nm.

to have an effect on EEO for a few classes of CSPs.[20] Therefore, all these parameters may need to be scrutinized during separation optimization. In addition, derivatization is also an effective way to control the EEO of chiral analytes through changing the interacting groups and thus the chiral recognition mechanism.[23] To facilitate the identification of individual enantiomer peaks, a good practice is to prepare sample solutions with enantiomers and/or diastereomers in different ratios for chiral column screening. If a racemic sample has to be used, a chiral detector such as a polarimeter or circular dichroism (CD) detector can be coupled online with UV detection to help enantiomer peak tracking.[24]

Chiral column screening may require testing different sets of chiral columns, depending on the analytes. In the first step, CSPs with broad enantioselectivity and good stability are normally considered for initial screening. Polysaccharide-based CSPs containing phenylcarbamate and benzoate derivatives of cellulose and amylose represent the most versatile and popular class of CSPs for the enantioseparation of pharmaceutical compounds. The most popular polysaccharide-based CSPs that are widely evaluated for pharmaceutical compounds are listed in Tables 10.1 and Table 10.2. The enantioselectivity of these CSPs is heavily influenced by the properties and position of substituents on the phenyl group of derivatized cellulose and amylose. Cellulose tris(4-methylbenzoate) and tris(3,5-dimethyl or 3,5-dichloro phenylcarbamate) and tris(chloromethyl phenylcarbamate) derivatives of cellulose and amylose have a superior chiral recognition ability compared to other investigated amylose and cellulose derivatives.[20] Therefore, these CSPs are recommended for the first step column screening. If no promising enantioseparations are identified, the column screening can be extended to the next step by evaluating other polysaccharide-based CSPs, since different

Table 10.1 Structures and Trade Names of Commonly Used Cellulose-Based CSPs in Pharmaceutical Chiral Analysis

	Substituted Phenylcarbamate and Benzoate Derivatives of Cellulose*				
Manufacturer	**Tris(3,5-Dimethyl-Phenyl-Carbamate)**	**Tris(3-Chloro-4-Methyl-Phenylcarbamate)**	**Tris (4-Methylbenzoate)**	**Tris(4-Chloro-3-Methyl-Phenylcarbamate)**	**Tris(3,5-Dichloro-Phenylcarbamate)**
Chiral technologies (Daicel)	Chiralcel OD-H and OD-RH Chiralpak IB (immobilized)	Chiralpak OZ	Chiralcel OJ and OJ-RH	Chiralcel OX	Chiralpak IC (immobilized)
Phenomenex	Lux cellulose-1	Lux cellulose-2	Lux cellulose-3	Lux cellulose-4	
Sepaserve	Sepapak-1	Sepapak-2		Sepapak-4	Sepapak-5
Regis	RegisCell				
Kromasil	CelluCoat				
Sigma–Aldrich	Astec cellulose DMP				
Dr. Maisch GmbH	Reprosil Chiral-OM				

Note: all CSPs are coated on a silica-gel support, unless specified.

Table 10.2 Structures and Trade Names of Commonly Used Amylose-Based CSPs in Pharmaceutical Chiral Analysis

	Substituted Phenylcarbamate Derivatives of Amylose*					
Manufacturer	**Tris(3,5-Dimethyl-Phenyl-Carbamate)**	**Tris(3-Chloro-4-Methylphenyl-Carbamate)**	**Tris(5-Chloro-2-Methylphenyl-Carbamate)**	**Tris [(S)-1-Phenyl-Ethylcarbamate]**	**Tris(3-Chloro-Phenyl-Carbamate)**	**Tris(3,5-dichloro-phenylcarbamate)**
Chiral technologies (Daicel)	Chiralpak AD-H and AD-RH Chiralpak IA (immobilized)	Chiralcel AZ Chiralpak IF (immobilized)	Chiralpak AY	Chiralpak AS-H and AS-RH	Chiralpak ID (immobilized)	Chiralpak IE (immobilized)
Phenomenex			Lux amylose-2			
Sepaserve			Sepapak-3			
Regis	RegisPack		RegisPack CLA-1			
Kromasil	AmyCoat					
Dr. Maisch GmbH	Reprosil Chiral-AM					

Note: all CSPs are coated on a silica-gel support, unless specified.

polysaccharide-based CSPs often show complementary enantiorecognition.[25] There is also a significant overlap in the chiral analytes separated on these CSPs.

There are a few points that are worth mentioning here:

- A number of chiral selectors are used to prepare coated and covalently immobilized CSPs. Due to notable differences in enantioselectivity, the coated and covalently attached versions should be treated as different CSPs. Generally, coated CSPs have a higher enantioselectivity than immobilized ones under the same chromatographic conditions.[26] In the worst case, an opposite elution order of enantiomers has been observed between coated and immobilized versions of cellulose and amylose tris(3,5-dimethylphenyl-carbamate) CSPs.[20] The advantage of covalently attached CSPs over coated CSPs is that a much broader range of organic solvents other than standard alcohol/alkane mixtures can be used in the MP, a feature that greatly expands the application scope of polysaccharide-based CSPs.[27]
- Some CSPs are available from different column manufacturers. The overall enantioselectivity of CSPs from different manufacturers is similar; however, the differences in column performance in terms of retention time, separation efficiency, and chemoselectivity and enantioselectivity is quite significant for some analytes.
- Considering that the backpressure limit is relatively low for certain types of polysaccharide-based CSPs, it is helpful to extend column life by running at either elevated temperature (thus reducing solvent viscosity) or low flow rate, or using a shorter column (such as 150 vs 250 mm).
- Polysaccharide-based CSPs can be operated under RP, normal phase (NP), and polar organic phase (POP) conditions; however, it is not recommended to switch the MP from the RP to the NP on the same column since the enantioselectivity in the NP might be affected by the residual polar component left in the column.

In many cases, chiral column screening must go beyond polysaccharide-based CSPs. Depending on the functional groups surrounding the chiral center(s) of a target compound, CSPs targeting specific enantioselective interactions can be selectively chosen for the next step of column screening. For example, if a primary amine group is present in the structure of a chiral compound, CSPs based on vancomycin, teicoplanin, α_1-acid glycoprotein (AGP), cellobiohydrolase (CBH), ovomucoid (OVM), chiral crown ethers, and cyclofructans all may be suitable for enantioseparation. For chiral compounds containing acidic functional groups, teicoplanin-, ristocetin-, AGP-, human serum albumin (HSA-), OVM-, quinine-, and quinidine-based CSPs can be considered for the column screen. In the case of underivatized amino acids, CSPs that are based on macrocyclic glycopeptides, chiral crown ethers, zwitterionic derivatives of quinine and quinidine, and chiral ligand exchange phases are quite successful. Pirkle-type CSPs and cyclodextrin-based CSPs are suitable for compounds with functional groups that engage in $\pi-\pi$ and hydrogen-bonding interactions in the proximity of chiral center(s).[28] Clearly, understanding the scope of the application of each CSP helps to avoid the screening of unsuitable chiral columns.

10.3.2.2 Mobile phase (MP) design

The HPLC MP plays an essential and active role in the chiral recognition process. The components of the MP not only regulate the solvation, ionization (for ionizable compounds), and conformation of CSPs and analytes, but they also promote noncovalent intermolecular interactions that support the chiral recognition of enantiomers.[21] For example, it is well known that hydrophobic interactions prevail in RP separations. Highly polar MP molecules (water, alcohols, acetonitrile (ACN), and tetrahydrofuran [THF]) easily disrupt polar and dipolar intermolecular interactions, while long-range electrostatic interactions may be diminished but are still well preserved under these conditions. On the contrary, NP conditions, which provide a low polarity environment, are more in favor of hydrogen bonding, dipole–dipole, $\pi-\pi$ stacking, and electrostatic interactions. In the meantime, hydrophobic interactions are largely suppressed in NP separations. Compared to RP and NP, POP, whose MP largely consists of polar organic solvents, such as ACN and alcohols, stimulates (or restrains) all these intermolecular interactions to moderate levels, thus yielding a relatively short retention time for most analytes. Furthermore, MP additives and polar components also compete with analytes for binding sites on CSPs, which may minimize nonenantioselective interactions or even change chiral recognition mechanisms.

Each CSP is rationally designed to take advantage of certain types of intermolecular interactions for enantioseparation. For instance, polysaccharide-based and Pirkle-type CSPs have aromatic, carbamate, or ester moieties incorporated into their enantioselective binding sites; therefore, they favor hydrogen-bonding, $\pi-\pi$ stacking, and dipole–dipole interactions. Since every entity of these two classes of CSPs is neutral in nature, it is important to keep analyte molecules in a neutral form in order to enhance the enantioselective interactions between CSPs and analytes. Consequently, small amounts of acidic or basic additives in the MP (usually $<0.5\%$ v/v) may be required for highly efficient enantioseparation of acidic or basic analytes, respectively. For CSPs that contain ion-exchange binding sites, such as macrocyclic glycopeptides, cinchona carbamate, and protein CSPs, electrostatic (or ionic) interaction is the driving force in the chiral separations of ionizable analytes in RP and POP conditions. To promote ionic interactions, small amounts of MP additives are often used to keep analytes in the desired ionization state. Therefore, the selected MP should have relatively weaker interactions with solutes so that stronger enantioselective interactions can take place between solutes and CSPs. The starting MP compositions for a number of important CSPs are listed in Table 10.3 for a generic chiral column screen. These conditions are chosen to ensure that analytes elute out of the chiral column within a reasonable time ($\sim$30 min). If the retention time of analytes turns out to be too short or too long, the MP solvent strength can be adjusted to initiate the next round of column screening.

In a chiral column screen, it is valuable to evaluate selected chiral CSPs in all applicable separation modes in order to gain a knowledge of column enantioselectivity under different MP polarities. Certain parameters may have a significant influence on the chiral separation on some types of CSPs. For instance, it is known that ethanol and isopropanol often give remarkably different enantioselectivities or even opposite EEO in the NP enantioseparations on derivatized polysaccharide phases. Therefore, a screen of different alcohols should be integrated into the column screen for polysaccharide-based CSPs in the NP. For covalently attached polysaccharide-based CSPs, if no satisfactory separation is achieved with conventional alcohol/alkane MPs, other organic solvent combinations, such as methylene chloride (CH_2Cl_2)/alkanes, methyl-*tert*-butyl ether (MtBE)/alcohol/alkane, or THF/alkanes, can be further explored.[25]

It should be kept in mind that a chiral method is often a living document during drug development since synthetic chemistry may continue evolving to meet the clinical demands for API. Besides impurity profile changes, issues with an existing method may also result in the redevelopment of a chiral method. For example, when this group developed a chiral method for a chiral drug with a carboxylic acid and a quaternary amine functional group around its chiral center, the initial column screening was focused on polysaccharide-based CSPs (10 coated and immobilized CSPs in three separation modes). Lux

Table 10.3 Starting MP Conditions for Chiral Column Screening

CSP Type	Compound Type	RP*	POP*	NP*	Reference
Derivatized polysaccharides	Acidic	50/50/0.1 ACN/Water/FA	100/0.1 ACN/TFA 100/0.1 MeOH/TFA	80/20/0.1 Hep/IPA/TFA 80/20/0.1 Hep/EtOH/TFA	
	Neutral and basic	50/50 ACN/20 mM NH_4OAc	ACN/0.1% DEA and MeOH/0.1% DEA	80/20/0.1 Hep/IPA/DEA 80/20/0.1 Hep/EtOH/DEA	
Macrocyclic glycopeptides	Acidic, neutral, basic	30/70 ACN (or MeOH)/20 mM NH_4OAc, pH 5.0	100/0.1/0.1 MeOH/HOAc/TEA	70/30 Hep/EtOH	29
AGP	Hydrophobic amine	10 mM NH_4OAc, pH 4.5	NA	NA	30
	Hydrophilic amine, weak acid, and neutral	5/95 IPA/10 mM phosphate buffer, pH 7.0	NA	NA	
	Strong acid (containing −COOH)	10 mM Phosphate buffer, pH 7.0	NA	NA	
CBH	Basic and neutral	5/95 IPA/10 mM phosphate buffer (pH 6.0) + 50 μM EDTA	NA	NA	
HSA	Acidic and neutral	5/95 IPA/10 mM phosphate buffer, pH 7.0	NA	NA	
OVM	Acidic	10/90 ACN/20 mM phosphate buffer, pH 3.0	NA	NA	31
	Neutral and basic	10/90 ACN/20 mM NH_4OAc, pH 4.6	NA	NA	
Cyclodextrins	Acidic, basic, and neutral	30/70 ACN (or MeOH)/20 mM NH_4OAc, pH 5.0	95/5/0.1/0.1 ACN/MeOH/HOAc/TEA	70/30 Hep/EtOH	32
Pirkle type	Acidic/neutral	Less enantioselective	Less enantioselective	70/30/0.1 Hep/EtOH/HOAc	33
	Basic/neutral			70/30/0.1 Hep/EtOH/DEA	
Chiral crown ethers	Amino containing	70/30 MeOH/10 mM HOAc in H_2O	Less enantioselective	NA	34
Cyclofructans	Amino containing	Less enantioselective	70/30/0.3/0.2 MeOH/ACN/HOAc/TEA	Less enantioselective	35
Zwitterionic quinine and quinidine derivatives	Acidic, basic, and zwitterionic	Less enantioselective	50 mM FA + 25 mM DEA in 1) 49/49/2 MeOH/ACN/H_2O 2) 49/49/2 MeOH/THF/H_2O	NA	36

Abbreviations: NA—not applicable; EtOH—ethanol; IPA—isopropanol; Hep—n-heptane; FA—formic acid; HOAc—acetic acid; TEA—triethylamine; EDTA—Ethylenediaminetetraacetic acid; DEA—diethylamine; TFA—trifluoroacetic acid.
**The MP composition is based on the volume ratio of each component.*

Cellulose-4 (previously Sepapak-4) was the best column to separate the wrong enantiomer from the desired enantiomer (Rs > 4.0) and other potential synthetic impurities. However, it was noticed later that the chiral column had robustness issues, evidenced by retention time shifts and a peak efficiency decline after a number of injections. To develop a more robust method, chiral column screening was resumed to include macrocyclic glycopeptide CSPs. As a result, a better chiral separation was achieved on a Chirobiotic T2 column with the preferred EEO and a more reproducible retention time.

10.3.3 Enantioseparation optimization

When an appropriate chiral column is identified through column screening, chromatographic conditions may need further optimization for better resolution, higher peak efficiency, better detection sensitivity, or shorter retention time. If a chiral drug is in late development stages, enantioseparation optimization should be focused on the method robustness, i.e. understanding how method parameters relate to analytical performance and defining the method operating space (see Chapter 3). At the optimization step, several parameters are recommended for further evaluation, including the particle size of chiral column packing material, MP conditions (such as organic content, pH, and additive), column temperature, column equivalence, and flow rate. The first three factors are discussed in more detail in the following sections. As with any method where multiple factors may need to be simultaneously optimized, a systematic approach to identification of critical factors, and their systematic optimization is preferred (see Chapter 3 of this volume on Quality by Design (QbD) for analytical methods). Modeling and QbD approaches for chiral separations are described in the literature,[37,38] and probably deserve broader application.

10.3.3.1 Particle size of chiral column packing material

Similar to achiral separations, the quality of enantioseparation is primarily related to resolution via the selectivity (α) of the CSP and column efficiency (N) as defined in Eqn (10.3). Reduction of the particle size of packing material is an effective way to improve the value of N given the reciprocal relationship to particle size (Eqn (10.4)):

$$R_s = \frac{\sqrt{N}}{4}(\alpha - 1)\frac{k}{k+1} \tag{10.3}$$

$$N = \frac{L}{d_p \cdot H} \tag{10.4}$$

where R_s is the resolution between two peaks, k is the retention factor, L is the column length, H is the reduced plate height, and d_p is the particle diameter of the packing material.

Analytical chiral columns on the market today are typically available in a 5-μm particle size; however, there is a clear trend that chiral columns are moving toward smaller packing materials.[39] If a CSP gives a high-enough R_s value (>2) and peak efficiency, a chiral column with a 5-μm particle size is preferred since it is more cost effective. In the situation where only marginal enantioseparation is achieved after extensive column screening, it is beneficial to use a chiral column with a smaller particle size. An example is shown Fig. 10.3. Resolution was improved significantly by simply switching the particle size of the packing material from 5 to 3 μm, while maintaining the same column length and chromatographic conditions. It is notable that method sensitivity was also improved by using a smaller particle size.

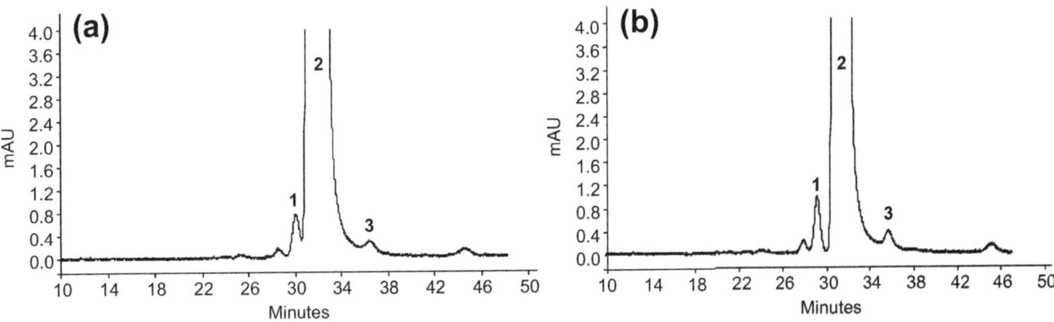

FIGURE 10.3

Comparison of enantioseparation of a chiral compound with four centers on the Chiralpack IB column (250 × 4.6 mm) with (a) 5 μm and (b) 3 μm packing material. Peak identification: (1) Diastereomer; (2) desired enantiomer; (3) wrong enantiomer. MP: 80/10/10/0.05 (v/v/v/v) Hep/MtBE/MeOH/DEA; flow rate: 1 mL/min; temperature: 38 °C.

10.3.3.2 Mobile phase optimization

As discussed in Section 10.3.2.2, the MP provides an environment in which a desirable chiral interaction may occur, leading to a separation. Many MP parameters may influence an enantioseparation, such as composition, polar component, additive, pH, and ionic strength of the buffer (for ionizable analytes in RP). It is important to identify the critical factors that are truly impactful on the chiral separation so that MP optimization can focus on these parameters and help define a meaningful method operating space.

Case study 1: A Chiralpak IB column was selected for the NP enantioseparation of a chiral drug with four chiral centers. Better resolution was achieved for the three peaks of interest by using conventional solvents for the NP separation. Since the primary intermolecular interactions are H-bonding and dipole–dipole interactions in such a system, the addition of polar organic solvents (e.g. MeOH and MtBE) in the MP plays a critical role in controlling the retention and resolution between the analytes. Separation of a main diastereomeric impurity, the desired enantiomer, and the wrong enantiomer is depicted in Fig. 10.4. The nature and the concentration of MP additives were also found to be influential. The MP conditions were optimized after a systematic evaluation of these key elements.

MP additives are extensively used in chiral separations to control the ionization state of ionizable chiral analytes and suppress the nonenantioselective interactions with residual silanols on the silica surface or with the CSP. The impact of MP additives on a chiral separation is complicated and should be carefully examined in method optimization to ensure method robustness.[20] For enantioseparation on polysaccharide-based CSPs, trifluoroacetic acid (TFA) and diethylamine (DEA) (typically in the range of 0.05–0.5%) are the most commonly used additives for acidic and basic chiral analytes, respectively. It was reported that the nature of additives, such as DEA, *n*-butylamine, ethanolamine, and ethylenediamine (EDA), had a remarkable influence on the chiral separation of studied basic compounds in terms of plate number and peak symmetry, as illustrated in Fig. 10.5.[27] This observation may be beneficial to chiral analysis when the resolution between a critical pair is important to impurity determination.

Case study 2: A chiral method was developed to separate the desired enantiomer, wrong enantiomer, and a major diastereomer of a chiral drug. However, only partial separation was achieved with

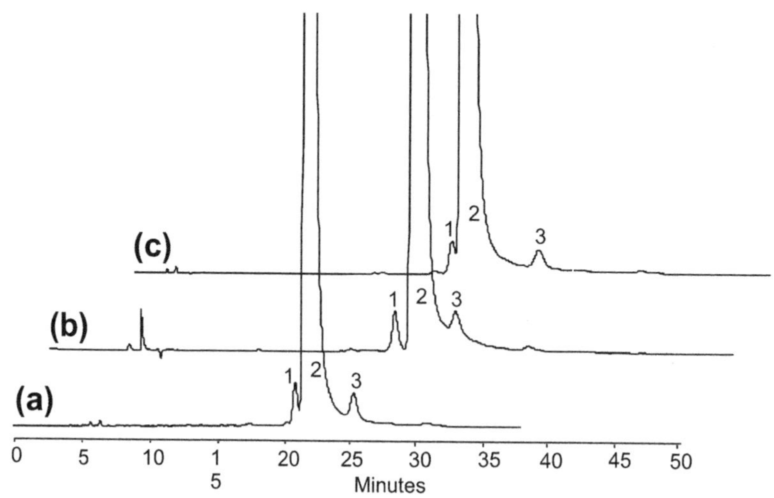

FIGURE 10.4

Effect of polar organic solvents on the NP chiral separation of a pharmaceutical compound on a Chiralpak IB column (250 × 4.6 mm, 3 μm). Peak identification: (1) Diastereomeric impurity; (2) desired enantiomer; (3) wrong enantiomer. MP: (a) 15/20/65/0.1 (v/v/v/v) MeOH/MtBE/Hep/DEA; (b) 15/10/75/0.1 (v/v/v/v) MeOH/MtBE/Hep/DEA; (c) 13/20/67/0.1 (v/v/v/v) MeOH/MtBE/Hep/DEA. Other conditions: flow rate 0.5 mL/min; temperature: 38 °C; UV detection: 306 nm.

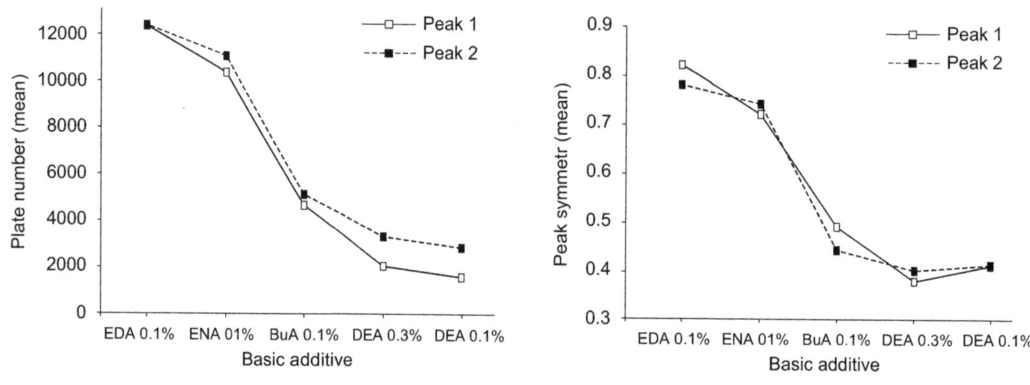

FIGURE 10.5

Effects of different basic additives on the enantioseparation of basic chiral compounds in terms of plate number and peak symmetry (mean values of five analytes).

Reproduced with permission from Ref. 27.

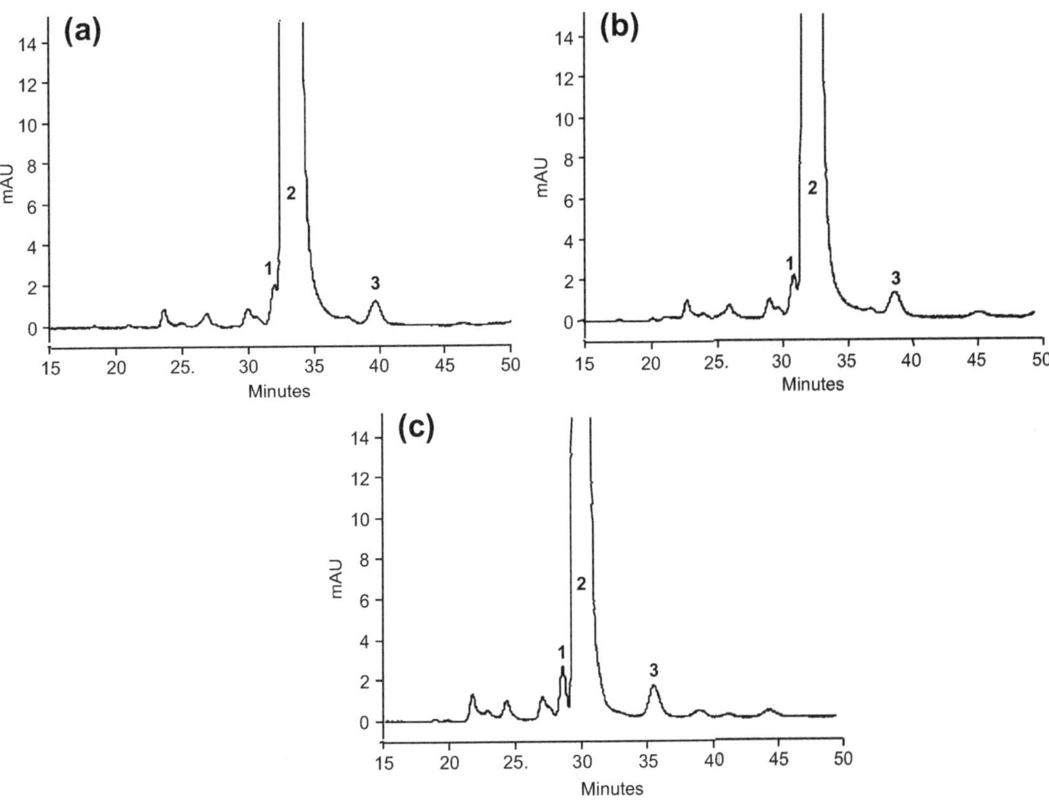

FIGURE 10.6

Effect of MP additives on the enantioseparation of a chiral drug on Chiralpack IB column (250 × 4.6 mm, 3 μm). MP consisting of 77/13/10 (v/v/v) Heptane/MtBE/MeOH with (a) 0.1% DEA, (b) 0.1% TEA and (c) 0.1% EDA, respectively. Flow rate: 0.8 mL/min; temperature: 30 °C; UV detection: 306 nm. Peak identification: (1) Diastereomer; (2) API; (3) enantiomer.

DEA as the MP additive as shown in Fig. 10.6(a). By changing the additive nature, the resolution between the diastereomer and desired enantiomer was significantly improved and allowed an accurate determination of both diastereomer and enantiomer at the 0.05% level (Fig. 10.6(c)).

It is well recognized that mobile phase pH and ionic strength play important roles in the RP enantioseparation of ionizable chiral compounds.[40] An analyte's retention time and resolution may be highly sensitive to changes in both of these parameters. An appropriate buffer system should be carefully selected to precisely control mobile phase pH and provide adequate buffer capacity.

Case study 3: Simultaneous separation of a chiral amine-containing drug and its three stereoisomers was obtained on a polysaccharide-based column (Chiralcel OD-RH) under RP conditions. Mobile phase pH was shown to have a profound impact on the chiral separation due to the ionizable nature of the compound (Fig. 10.7). The operating mobile phase pH was determined to be in the range

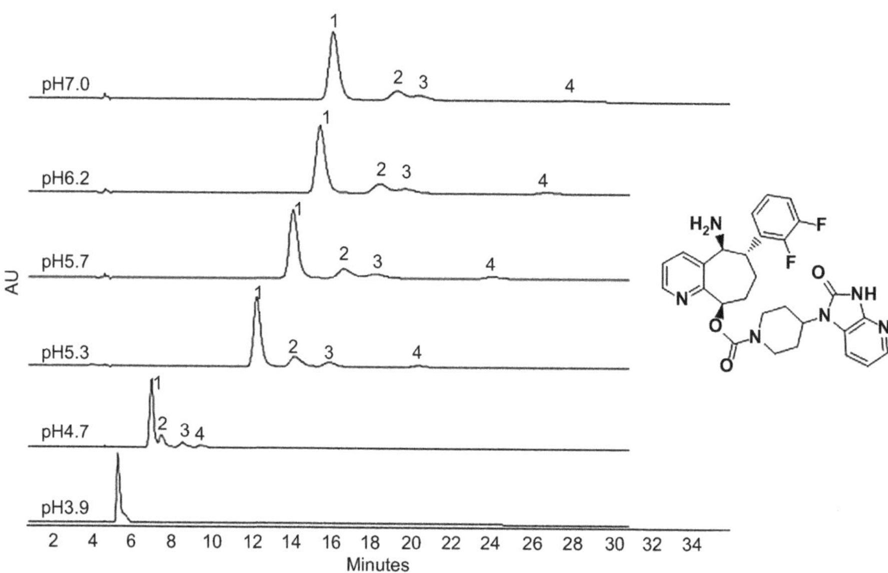

FIGURE 10.7

The pH effect on the separation of a chiral amine-containing compound and its three stereoisomers under the RP condition on a Chiracel OR-RH column (150 mm × 4.6 mm, 5 μm). MP: 35/65 (v/v) ACN/50 mM citrate buffer, flow rate: 0.5 mL/min; temperature: 30 °C; UV detection: 290 nm. Peak identification: (1) Desired enantiomer; (2) wrong enantiomer; (3) diastereomer 1; (4) diastereomer 2.

of 5.3–5.8. To maintain the pH in such a narrow range, a citric acid/sodium citrate buffer system was selected. A relatively high buffer concentration (50 mM) was used to control the mobile phase pH within the operating range and improve peak efficiency.

10.3.3.3 Column temperature

The effect of column temperature on enantioseparation is described by Eqn (10.5). Most enantioseparations are enthalpy-driven processes, where the enthalpy term is much larger than that of entropy; thus, lower temperatures are preferred for better enantioselectivity.[28]

$$\ln \alpha = -\frac{\Delta(\Delta H^\circ)}{RT} + \frac{\Delta(\Delta S^\circ)}{R} \tag{10.5}$$

where α is the enantioselectivity; $\Delta(\Delta H^\circ)$ and $\Delta(\Delta S^\circ)$ are the differences between two enantiomers in the enthalpy and entropy of adsorption, respectively, onto the stationary phase, R is the gas constant, and T the absolute temperature.

However, higher temperatures increase the rates of mass transfer in a chromatographic process and lower the viscosity of MP, which results in a higher peak efficiency and a lower column back pressure. In the example shown in Fig. 10.8, peak efficiency was greatly improved by increasing column temperature. Therefore, it is recommended to use elevated column temperatures for chiral separations when there is a large enough resolution and on-column racemization does not occur. In cases where the

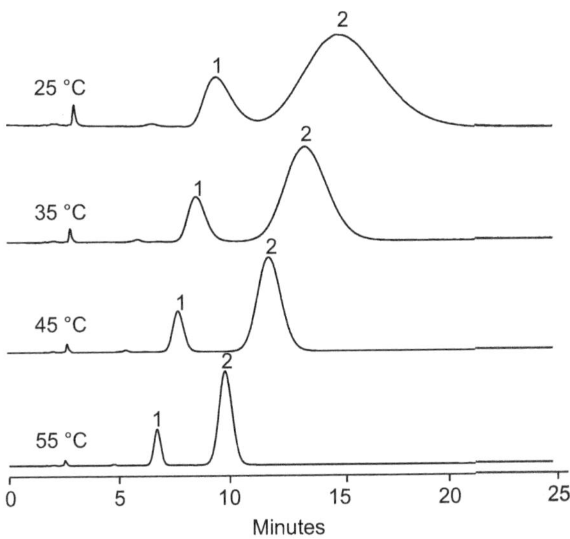

FIGURE 10.8

Effect of column temperature on the enantioseparation of a chiral pharmaceutical compound on Chiralcel OJ-RH (150 × 4.6 mm, 5 μm). MP: 50:50/0.1 (v/v/v) water/ACN/TFA; Flow rate: 1.0 mL/min, UV detection: 260 nm.

entropy contribution $\Delta(\Delta S^{\circ})$ dominates the separation process, enantioseparation is favored at higher temperatures. At the isoenantioselective temperature (T_{iso}), enthalpy and entropy contributions cancel each other out. If T_{iso} is within the allowed temperature range of the chiral column, temperature-induced reversal of EEO can be observed.[20] Hence, it is important to make sure that the operating temperature range of a chiral method is well removed from T_{iso}.

10.3.3.4 Sample diluent

Sample diluent is a factor that is often overlooked, but which may deserve attention at the stage of method optimization. If the diluent has a stronger eluting strength than the MP, sample peaks (especially early-eluting peaks) may be severely distorted due to solvent mismatch, which may cause problems in impurity determination. Such a situation can be encountered in both RP and NP when a sample has a limited solubility in diluents with a composition close to the MP. A higher content of a polar organic solvent is often added to the diluent in order to achieve adequate sample concentration as illustrated in Fig. 10.9(a-1) and (b-1).[25] To prevent this from occurring, other solutions to increase the analyte's solubility in the diluent with a low polarity should be considered, depending on the analyte's physicochemical properties.

10.3.3.5 Column equivalence

As mentioned in Section 10.3.2.1, several CSPs based on derivatized polysaccharides are commercially available from different manufacturers. Notable discrepancies in their analytical performance, such as retention time, enantioselectivity, and peak efficiency, have been observed for a number of applications. In the example shown in Fig. 10.10, the detection sensitivity for impurities was

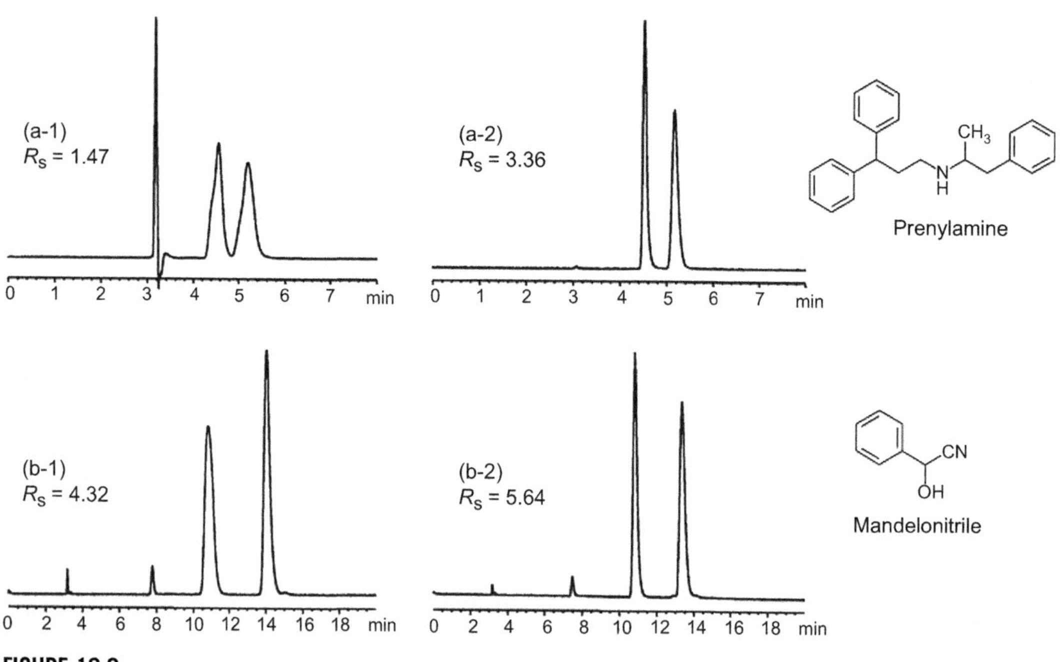

FIGURE 10.9

Effect of the sample solvent. (a) Prenylamine on Chiralpak IA. MP: 85/15/0.1 (v/v/v) Hexane/THF/DEA sample solvent: (a-1) 100% THF, (a-2) 85/15 (v/v) Hexane/THF. (b) Mandelonitrile on CHIRALPAK IC. MP: 70/30 (v/v) Hexane/MtBE; sample solvent: (b-1) 98/2 (v/v) MtBE/EtOH, (b-2) 70/30 (v/v) Hexane/MtBE.

Reproduced with permission from Ref. 25.

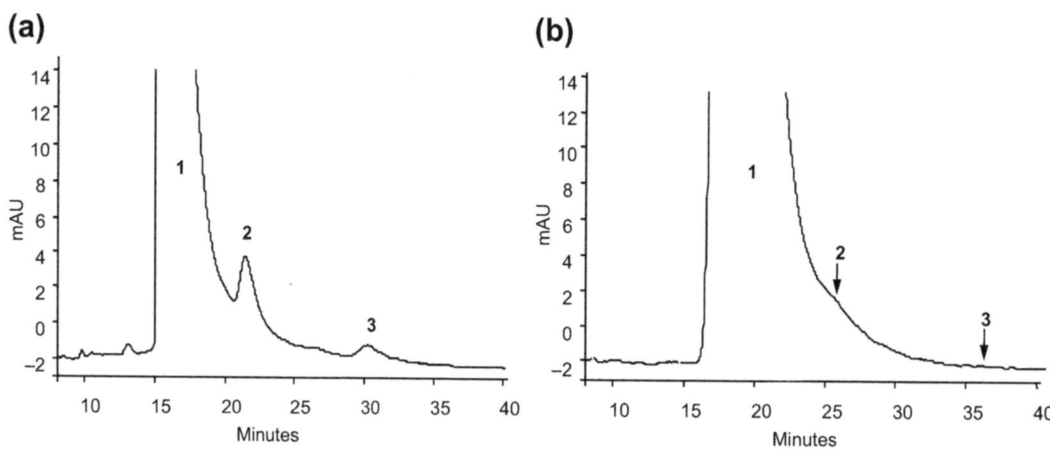

FIGURE 10.10

Comparison of the chiral separation of a chiral drug compound, enantiomeric and meso impurities on (a) Lux Cellulose-1 and (b) Chiralcel OD-H column. Column dimensions: 250 × 4.6 mm, 5 μm; MP: 50/40/10/0.02/0.005 (v/v/v/v/v) Hep/ACN/IPA/ethylsulfonic acid/DEA; flow rate: 0.4 mL/min; temperature: 40 °C; UV detection: 306 nm. Peak identification: (1) Desired enantiomer, (2) meso impurity, (3) wrong enantiomer.

significantly less in the separation on an Chiralcel OD-H column. A few factors may contribute to the observed column performance differences, including chemical composition of chiral selector, column packing, chiral selector loading, and the nature of the silica support.[20] Precautions should be taken if an analyst considers switching columns from one vendor to another one.

10.4 OTHER CHIRAL SEPARATION METHODS

10.4.1 Capillary gas chromatography

Chiral capillary GC is an important technique for enantiomeric separation. Enantioselective GC based on CSPs offers a high separation efficiency, detection sensitivity, and fast speed for chiral impurity analysis, but is constrained in its application to APIs since they need to be volatile (or made volatile with derivatization).[41] The application of chiral GC in pharmaceutical analysis is thus often limited to volatile or semivolatile starting materials and intermediates that are thermally stable. Compared to LC, SFC, and CE method development, it is much more straightforward to develop a chiral GC method. Given that the typical GC carrier gas is helium, CSP and column temperature are the two major parameters that need to be evaluated in chiral GC method development. Chemically modified cyclodextrins represent the most versatile CSPs for GC enantiomeric separation. As a result, GC chiral column screening can be simply focused on cyclodextrin-derived CSPs.[42] The availability of universal detectors, such as flame ionization, makes GC an attractive tool for the analysis of small chiral analytes without good UV chromophores. For polar analytes, precolumn derivatization might be needed to increase the volatility or improve the peak efficiency of samples.

10.4.2 Subcritical and supercritical fluid chromatography

SFC is a chiral separation technique which, on both the analytical and preparative scales, plays an increasingly important role in pharmaceutical development,[43] sharing many of the column technologies with HPLC.[44] It is routinely used in research laboratories, but historically the instrumentation has had marginally acceptable sensitivity and reproducibility. Innovations and improvements in SFC instrument design in recent years have significantly improved the performance of analytical SFC and make it possible to conduct impurity analysis with the levels of sensitivity and reproducibility comparable to contemporary HPLC.[45] However, there is currently a very low installed base of GMP-qualified analytical instruments, so SFC would typically not be the technology of choice for QC analyses at this time. Nevertheless, SFC has an excellent ability for separation of drug-like molecules; a >95% success rate in the resolution of small organic molecules and drugs has been reported.[43] Thus, in an environment such as discovery analytics, SFC excels in providing a quick answer to chiral separation problems, using a relatively limited number of stationary phases. Another area of pharmaceutical development where SFC has great application is in preparative separations of enantiomers.[46] The most obvious advantages for such separations are the ease of removal of the MP from the isolated compounds of interest, and the relative greenness of the process (compared to NP separations based on hexane-containing MPs). Furthermore, high selectivity and efficiency in chiral SFC lead to more productive preparative separations.

10.4.3 Capillary electrophoresis

CE is a term that encompasses a variety of related, electrically driven liquid-phase separation techniques that occur in a capillary tube. Operation in capillaries results in low heat generation because of the low current flow through the liquid (background electrolyte, BGE) in the capillary, and effective dissipation of the heat that does arise. Enantiomers cannot be resolved in capillary zone electrophoresis where the separation medium is a simple buffer, but resolution may be achieved by the addition of a chiral compound to the separation medium (Mikuš provides a recent comprehensive review of chiral CE[47]). The resulting system is a form of capillary electrokinetic chromatography, in which the analyte and chiral additive form transient complexes. A separation may be achieved as long as:

- There is some difference in the binding of the analyte stereoisomers to the single stereoisomer additive and
- The complexes have a different velocity from that of the free analyte when an electric field is applied

The principle of separation, differential complexation, is thus similar between CE and LC,[48] and so in principle many of the same selectors may be used. However, the instrumental setup in CE introduces certain constraints, e.g. the chiral selector passes through the detector, which should therefore preferably not respond to this additive. CE separations are frequently characterized by very high efficiency—even chiral CE separations—and so this confers an important potential advantage when compared to LC. Resolution may be achieved in a high-efficiency CE separation system based on relatively minor differences in binding strength; in a low-efficiency system, much greater selectivity is needed to achieve baseline resolution.[48] Cyclodextrins have become the first-choice chiral selector in CE for a variety of reasons, primarily their lack of interference with UV detection, and easy solubilization in aqueous CE buffers at concentrations appropriate for achieving separations, although other chiral selectors such as macrocyclic antibiotics, proteins, and chiral surfactants are occasionally used.

10.4.3.1 Generic CE methods

One approach to chiral CE separations is to start with a generic method or methods that have proven to be suitable for a large variety of compounds. This can work because of the high efficiency of CE and the relatively low difference in binding strength needed to achieve a usable separation. The generic method is unlikely to be optimal, but may well be sufficient. A single selector may be adequate within a class of compounds, e.g.

- neutral CD with neutral-low pH BGE for bases
- neutral CD with neutral-high pH BGE for acids
- charged CD for neutral compounds

In each of these combinations, the separation is designed to meet the necessary condition that there be a difference in the mobility between the free and complexed species (e.g. a neutral analyte will have no electrophoretic mobility, but the complex with a charged CD will move under the influence of an electric field).[49] For separations of a broader range of analytes, a BGE containing multiple CDs for broader selectivity may be required.[50]

10.4.3.2 Screening approaches to method development

Taking a screening approach, one or more sets of standard conditions are set up using a variety of selectors, and screening analyses are made. Upon review of results, one may either

- accept one of the existing results
- optimize based on one or more conditions offering some separation
- continue screening

An example protocol for basic analytes would be to first select the CD size (α, β, γ) according to the number of aromatic rings in the analyte, then select types of CD to screen, e.g. native CD plus derivatives such as HP-β-CD and DM-β-CD (the derivatized CDs often show a markedly different selectivity than the native CD). An acidic BGE (e.g. phosphate buffer) will ensure that the analytes are charged, and native and derivatized CDs should be added to this at a variety of concentrations, e.g. at 1, 5, and 25 mM (if soluble). Having performed these runs, hopefully one or more will show some selectivity. To optimize further, the CD concentration can be systematically varied, increasing the selector concentration if the interaction is too weak. Alternatively, if binding is strong, addition of a modifier such as an alcohol to weaken the binding may be proposed. El Deeb et al.[51] provide a clear description of a screening and optimization approach.

Despite generating extensive research literature,[47] chiral CE has found relatively limited acceptance in regulated bioanalysis or pharmaceutical analysis. This is not necessarily because of limitations of the technique—the large installed base of LC equipment and the relatively high level of knowledge of LC among industry analysts makes LC the approach of choice. Even when CE might provide a demonstrably better separation, it will unlikely be used unless LC is unable to provide even a barely-adequate analysis because of the large LC infrastructure and experience base. Still, if a suitable LC analysis has been proven through extensive experimentation to be impossible, CE provides a viable alternative (Fig. 10.1).

10.5 NONSEPARATION TECHNIQUES

10.5.1 Optical rotation and circular dichroism

Chiral molecules exhibit the property of optical activity, which can be described as a differential interaction with left circularly polarized (LCP) and right circularly polarized (RCP) light. This may be manifested as optical rotation, the rotation of the axis of polarization of plane polarized light due to a difference in the velocity of LCP or RCP light when passing through a chiral solid or solution of chiral molecules, or as circular dichroism due to the preferential absorption of either LCP or RCP light.[52] When this absorption occurs in the UV–visible region, this is referred to as electronic circular dichroism (ECD, referring to the absorption being related to an electronic transition within a molecule). The two enantiomers of a chiral compound have equal but opposite interactions with polarized light. At a given temperature and wavelength, the magnitude of the rotation experienced by polarized light passing through a solution containing a pure chiral compound is proportional to the concentration of the chiral compound and the optical path length. The *specific rotation* $[\alpha]_{\lambda}^{T}$ of that compound at a given temperature, T, and wavelength, λ, is given by

$$[\alpha]_{\lambda}^{T} = \alpha / lc \tag{10.6}$$

where α is the observed rotation in degrees, l is the optical path length in decimeters and c is the concentration in grams per milliliter.[53] A racemate will exhibit no rotation (the effects of the two enantiomers canceling each other out) while a sample which is enantiomerically enriched but not enantiopure will show a rotation which is lower than the specified specific rotation. A measurement of optical rotation under specified standard conditions may be an acceptable chiral identity test; there is a USP General Chapter <781>, *Optical Rotation*, and this test is reported in some drug monographs. Standard test conditions are typically a wavelength of 589 nm (the mercury D-line) and a temperature of 20 or 25 °C. When measuring specific rotation, the analyte has to be reasonably pure, so that the measurement is not influenced by the presence of chiral impurities. Circular dichroism has found more use in drug discovery and development rather than as a QC tool for small molecules[54] although validated assay methods using circular dichroism have been reported.[55] In contrast to small molecules, circular dichroism is widely used for large molecules.

While ECD is a measure of differential absorption in the UV or visible parts of the spectrum due to electronic transitions, vibrational circular dichroism (VCD) is an analogous measurement due to differences in absorbance of RCP and LCP light in the vibrational spectral region.[56] VCD is most commonly used in structure elucidation rather than as a quantitative method in drug analysis, and like X-ray crystallography, is capable of determining the absolute configuration of a compound.[56] Unlike crystallography, a liquid sample is typically used.

Chiroptical detection may also be used in combination with LC.[57] Qualitatively, this is useful to determine the order of elution of enantiomers. Coupled chiral detection with achiral chromatography can also be used quantitatively for chiral assay if a pure enantiomer standard is available, by taking the ratio of the output of the chiral detector to that of a nonchiral detector (e.g. UV).[58] Although the advances in chiral chromatography have largely supplanted this approach for finished product testing, it may still be a useful way to monitor reactions on- or at-line.

10.5.2 X-ray crystallography

Given that Pasteur identified chirality via the observation of enantiomorphous crystals,[4] it is no surprise that X-ray crystallography is a powerful technique for the determination of chirality.[59] X-ray crystallography can be used as a relative method, wherein an internal reference compound of known stereochemical configuration is crystallized together with the analyte of interest.[60] Absolute determination may also be performed without any external reference.[61] As long as a suitable crystal of an enantiopure small molecule can be grown, the determination of the absolute configuration of that material by X-ray crystallography is considered to be possible in most cases.[59]

10.5.3 Nuclear Magnetic Resonance

NMR may be used either to measure the level of an enantiomeric impurity in a sample or to perform the determination of absolute configuration of a single enantiomer.[62] In an achiral environment, a single enantiomer will produce individual resonances, however, in a chiral environment, e.g. due to a chiral solvent or chiral additive (a "shift reagent"), a mixture of enantiomers will produce pairs of peaks whose area is in proportion to the composition of the mixture. The peaks arise because of the transient diastereomeric complexes that are formed in solution. Some of the chiral agents used are familiar from other chiral analysis techniques, e.g. CDs or crown ethers,[63] while others are more specifically useful in NMR, with complexing agents containing paramagnetic lanthanide metals being a broadly used example.[63]

Determination of the absolute configuration is typically performed using chiral derivatizing agents, where a covalently bound diastereomeric complex is formed.[62,63]

10.6 CHIRAL METHOD VALIDATION

The general principles outlined in Chapters 4 and 6 apply to the validation of chiral methods, whether they be separations or spectroscopic[55] analyses. The usual parameters need to be studied, as appropriate: specificity, linearity, sensitivity, precision, accuracy, and various aspects of ruggedness and robustness. System suitability tests may be added to the method where needed to ensure that the method performance criteria are met.

Demonstrating adequate selectivity is important, since with many chiral chromatographic methods the peak width can be fairly broad, and baseline resolution may not be achieved. This can be particularly problematic if the wrong enantiomer elutes after the drug peak. In such a case, tailing of the main peak can make it very difficult to quantitate the minor enantiomer. As discussed in Section 10.3.2.1, it may well be possible with the array of chiral columns available today to find an alternative separation with the preferred elution order. However, if it is not, broad peaks and partial resolution are two factors that can lead to relatively poor lower limits of quantitation for chiral chromatographic methods, due to low signal-to-noise ratios and challenges in integration. This is recognized in regulatory guidance documents, and enantiomer analysis is specifically excluded from the reporting limit guidance in ICH Q3A, and levels >0.05% may be acceptable. Partial resolution may also require the analyst to carefully consider how to integrate samples to avoid introducing a bias at low concentrations: is drop-line or a skim more appropriate? It may be necessary to carefully specify such considerations in the method to obtain good quantitation[64] and to put in place appropriate system suitability tests to ensure adequate performance.

A variety of literature references[51] illustrate the use of fully validated chiral CE methods in pharmaceutical analysis. One point to consider for validating chiral CE methods is that if two enantiomers are separated, they will actually travel through the on-capillary detector at different speeds, resulting, for example, in different integrated peak areas for two peaks from a racemic mixture. The solution is to report the time-normalized peak area.[65] If this is done, the unwanted enantiomer can be accurately quantitated against a standard of the drug itself.

Chiral stability during the analytical process (e.g. during extraction and for storage of samples) should be determined to ensure that racemization does not occur. Ruggedness and robustness of chiral chromatographic methods may also be a concern. Unlike simple and relatively inexpensive C18 columns, laboratories may be hesitant to keep an expensive chiral column just for one assay, and instead use it for method development with multiple MPs and additives, which can lead to modification of the column selectivity.

10.7 CONCLUSIONS

A variety of analytical techniques are available today to meet the regulatory requirements for drug analysis during development and commercialization. Developments in chiral LC have made this the most used separation technique, especially for later-stage development and at commercialization. Crystallography or spectroscopic techniques play an important role in determining the absolute configuration of chiral compounds.

Acknowledgments

The authors wish to Qinggang Wang, Yingru Zhang Li Li, Himanshu Gandhi, and Meng Chai for the analytical work that generated several of the chromatograms shown here.

References

1. Mislow, K. *Introduction to Stereochemistry;* W. A. Benjamin: New York, 1965.
2. Eliel, E. L.; Wilen, S. H.; Doyle, M. P. *Basic Organic Stereochemistry;* Wiley-Interscience: New York, 2001.
3. Jozwiak, K. Stereochemistry—Basic Terms and Concepts. In *Drug Stereochemistry: Analytical Methods and Pharmacology;* Jozwiak, K., Lough, W. J., Wainer, I. W., Eds.; Informa Healthcare: Independence, 2012; pp 17–29.
4. Pasteur, L. The Asymmetry of Naturally Occurring Organic Compounds (lectures delivered before the Chemical Society of Paris, 1860). In *The Foundations of Stereo Chemistry: Memoirs by Pasteur, van't Hoff, Lebel and Wislicenus;* Richardson, G. M., Ed.; American Book Company: New York, 1901.
5. Ariens, E. J. Stereoselectivity of Bioactive Agents: General Aspects. In *Stereochemistry and Biological Activity of Drugs;* Ariens, E. J., Soudjin, W., Timmermanns, T. M. W. M., Eds.; Blackwell: Oxford, 1983.
6. Smith, S. W. Chiral toxicology: It's the same thing only different. *Toxicol. Sci.* **2009,** *110* (1), 4–30.
7. Jóźwiak, K.; Lough, W. J.; Wainer, I. W. *Drug Stereochemistry: Analytical Methods and Pharmacology,* 3rd ed; Informa healthcare: London, 2012.
8. Ariens, E. J. Stereochemistry, a Basis for Sophisticated Nonsense in Pharmacokinetics and Clinical Pharmacology. *Eur. J. Clin. Pharmacol.* **1984,** *26* (6), 663–668.
9. Branch, S. K.; Hutt, A. J. Regulatory Perspective on the Development of New Stereoisomeric Drugs. In *Drug Stereochemistry: Analytical Methods and Pharmacology;* Krzysztof, J., Lough, W. J., Wainer, I. W., Eds.; Informa Healthcare, 2012.
10. Development of new stereoisomeric drugs. 1992, updated April 27, 2011; January 29, 2013; Available from: http://www.fda.gov/drugs/GuidanceComplianceRegulatoryInformation/Guidances/ucm122883.htm.
11. *Investigation of Chiral Active Substances,* January 29, 2013. Available from: http://www.ema.europa.eu/docs/en_GB/document_library/Scientific_guideline/2009/09/WC500002816.pdf.
12. Sahajwalla, C. Regulatory Considerations in Drug Development of Stereoisomers. In *Chirality in Drug Design and Development;* Reddy, I. K., Mehvar, R., Eds.; Dekker: New York, 2004; pp 369–379.
13. Zeid, R. L. Regulatory and Development Considerations of Chiral Compounds. In *Chiral Separation Methods for Pharmaceutical and Biotechnological Products;* Ahuja, S., Ed.; John Wiley and Sons: Hoboken, 2011; pp 9–34.
14. Specifications: Test Procedures and Acceptance Criteria for New Drug Substances and New Drug Products: Chemical Substances Q6A. In *International Conference on Harmonization of Technical Requirements for Registration of Pharmaceuticals for Human Use;* 1999.
15. Williams, R. C.; Riley, C. M.; Sigvardson, K. W.; Fortunak, J.; Ma, P.; Nicolas, E. C.; Unger, S. E.; Krahn, D. F.; Bremner, S. L. Pharmaceutical development and specification of stereoisomers. *J. Pharm. Biomed. Anal.* **1998,** *17* (6–7), 917–924.

16. Eliel, E. L.; Wilen, S. H.; Mander, L. N. Determination of Enantiomer and Diastereomer Composition. In *Stereochemistry of Organic Compounds;* John Wiley & Sons, Inc: New York, 1994; pp 214–295.
17. Zhang, T.; Nguyen, D.; Franco, P. Use of Evaporative Light Scattering Detector in the Detection and Quantification of Enantiomeric Mixtures by HPLC. *J. Sep. Sci.* **2006,** *29* (10), 8.
18. Ahuja, S., Ed. *Chiral Separation Methods for Pharmaceutical and Biotechnological Products;* John Wiley & Sons, Inc.: Hoboken, 2011; pp 491.
19. Maier, N. M.; Lindner, W. Stereoselective Chromatographic Methods for Drug Analysis. In *Chirality in Drug Research;* Francotte, E., Lindner, W., Eds.; Wiley-VCH Verlag GmbH & Co. KGaA: Weinheim, 2006; pp 189–259.
20. Chankvetadze, B. Recent Developments on Polysaccharide-Based Chiral Stationary Phases for Liquid-Phase Separation of Enantiomers. *J. Chromatogr. A* **2012,** *1269,* 26–51.
21. Lämmerhofer, M. Chiral Recognition by Enantioselective Liquid Chromatography: Mechanisms and Modern Chiral Stationary Phases. *J. Chromatogr. A* **2010,** *1217,* 814–856.
22. Berthod, A., Ed. *Chiral Recognition in Separation Methods: Mechanisms and Applications;* Springer-Verlag: Berlin Heidelberg, 2010.
23. He, B. L.; Shi, Y.; Kleintop, B.; Raglione, T. Direct and Indirect Separations of Five Isomers of Brivanib Alaninate Using Chiral High-Performance Liquid Chromatography. *J. Chromatogr. B* **2008,** *875,* 122–135.
24. Zhang, Y.; Watts, W.; Nogle, L.; McConnell, O. Rapid Method Development for Chiral Separation in Drug Discovery Using Multi-Column Parallel Screening and Circular Dichroism Signal Pooling. *J. Chromatogr. A* **2004,** *1049,* 75–84.
25. Zhang, T.; Franco, P.; Nguyen, D.; Hamasaki, R.; Miyamoto, S.; Ohnishi, A.; Murakami, T. Complementary Enantiorecognition Patterns and Specific Method Optimization Aspects on Immobilized Polysaccharide-Derived Chiral Stationary Phases. *J. Chromatogr. A* **2012,** *1269,* 178–188.
26. Thunberg, L.; Hashemi, J.; Andersson, S. Comparative Study of Coated and Immobilized Polysaccharide-Based Chiral Stationary Phases and Their Applicability in the Resolution of Enantiomers. *J. Chromatogr. B* **2008,** *875,* 72–80.
27. Zhang, T.; Nguyen, D.; Franco, P.; Murakami, T.; Ohnishi, A.; Kurosawa, H. Cellulose 3,5-Dimethylphenylcarbamate Immobilized on Silica: A New Chiral Stationary Phase for the Analysis of Enantiomers. *Anal. Chim. Acta* **2006,** *557,* 221–228.
28. He, B. L. Chiral Recognition Mechanism: Practical Considerations for Pharmaceutical Analysis of Chiral Compounds. In *Chiral Recognition in Separation Methods – Mechanisms and Applications;* Berthod, A., Ed.; Springer: Lyon, 2010; pp 153–201.
29. Astec Chirobiotic: Macrocyclic Glycopeptide-Based Chiral HPLC Phases, http://www.sigmaaldrich.com, Supelco (Editor).
30. Operating Guidelines for ChromTech Chiral-AGP, Chiral-HSA, and Chiral-CBH HPLC Columns. Available from: http://www.sigmaaldrich.com/etc/medialib/docs/Supelco/Product_Information_Sheet/t709074.Par.0001.File.tmp/t709074.pdf. (accessed March 5, 2013).
31. *Ultron ES-OVM Column: A Bonded Ovomucoid Protein Column For Direct HPLC Separations of Chiral Compounds - Technical Overview;* Agilent Technologies, Inc, 2008.
32. Astec Cyclobond: Bonded Cyclodextrin-based Chiral HPLC Phases, http://www.sigmaaldrich.com, Supelco, (Editor).
33. *Regis Technologies Chiral Application Guide VI,* 2007; pp 83–85.
34. Jin, J. Y.; Lee, W.; Hyun, M. H. Development of the Antipode of the Covalently Bonded Crown Ether Type Chiral Stationary Phase for the Advantage of the Reversal of Elution Order. *J. Liq. Chromatogr. Relat. Technol.* **2006,** *29,* 841–848.
35. Sun, P.; Armstrong, D. W. Effective Enantiomeric Separations of Racemic Primary Amines by the Isopropyl Carbamate-Cyclofructan6 Chiral Stationary Phase. *J. Chromatogr. A* **2010,** *1217,* 4904–4918.

36. Hoffmann, C. V.; Reischl, R.; Maier, N. M.; Lämmerhofer, M.; Lindner, W. Investigations of Mobile Phase Contributions to Enantioselective Anion- and Zwitterion-Exchange Modes on Quinine-Based Zwitterionic Chiral Stationary Phases. *J. Chromatogr. A* **2009,** *1216,* 1157–1166.
37. Lämmerhofer, M.; Di Eugenio, P.; Molnar, I.; Lindner, W. Computerized Optimization of the High-Performance Liquid Chromatographic Enantioseparation of a Mixture of 4-Dinitrophenyl Amino Acids on a Quinine Carbamate-Type Chiral Stationary Phase Using Drylab. *J. Chromatogr. Biomed. Appl.* **1997,** *689* (1), 123–135.
38. Nistor, I.; Lebrun, P.; Ceccato, A.; Lecomte, F.; Slama, I.; Oprean, R.; Badarau, E.; Dufour, F.; Dossou, K. S. S.; Fillet, M.; Liégeois, J.-F.; Hubert, P.; Rozet, E. Implementation of a Design Space Approach for Enantiomeric Separations in Polar Organic Solvent Chromatography. *J. Pharm. Biomed. Anal.* **2013,** *74* (0), 273–283.
39. Kotoni, D.; Ciogli, A.; D'Acquarica, I.; Kocergin, J.; Szczerba, T.; Ritchie, H.; Villani, C.; Gasparrini, F. Enantioselective Ultra-High and High Performance Liquid Chromatography: A Comparative Study of Columns Based on the Whelk-O1 Selector. *J. Chromatogr. A* **2012,** *1269,* 226–241.
40. Tachibana, K.; Ohnishi, A. Reversed-Phase Liquid Chromatographic Separation of Enantiomers on Polysaccharide Type Chiral Stationary Phases. *J. Chromatogr. A* **2001,** *906,* 127–154.
41. Schrig, V. Separation of Enantiomers by Gas Chromatography on Chiral Stationary Phases. In *Chiral Separation Methods for Pharmaceutical and Biotechnological Products;* Ahuja, S., Ed.; John Wiley & Sons, Inc.: Hoboken, 2011; pp 251–297.
42. He, L.; Beesley, T. E. Applications of Enantiomeric Gas Chromatography: A Review. *J. Liq. Chromatogr. Relat. Technol.* **2005,** *28,* 1075–1114.
43. Ren-Qi, W.; Teng-Teng, O.; Siu-Choon, N.; Weihua, T. Recent Advances in Pharmaceutical Separations with Supercritical Fluid Chromatography Using Chiral Stationary Phases. *TrAC, Trends Anal. Chem.* **2012,** *37,* 83–100.
44. De Klerck, K.; Mangelings, D.; Vander Heyden, Y. Supercritical Fluid Chromatography for the Enantioseparation of Pharmaceuticals. *J. Pharm. Biomed. Anal.* **2012,** *69,* 77–92.
45. Alexander, A. J.; Hooker, T. F.; Tomasella, F. P. Evaluation of Mobile Phase Gradient Supercritical Fluid Chromatography for Impurity Profiling of Pharmaceutical Compounds. *J. Pharm. Biomed. Anal.* **2012,** *70,* 77–86.
46. Miller, L. Preparative Enantioseparations Using Supercritical Fluid Chromatography. *J. Chromatogr. A* **2012,** *1250,* 250–255.
47. Mikuš, P. *Chiral Capillary Electrophoresis in Current Pharmaceutical and Biomedical Analysis;* In Tech: Rijeka, 2012.
48. Ahmed, A.; Ibrahim, H.; Pastoré, F.; Lloyd, D. K. Relationship Between Retention and Effective Selector Concentration in Affinity Capillary Electrophoresis and High-Performance Liquid Chromatography. *Anal. Chem.* **1996,** *68,* 3270–3273.
49. Rocheleau, M. J. Generic Capillary Electrophoresis Conditions for Chiral Assay in Early Pharmaceutical Development. *Electrophoresis* **2005,** *26* (12), 2320–2329.
50. Szucs, R.; Caron, I.; Taylor, K. A.; Gee, S. P.; Ferguson, P. D.; Kelly, M. A.; Beaman, J. V.; Lipczynski, A. M.; Hailey, P. A. Generic Approach to Chiral Separations: Chiral Capillary Electrophoresis with Ternary Cyclodextrin Mixtures. *J. Microcolumn Sep.* **2000,** *12* (11), 568–576.
51. El Deeb, S.; Hasemann, P.; Wätzig, H. Strategies in Method Development to Quantify Enantiomeric Impurities Using CE. *Electrophoresis* **2008,** *29* (17), 3552–3562.
52. Chiroptical Properties; Chapter 12. In *Basic Organic Stereochemistry;* Eliel, E. L., Wilen, S. H., Doyle, M. P., Eds.; Wiley: New York, 2001; pp 534–607.
53. Mason, S. F. *Molecular Optical Activity and the Chiral Discriminations;* Cambridge University Press: Cambridge, 2009.

54. Bertucci, C.; Pistolozzi, M.; De Simone, A. Circular Dichroism in Drug Discovery and Development: An Abridged Review. *Anal. Bioanal. Chem.* **2010,** *398* (1), 155–166.
55. Do Monte, Z. S.; Ramos, C. S. Development and Validation of a Method for the Analysis of Paroxetine HCl by Circular Dichroism. *Chirality* **2013,** *25* (4), 211–214.
56. He, Y.; Wang, B.; Dukor, R. K.; Nafie, L. A. Determination of Absolute Configuration of Chiral Molecules Using Vibrational Optical Activity: A Review. *Appl. Spectrosc.* **2011,** *65* (7), 699–723.
57. Lloyd, D. K.; Goodall, D. M. Polarimetric Detection in High Performance Liquid Chromatography. *Chirality* **1989,** *1,* 251–264.
58. Wu, Z.; Goodall, D. M.; Lloyd, D. K. Determination of Enantiomeric Purity of Ephedrine and Pseudoephedrine by High-Performance Liquid Chromatography with Dual Optical Rotation/UV Absorbance Detection. *J. Pharm. Biomed. Anal.* **1990,** *8* (4), 357–364.
59. Thompson, A. L.; Watkin, D. J. X-ray Crystallography and Chirality: Understanding the Limitations. *Tetrahedron Asymmetry* **2009,** *20* (6–8), 712–717.
60. Harada, N. Determination of Absolute Configurations by X-ray Crystallography and 1H NMR Anisotropy. *Chirality* **2008,** *20* (5), 691–723.
61. Flack, H. D.; Bernardinelli, G. The Use of X-ray Crystallography to Determine Absolute Configuration. *Chirality* **2008,** *20* (5), 681–690.
62. Seco, J. M.; Quiñoá, E.; Riguera, R. The Assignment of Absolute Configuration by NMR. *Chem. Rev.* **2004,** *104* (1), 17–117.
63. Wenzel, T. J.; Wilcox, J. D. Chiral Reagents for the Determination of Enantiomeric Excess and Absolute Configuration Using NMR Spectroscopy. *Chirality* **2003,** *15* (3), 256–270.
64. Meyer, V. R. Accuracy in the Chromatographic Determination of Extreme Enantiomeric Ratios: A Critical Reflection. *Chirality* **1995,** *7* (8), 567–571.
65. Altria, K. D. Essential Peak Area Normalisation for Quantitative Impurity Content Determination by Capillary Electrophoresis. *Chromatographia* **1993,** *35* (3–4), 177–182.

CHAPTER

Water determination

11

Nilusha L.T. Padivitage, Jonathan P. Smuts, Daniel W. Armstrong*

Department of Chemistry and Biochemistry, University of Texas at Arlington, Arlington, TX, USA

**Corresponding author*

CHAPTER OUTLINE

11.1 INTRODUCTION

The determination of water content is an important aspect of the pharmaceutical industry. It is of great significance in pharmaceutical development, process, production and quality control. Knowing the water content and understanding the hygroscopic nature of a drug substance, as well as the final product in which it is contained is essential. Water can affect the physical and chemical stability of active pharmaceutical ingredients (APIs) and drug formulations,[1] the microbial sustainability and activity, potency, efficacy and shelf life.[2] Therefore, it is very important to set proper specifications and have accurate methods for the determination of water in pharmaceuticals while they are still in developmental stages.

Specification of Drug Substances and Products. http://dx.doi.org/10.1016/B978-0-08-098350-9.00011-4

The US Food and Drug Administration (FDA) quality by design initiative and International Conference on Harmonization (ICH) recognize the importance of a thorough understanding of the product and the manufacturing process, and they encourage a thoughtfully designed risk management program backed by scientifically sound testing methods. ICH makes recommendations toward achieving greater harmonization in the interpretation and application of technical guidelines and requirements for pharmaceutical product registration. These guidelines are intended to assist, to the extent possible, in the establishment of a single set of global specifications for new drug products. A specification includes a list of tests, references to analytical procedures, and appropriate acceptance criteria, which are numerical limits, or other criteria for the tests described. ICH Q6A outlines testing procedures and acceptance criteria for drug release programs.[3,4] According to ICH Q6A, the universal tests such as description, identification, assay and impurities, which are required for any new drug substance or drug product, are specified. In addition, specific tests, such as physicochemical properties, particle size, polymorphic forms, isomeric forms (chiral), water content, inorganic impurities, and microbial limits (determined on a case-by-case basis), are included.

Moreover, instructions on the best methods for determining microbiological attributes of drug substances and excipients and of non-sterile drug products are found in decision trees #6 and #8 in ICH Q6A. In both decision trees, the need for microbial limits testing is based on whether the product is inherently "dry" enough to prevent microbial growth. This dryness can be readily determined via moisture content analysis. Excessive moisture also can adversely affect the stability of the active ingredient. Hence, the accurate determination of water content is one of the most important and frequent analyses performed in the pharmaceutical industry.

A number of analytical techniques have been developed for this purpose, which vary in their accuracy, cost, speed, sensitivity, specificity, ease of operation, etc. However, the choice of an analytical procedure for a particular application depends on the nature of the substances being analyzed and the reason the information is needed. Traditionally, water content in pharmaceutical products is determined by one of two methods, loss on drying (LOD) or Karl Fischer Titration (KFT).[5–7] There have been a number of other techniques reported, such as gas chromatography (GC),[8–17] near-infrared (NIR),[8,18,19] solvatochromic sensing,[20,21] fluorine nuclear magnetic resonance (NMR),[22] isotope ratio mass spectrometry,[20] for the determination of water in different compounds. In this chapter, we focus on LOD and KFT, as they are specified in the leading regulatory documents such as the US and European Pharmacopeias, as well as in ICH guidelines.[5–7] Also, we shall discuss GC and NIR methods as they have great potential for determination of water content in pharmaceutical ingredients and will become increasingly important in the future.

11.2 LOSS ON DRYING (LOD)

LOD, expressed as percent (w/w), is the loss of weight under specific conditions (usually 105 °C for a given time period). In the pharmaceutical industry, the LOD procedure was introduced in the beginning of the last century[23] and still may be considered adequate in some cases. According to the United Standards Pharmacopeia (USP <731>), LOD[5] may still be used in those cases where the weight loss sustained on heating may be not entirely water. This method can be carried out when the sample material is abundant and will not decompose or melt at 110 °C. In this method, the prescribed quantity of the substance specified in the appropriate monograph is dried to constant mass or for the prescribed time. Drying can be carried out in a desiccator, in vacuo, or in an oven within a specified temperature range. This method is commonly used to determine water content in excipients, tablets and for stable APIs.[24] Also, accurate LOD values can be determined using thermogravimetric analysis.[25]

Although this is a common method for water determination, LOD suffers from several disadvantages. The difference in mass before and after drying is not necessarily the water content but rather mass loss on drying. This mass loss is sometimes called "moisture" but this term is problematic, as it is commonly used to mean water and any other volatile liquids. Some of the mass loss may result from intrinsic gaseous and volatile compounds and/or decomposition products. A further complication is posed by the existence of free water molecules and different forms of bound water that are associated more or less strongly to components, often evading determination. Therefore, this is not a specific method for the water determination. Furthermore, the time required for loss on drying may be several hours. When the sample contains low melting ingredients, it may not be applicable for detection of the entire amount of water.

11.3 KARL FISCHER TITRATION (KFT)

Currently, the preferred and most widely used method for water determination is KFT,[26–30] which was first reported in 1935.[31] It has been a well-established standard method for water analysis for over 70 years, and is specified in the leading pharmacopeias such as the USP[5,6] and European Pharmacopoeia (EP).[7] It is estimated that nearly 500,000 KF determinations are performed daily around the world. The KF method is based on the modified Bunsen reaction, which is used for the determination of sulfur dioxide in aqueous solutions.

11.3.1 KFT Theory

The KF reaction is a specific quantitative reaction of water with an anhydrous solution of sulfur dioxide and iodine in the presence of a buffer that reacts with hydrogen ions. A variety of modifications have been introduced since the original method was published, and the conditions and reagent composition can vary greatly depending on the type of sample to be analyzed and the technique to be used. According to the original interpretation of this method, the KF reaction is based on the well-known Bunsen reaction in aqueous media (Eqn (11.1)).

$$I_2 + 2H_2O + SO_2 \rightarrow H_2SO_4 + 2HI \qquad (11.1)$$

The reagent is buffered with pyridine (Py) in order to neutralize liberated protons and ensure complete reaction (Eqn (11.2)).

$$2H_2O + (Py)_2 \cdot SO_2 + I_2 + 2Py \rightarrow (Py)_2 \cdot H_2SO_4 + 2Py \cdot HI \qquad (11.2)$$

Fischer proposed the above equation (Eqn (11.2)) for water determination, which requires a 2:1 molar ratio of H_2O to iodine. This reaction led to the establishment of the classical KF reagent, which is a solution of iodine and sulfur dioxide in a mixture of pyridine and methanol.

Later on, Smith et al. showed that methanol plays an important role in the KF reaction.[32] They showed that water forms a pyridine sulfur trioxide complex in the first partial reaction (Eqn (11.3)) that preferably reacts with methanol (Eqn (11.4)), or in lack of alcohol, with a second molecule of water.

$$H_2O + SO_2 + I_2 + 3C_5H_5N \rightarrow 2C_5H_5N \cdot HI + C_5H_5N \cdot SO_3 \qquad (11.3)$$

$$C_5H_5N \cdot SO_3 + CH_3OH \rightarrow C_5H_5N \cdot HSO_4CH_3 \qquad (11.4)$$

From the reaction scheme, it is clear that the stoichiometry between water and iodine is 1:1 in this solvent. Hence, the amount of water can be determined by measuring consumption of iodine. Although this theory could explain the original KF 2:1 stoichiometry between water and iodine in the absence of methanol, this was questioned during later years. For example, Scoholz[33] and Eberius[34] pointed out that pyridine sulfur trioxide reacts very slowly with methanol or water.

Cerdergren investigated the kinetics of the KF reaction by using a platinum electrode to monitor the iodine concentration when varying the reagent composition.[35] The reaction was shown to follow a third order rate expression with a first order dependence of each of the three reactants-water, sulfur dioxide and iodine, as shown as Eqn (11.5). This reaction rate did not depend much on pyridine content.

$$\frac{d[I_2]}{dt} = -k_3[I_2][SO_2][H_2O] \qquad (11.5)$$

Verhoef and Barendrecht[36,37] confirmed the result that was observed by Cerdergren[35] and they also showed that similar reaction rates were obtained for different bases at constant pH. Also, they showed that the reaction rate was pH dependent up to pH 5. The rate remained constant between pH 5.5 and 8. However, the reaction rate increased only slightly above pH 8. From the pH dependence of the reaction rates, they suggested that neither pyridine sulfur trioxide nor SO_2 was the reactive component in the reaction. It was suggested that the monomethyl sulfite (that is formed from sulfur dioxide and methanol, Eqn (11.6)), was the reactive component since the activity of monomethyl sulfite would be expected to change with pH. Furthermore, they found that pyridine does not take part in the reaction, but only acts as a buffer.

$$2CH_3OH + SO_2 \rightleftharpoons CH_3OH_2^+ + SO_3CH_3^- \qquad (11.6)$$

The reaction rate also is strongly influenced by the concentration of iodide ions. At high concentrations of iodide, both iodide and triiodide ions will be present to some extent (Eqn (11.7)) and the triiodide ions are very stable in methanol.

$$I_2 + I^- \rightleftharpoons I_3^- \qquad (11.7)$$

Hence, Verhoef and Barendrecht considered that these ions are able to oxidize the sulfurous base (methyl sulfite). However, Wünsch and Seubert assumed that the free iodine reacts far more quickly than the large negatively charged triiodide complex.[38]

During the development of efficient KF reagents, several modifications of the original reagents have been proposed. Replacement of pyridine by other bases such as dimethanolamine, sodium acetate and sodium salicylate was tested due to the noxious odor and toxicity of pyridine, with varying degrees of success. Finally, odor-free imidazole was found as the ideal substance for the KF reaction as it is more basic than pyridine.[39] Methanol is the alcohol most commonly used for the KF reaction; however it can be replaced by other alcohols under certain conditions.[40] For example, Schöffski has shown that ethanol-based reagents can be used instead of methanol.[41] Sulfur dioxide is used for all types of KF reactions and no alternative substances have been reported. Wünsch and Seubert showed that a solvation product similar to methyl sulfite forms when formamide is used as a solvent,[38] but the evidence for this has not been published. Finally, other modifiers can be used to enhance such desired KF reagent properties as sample solubility, and reaction rate, suppression of side reactions.

Typical KFT systems consist of a closed vessel with a septum for sample introduction, a stirrer to ensure efficient mixing, an indication system to measure the iodine concentration and either a dispenser to add iodine solution (for volumetric KFT) or a generator electrode to produce iodine (for coulometric KFT). A KF titrator generates or introduces iodine to maintain a certain excess (the end point concentration) and the titration is controlled by the titrator. When a sample is injected, the water will react with iodine and the amount of water in the sample can be calculated from the amount of iodine consumed. However, it should be noted that KFT does not imply one universal method, but is rather a concept covering numerous methods. There are a number of reports for different KFT methods, including flow injection analysis,[42–44] direct potentiometric determination[45,46] and so on. However, volumetric and coulometric methods are by far the most common methods.

11.3.2 Volumetric KFT

The volumetric KFT method is simple, cheap and is preferably used to determine higher water contents. The typical analytical range from 10 ppm to 100% requiring sample sizes of ~15 and ~0.02 g, respectively.[47] In volumetric KFT, not only liquid samples, but also solid or paste samples, can be directly introduced into the titration vessel, and the analysis can be carried out with a variety of suitable organic solvents. According to the original KF method, the titration solution contains all the necessary reactants (I_2, SO_2 and base) in an appropriate solvent and the sample is dissolved (or dispersed) in methanol. In this method, the solvent must be pretitrated as the blank. Then, samples can be dissolved in the pretitrated solvent and subsequently titrated with iodine to determine the water content. However, due to the instability of the single volumetric component reagent, Johansson suggested dividing it into two components which consist of: (1) a methanolic solution of I_2 as a titrant and (2) the sample in a solution of the base and SO_2 in methanol.[48] This two-component reagent provides better shelf life and the greater buffer capacity of the solvent ensures a rapid KF reaction and a distinct end point. The disadvantages of the volumetric method are that the titer has to be determined regularly and large sample volumes have to be used for samples containing small amounts of moisture.

11.3.3 Coulometric KFT

In 1959, Meyer and Boyd presented a method for coulometric generation of iodine in the KF reagent as an alternative to the original volumetric addition of KF reagent.[49] Coulometric water determination is used primarily for the determination of small amounts of water, ranging from 1 ppm (sample size—~ 10 g or more) to 5% (sample size—~ 0.05 g) and is regarded as a trace method.[47] The sample must be introduced in liquid or gaseous form through a septum. For special applications, the water of solid samples is released in an oven and the vapor driven into the cell by a stream of dried air.[50] In this method, the current releases, stoichiometrically, corresponding amounts of iodine from

Table 11.1 Comparison of the Volumetric and Coulometric Karl Fisher Titration Methods

Method Details	Volumetric Karl Fischer Titration	Coulometric Karl Fischer Titration
Source of iodine	• Titration reagent.	• Anodic oxidation of iodide.
Calibration and standardization	• Required frequently. • Two-component reagents are more stable than one-component reagents, but titer has to be determined at time of analysis.	• Not required. • Quantitation is based on Faraday's law.
Range	10 ppm to 100%	1 ppm to 5%
Sample sizes	~15 and ~0.02 g	~10 g or more and ~0.05 g
Advantageous	• Simple. • Cheap. • Preferably used to determine higher water contents. • Applicable for liquid, solid or paste samples. • Different titers available.	• Primarily for the determination of small amounts of water. • An absolute method as no titrant is added. • Highly accurate.
Disadvantages	• The titer has to be determined regularly. • Large sample volumes have to be used for samples containing small amounts of moisture. • No side reactions can occur. • Frequent maintenance is required. • Sample should be soluble in the titration reagent.	• Sample must be introduced in liquid or gaseous form. • The process must take place in 100% current efficiency. • No side reactions can occur. • Replacement of reagent as required. • Sample should be soluble in the titration reagent.

the iodide-containing KF reagent by electrolysis. Exactly the same chemical processes take place as in a volumetric KFT, but I_2 is formed in the titration cell itself by anodic oxidation of iodide. In contrast to the volumetric KFT, the coulometric KFT is an absolute method as no titrant is added but I_2 is generated electrolytically from the iodide. The amount of iodine produced can be determined very accurately by measuring the generating current. It should be noted that the process must take place with 100% current efficiency, and no side reactions can occur during the titration process. The comparison of the volumetric and coulometric KFT methods is summarized in Table 11.1.

11.3.4 Indication systems for end point detection

The detection method is important for obtaining accurate and precise data. The onset of excess iodine indicates the end point of the titration. When excess iodine is present, the solution turns increasingly yellow to brown (with a large excess of iodine). It may be hard to determine the exact end point visually, which means that good reproducibility can be a problem. In addition, the coloration differs in polar solvents and nonpolar solvents, and "visual end points" cannot be automated or easily validated.

Therefore, the usual indication of the end point is based on an electrochemical effect using two polarized platinum electrodes. The advantage of this method over the visual indication method is that the titration is always carried out to the same (slight) excess of iodine and better reproducibility and accuracy can be achieved.

There are two types of indication methods commonly used, namely biamperometric and bivoltametric. Both methods are based on simultaneous oxidation of iodide and reduction of iodine at two platinum electrodes. The most common is the bivoltametric indication method. In this method, two platinum electrodes are polarized by a constant current (~50 μA) and the voltage is monitored. Initially, when the concentrations of iodine are low, a large voltage is necessary to maintain the polarization current. At the end of the titration, the voltage drops suddenly as the concentration of iodine increases. If the voltage remains below a certain value (usually 250 mV) for a certain time, which is "stop delay", the determination is completed. This stop delay is very important for the determination of water in samples which do not contain free water that immediately reacts. The arrival of delayed water in the working medium makes the voltage rise above the critical value again and further reagent is required until the voltage remains below the chosen end point value for the desired delay time. In the biamperometric indication method, a constant voltage (maximum 500 mv) is applied to the electrodes and the resulting current is measured. In this method, the measured current will be low at low iodine concentrations and then it increases at the end point. Hence, minimal current flows as long as excess water is present and the current increases to a few μA when excess iodine is present.[51]

11.3.5 Applications

The KFT method is a selective, quantitative, preferred and long-standing standard method in pharmacopeias for the determination of water for both drug substances and drug products.[3] For example, the EP, fourth edition, Chapter 2.5.12 "Water: Semi-Micro Determination" describes the volumetric method while Chapter 2.5.32 describes the coulometric method.[7] Chapter 2.5.12 "Water: Semi-Micro Determination" describes Method A (direct titration of water) and Method B (the indirect method of back titration). According to Method A, methanol or the solvent indicated in the monograph or recommended by the supplier of the titrant, is introduced into the titration. Then the sample is introduced rapidly and titrated. Method B has basically the same requirements as Method A, however in Method B, the titrant, which is not consumed by the water in the sample, has to be back titrated using a standard with a known amount of water. Hence, an excess of titrant has to be added after addition of the sample. In practice, the direct Method A is easier to carry out and more widely used.

According to USP <921>,[6] the KFT is specified as Method I (Titrimetric) for the determination of water and emphasizes Method Ia (Direct Titration), Method Ib (Residual Titration) and Method Ic (Coulometric Titration) unless otherwise specified in the individual monograph. According to Method I, the test specimen may be titrated with the KF reagent directly, or the analysis may be carried out by a residual titration procedure which is specified in Method Ib. The precision of Method Ia is governed largely by the extent to which atmospheric moisture is excluded from the system. Method Ib (Residual Titration) is generally applicable and avoids the difficulties that may be encountered in the direct titration of substances from which the bound water is released slowly. In this method, excess reagent is added to the test sample, sufficient time is allowed for the reaction to reach completion, and the unconsumed reagent is titrated with a standard solution of water in a solvent such as methanol.

In Method Ic (Coulometric Titration), the KF reaction is used with coulometric determination of water as described earlier. Both the EP and the USP describe sample size/sample weight, titration method (95% direct volumetric titration), the range of water content permitted, and the reagents to be used. The use of pyridine-containing reagents for KF volumetry is described in both standards. The USP has fewer restrictions, which permits the use of commercially available reagents containing other solvents than pyridine. Finally, KF titrations are effective at quantifying even tightly bound water and often are considered a better method than LOD. Also, KF is a standard method that is used for calibration of other indirect methods for water determination.[52]

11.3.6 Challenges posed by KFT

Although the KFT method has a number of advantages, pharmaceutical samples have posed certain challenges to conventional KFT methods and reagents. These can be categorized mainly as (1) reactivity issues, (2) solubility issues, and (3) pH issues.

11.3.6.1 Reactivity issues

A number of reactions can occur between the components of the sample to be analyzed and the various species present in the KF reagent. These undesired interfering reactions should be avoided or minimized in order to obtain accurate results. Specifically, no water should be released in side reactions, nor should the sample consume or release iodine during the titration. KFT has recognized limitations for molecules with certain functional groups such as aldehydes, ketones, and mercaptans. Those compounds will undergo interfering side reactions either with methanol or iodine. For example, the side reactions of aldehydes and ketones with the methanol in the KF reagents form acetals and ketals, respectively, with the release of water.[40] The water formed by these side reactions can result in an over estimation of the water content, and in some cases no end points are reached, making water determination impossible. Acetal and ketal formation is influenced by the presence of alcohol and the reactivity decreases as the chain length of the alcohol increases. Aldehydes react rapidly with methanol present in the KF reagent and the reactivity of the aldehydes decreases as the chain length of both the alcohol and the aldehyde increases. However, acetal formation is inhibited by phenyl groups (e.g. benzaldehyde, diphenylacetaldehyde) and halogen substitutions (e.g. chloral 2-bromobenzaldehyde). Further, it was indicated that the rates of water formation for pyridine or imidazole buffered KF reagents were nearly the same. All nonalcoholic solvents inhibit the formation of acetals and ketals while they promote bisulfite formation, which occurs when SO_2 reacts directly with aldehydes or ketones. This leads to an underestimation of water content. Hoogmartens et al. reported interlaboratory variations for the water content of erythromycin samples as determined by the KF method.[28] These variations were related to the reagents used. However, it is also known that Erythromycin A is transformed into Erythromycin A enol ether by acid catalyzed dehydration.[53] The composition of the KF reagents can affect the acid catalysis and the speed of degradation, which may be the reason for the interlaboratory variability. It was indicated that accurate titration of water in erythromycin samples is possible when a 10% (m/v) solution of imidazole in methanol is used when a compared to pyridine or a mixture of pyridine and methanol in the KF reagent.

Iodine-consuming side reactions also have to be considered and minimized for accurate results. Penicillin is one example of a compound that reacts with iodine in the presence of KF reagents. Penicillin always contains some penicilloic acid, which at least in aqueous solution consumes iodine.[26] The resulting error can be minimized by using commercially available fast-reacting KF reagents that provide shorter titration times, and therefore reduces the risk of interfering side reactions.

Compounds containing strong acids, silanol groups and boron compounds can esterify the methanol in the KF reagents and release water.[51] These issues can be somewhat overcome by using methanol-free KF reagents. Methanol can be replaced by a higher alcohol like 2-methoxyethanol. On the other hand, methyl sulfuric acid and hydroiodic acid are produced during the KF titration. Although the KF reagents were buffered with bases, they still form weak acids that react with some compounds containing carbonates, hydroxides and oxides, and thus release water. Also, reducing and oxidizing agents can directly or indirectly react with the iodine in the KF reagents and affect the water content of the original samples. For example, compounds containing reducing agents such as iron (III) salts, copper (II) salts, tin (II) salts, arsenites, arsenates and ascorbic acid (vitamin C) all react with iodine and produce hydrogen iodide which leads to an inaccurate water measurement. Kuselman et al. reported that water determination in vitamin C is not amenable to KF reagents due to the oxidation of the ene-diol group in vitamin C by iodine.[54] The use of a new reagent consisting of iodine, potassium iodide and sodium acetate in a nonaqueous medium provided somewhat reproducible and accurate results for the tablets containing vitamin C. Also, compounds containing oxidizing agents, such as dialkyl peroxides, can release elemental iodine from the iodide in the KF reaction which results in a low estimation of the water content. These types of reactivity issues sometimes can be resolved by performing titrations at lower temperatures.

A number of new KF reagents and methods have been developed for different types of samples.[54–58] Suitable solvents and methods should be carefully selected depending on the functional groups of the samples being analyzed. Basically, these methods and reagents are designed to suppress interfering side reactions. In addition, substances that release their water slowly or at high temperatures are not suitable for a direct KFT. In such cases, the sample is first heated in an oven and a carrier gas transfers the released water to the titration cell, where it is then determined by KF titration. This method is called the KF oven method. In this method, only the water enters the cell and the sample itself does not come into contact with the KF reagent. Therefore, unwanted side reactions and matrix effects are eliminated. For example, 5-aminolevulinic acid-HCl (lyophilisate) shows a strong side reaction in methanol and an end point cannot be reached with direct KF titration methods. For this compound, the use of a KF oven in combination with the coulometric titration technique (due to the low water content of this substance) is recommended.[59]

11.3.6.2 Solubility issues

Some pharmaceutical compounds have limited solubility in methanol. For example, atropine sulfate and calcium folinate are only partially soluble in methanol, and riboflavin phosphate sodium (biochemical cofactor; also used as food dye) is insoluble in the alcohol-based KF media.[59] In such cases sample preparation procedures are complicated. Hexanol, decanol, chloroform and xylene are frequently used as solvents to increase the solubility of the samples. However, samples that will not dissolve in any cosolvents will release their water in the presence of formamide. Also, solubility issues may be overcome by using different specialty reagents and performing titrations at elevated temperatures. Samples containing water that is only released at high temperatures and which does not allow for effective extraction using formamide, may be determined by using a KF oven coupled with a titrator.

Sampling also is critical, particularly with heterogenous samples. This is because only a small amount of sample is used for the titration. Many standardized methods provide detailed instructions for sample preparation. In general, it is very important that the sample preparation technique used does not introduce any additional water into the sample and that no water loss occurs by heating the sample.

11.3.6.3 Issues related to pH

The KF titration is a pH-dependent reaction and the optimum pH range is 5.5–8. The titration proceeds very slowly below that range, and reaction rates due to interfering side reactions increase above that range. Side reactions occur during the KF titration, possibly leading to coated electrodes, fading end points or no end points, and erroneous results. For example, compounds containing nitrogen such as benserazide hydrochloride (used in the treatment of Parkinson's disease) and proflavine hemisulfate (a topical antiseptic) may alter the pH value of the working medium.[59] Also, the amount of water in penicillins can be changed by pH influences due to the oxidation of penicillin derivatives such as penicilloic acid and other hydrolysis products by iodine. By conducting the titration in weakly acidic conditions, this side reaction can be suppressed. In general, samples that create highly acidic or basic conditions for the KF reactions must be buffered to ensure accurate water quantification. In the case of acidic samples, weak bases such as imidazole have proven to be the most effective titration aids while in the case of basic samples, buffering using a weak acid, such as salicylic acid is typically recommended. Although many variations on the basic KFT methodologies have been developed to overcome the aforementioned issues,[13,40,60] a number of questions still remain, such as the degradation of reagents with time, residual water in KF reagents and so on. The bottom line is that although the most commonly used method for water analysis is KFT, interference of side reactions,[13] regular instability,[61] sample insolubility,[60] pH issues[8] and the complexity of the analysis prevent it from being accepted as a universal method.

11.4 GAS CHROMATOGRAPHIC METHODS

GC methods for water quantification can play an important role and this technique greatly simplifies the analysis of water in a variety of applications. Specifically, GC methods can be used to determine water content in pharmaceutical samples when limited amounts are available. Early reports have been published for the determination of water using GC mainly based on packed columns (molecular sieves), involving both direct detection by thermal conductivity detection (TCD) and indirect detection with flame ionization detection (FID).[8,11–14,16] The early indirect attempts at GC analysis of water were hindered by the need to pass sample through reactors that converted the water to compounds that are more compatible with existing chromatographic systems. For example, Knight et al. reported a simple GC method for water determination by passing the analyte through a calcium carbide bed where the water reacted to form acetylene, which was then measured by FID.[9] The accuracy has been studied for the samples containing small amounts of water (3 ppm). This method was presumably applicable to all materials that are inert toward calcium carbide. Hogan et al. reported a GC technique employing methanol as an internal standard and polyaromatic beads as column packing for determining the amounts of water in a variety of solvents.[10] They studied the accuracy based on standard additions in order to obtain absolute values and reported the lower limit of detection as 0.1 ppm. Cook et al. also reported the use of Porapak Q (porous polymer) as the solid support for GC for the separation of water.[62] However, analysis of water below 1000 ppm using a column packed with Porapak Q is very difficult because of its poor resolution. Also, the utility of GC methods for the water analysis has been limited by the excessive tailing of the water peak due to its strong adsorption on the stationary phase.[9] Consequently, there have been a few special techniques developed to minimize peak tailing for specific systems. Bennett reported water determination in organic systems using commercially available columns packed with Teflon powder impregnated with 5% Carbowax 20-M.[63] Thermal conductivity detection was used for the analyses and the water peaks were nearly symmetrical even at low concentration (0.2%). Retention times of water showed no variation with changes of the concentration in the organic substance. Quiram has reported the use of wide-diameter open tubular columns for the analysis of organic compounds[10] (aromatic compounds and aliphatic alcohols) and also observed a well-resolved water peak with relatively improved symmetry. It was demonstrated that these columns can give better resolution and perform analyses faster than conventional packed columns. However, it is not possible to attain the higher efficiency (number of theoretical plates) using these columns when compared to the small diameter capillary columns that are standard today. Andrews has reported the analysis of trace amounts of water by capillary gas chromatography with a helium ionization detector.[64] The water content of various solvents and reagents was determined using a split injection technique and fused-silica capillary column coated with Carbowax. Utilization of a helium ionization detector showed improved sensitivity when compared to the TCD, which is commonly used for water analysis by GC. The lower limit of detection was reported as 2 ppm for water. However, the sample capacity of a capillary column is limited and a solvent peak tends to overload the column when a large quantity of sample is injected. In 1988, Oguchi et al. reported the use of a polyethylene glycol-based megabore column (0.53 mm i.d. DBWAX fused-silica), which has a larger sample capacity than a standard capillary, to determine the water content of common organic solvents.[62] They reported that the water content of commercially obtained spectro-grade organic solvents varied from 10 ppm in pentane to 230 ppm in methanol.

In 1998, Zhou et al. explored GC as a reference method for moisture determination by near-infrared spectroscopy (NIRS).[19] They used direct detection of water with TCD and a DBWAX column to determine water in freeze-dried drug products. The amount of water in each sample was calculated from the calibration curve based on peak height. The peak height was used because the use of peak area integration was less reproducible due to peak tailing. All samples spanned a water range of 0.14–5.69% (w/w). They emphasized the applicability of GC as a primary method for cases where thermal and titration methods cannot be used due to sample chemistry. Also, Nußbaum et al. have reported quantitative GC determination of water in small solid samples.[14] Quantification was carried out using ethanol as an internal standard. Solid samples such as drugs were completely dissolved in acetone and isopropanol depending on the solubility of the sample. Complete dissolution of the substances was important to destroy the crystal lattice and release any bound water. The dissolved samples were injected into a fused-silica Porous Layer Open Tubular (PLOT) column and TCD was used to detect the water peak. The authors demonstrated that the GC method was more sensitive and was not affected by chemical reactions. Although there were many developments of GC methods for water determination, concerns about peak asymmetry, poor sensitivity, poor efficiency, strong adsorption of water and many solvents by the stationary phase,[8,11] and instability of some stationary phases to steam, remained as drawbacks.

In 2012, Armstrong et al. introduced a rapid and efficient capillary GC method for quantification of water using ionic liquid-based GC columns with TCD.[17] Ionic liquid stationary phases have high thermal stabilities,[65] variable polarities,[66–74] and exceptional stability to water and oxygen. This makes them excellent choices as GC stationary phases. In the study of water determination done by Armstrong et al.,[17] three open tubular capillary columns, coated with specific ionic liquids, bis-3-hydroxyalkylimidazolium-polyethelene glycol (HAIM-PEG) triflate, trigonal tripropylphosphonium (TTIP) triflate or bis-2,3-dimethyl imidazolium-PEG (DMIM-PEG) triflate, were developed.[17,75] Water was detected in 50 different solvents using the ionic liquid GC columns and the results were compared to those obtained using a commercial PEG GC column as well as by KFT. The accurate quantification of water was achieved using one of two internal standards, either acetone or acetonitrile. Two different internal standards were used in case one coeluted with the analyte solvent under the conditions of the experiment. Water quantification was achieved using a standard curve by integration of the internal standard peak and the water peak. The concentration of internal standard in milligrams per kilogram was multiplied by the ratio of water peak to internal standard peak and the result was divided by the response factor (RF). The accuracy of the results was compared with 10 different solvents and a National Institute of Standards and Technology (NIST) methanol standard (97 ± 13 ppm water).[17,75] The ionic liquid-based stationary phases produced the most precise and accurate results. The limit of detection and the limit of quantification were calculated according to the guidelines of the FDA.[76,77] The ionic liquid-based GC method was shown to produce much lower limits of detection ($\sim$2.0 ng) when compared to coulometric KFT (10 μg). Moreover, only a very small amount of sample (1 μL) was used for GC methods while relatively large amounts of sample (0.5–15 g) are used for the KFT method. Water did not degrade or alter the ionic liquid-based stationary phases and therefore retention times were unchanged.

Figure 11.1 shows the GC separations of small amounts of water from t-butanol and dimethyl formamide. The water peak can be eluted before or after the solvent peak depending on the relative elution order of both water and the solvent on the ionic liquid-based stationary phase. The separation window between the water and solvent peaks was sufficiently large even though the solvent peaks were very broad relative to the water peak (Fig. 11.1(a)). However, a thermal gradient was used to further narrow the peak width of water and reduce analysis times (Fig. 11.1(b)). Generally, peak shapes for the water were symmetrical using the ionic liquid-based stationary phases when compared to the commercial PEG column. The authors pointed out that the ionic liquid stationary phases containing trifluoromethylsulfonate (TfO^-) anions resulted in more symmetrical water peaks than those that contained PF_6^-, BF_4^- or bis[(trifluoromethyl)sulfonyl]imide (NTf_2^-). The typical analysis time ranged from less than 3 min to 7 min on ionic liquid-based stationary phases and they possessed superior selectivity for water with no degradation or chromatographic changes with time. In this method, it was also possible to analyze samples containing high levels of water without pretreatment. The bottom line was that the ionic liquid GC columns increased the sensitivity and the ruggedness of GC technique. This method was highly sensitive and fast. Also, the ionic liquid-based GC method can be used regardless of the chemical nature of the solvent. Hence, this approach greatly simplifies the analysis of water and there is a great potential to use this technique in the pharmaceutical field. The ionic liquid capillary columns for water analysis have just recently become commercially available.[17,75]

11.5 NEAR-INFRARED SPECTROSCOPY

NIRS is a rapid, noninvasive, nondestructive analytical technique.[78–80] The NIR range has been defined by American Standard for Testing and Materials (ASTM) as including the wavelengths from 780 to 2526 nm (or 12,820–3959 cm^{-1}). It is that range of wavelengths lying between the

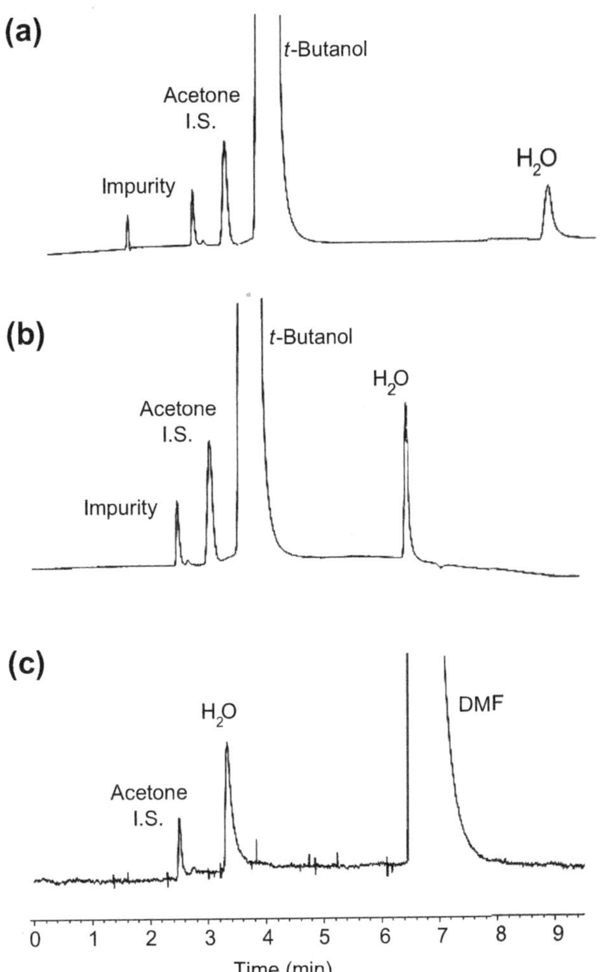

FIGURE 11.1

Chromatograms illustrating the relative retention orders of water and different organic solvents. Chromatograms (a) and (c) are isothermal separations. Chromatogram (b) is for the same sample as in (a), however a temperature gradient was used to decrease the analysis time and further "sharpen" the water peak. This enhanced the sensitivity and precision of the method. Column: HAIM-PEG (a) 1 μL injection, 50 °C, analysis time: 9 min, Internal Standard: acetone (0.4%) (b) 1 μL injection, 50 °C (hold 2 min), ramp 10 °C/min to 80 °C, analysis time: 6 min, Internal Standard: acetone (0.4%). (c) 0.2 μL injection; 110 °C, analysis time: 8 min, Internal Standard: acetone (0.2%).

Reprinted with permission from Ref. 75.

mid-IR and visible wavelength regions. The near-infrared region contains the same overtones and combinations of the fundamental vibrations of –CH, –NH, –OH (and –SH) functional groups as in the mid-IR region, but approximately 10–100 times weaker. The absorption bands are typically broad, and extensive overlap is a characteristic feature of NIRS.

11.5.1 Water determination by NIR

The determination of water, or moisture, by NIRS is attractive because it is one of the few analytes that displays strong absorption bands, which are often well resolved.[81] There are five bands of importance, namely 760, 970, 1190, 1450 and 1940 nm. The position of these bands may vary depending on the chemical and physical matrix of the sample.[78] Most publications, however, make use of ranges that include, at least, the 1450 nm (overtone) and 1940 nm (combination) bands.[18,19,77,82–90] In the pharmaceutical industry, moisture determination by NIRS has found application in the following areas/products: granulation,[83] lyophilization,[19,81,85,88,89] capsules (both hard and soft)[82,87], differentiation between surface and bound water[84,90] and other drying processes.[83] NIRS is particularly powerful in that it is readily amenable to *at-line*, *in-line* and *on-line* analysis.

NIRS, however, cannot be considered a primary method and in order to function quantitatively it requires a reference method.[79] Examples of (primary) reference methods are those that have already been discussed in this chapter, namely KFT, GC and LOD. The values generated by the reference method for a given set of samples are then used to create a calibration model for the NIR instrument. The goal of this calibration model is to predict the moisture content of unknown samples from their NIR spectra.[69,70] Firstly, a model is generated by using a reference sample set, called the calibration set, which is analyzed with the reference method and mathematically correlated to its NIR spectra. Typically, an NIR spectrum needs to go through a data pretreatment and regression step as well (Section 11.5.2). Secondly, the calibration model needs to be evaluated with a validation sample set. The validation sample set assesses the NIR calibration model's ability to predict the moisture content in unknown samples. It is important to ensure that the calibration and validation sample sets are independent of each other and span the necessary concentration range for the desired application. Once the calibration model has been validated it can operate on a routine basis. Ideally the calibration model is used with the instrument that was used to construct it since transferability of models between instruments often requires further adjustments.

Analysts need to ensure that the calibration set is updated periodically and that the primary reference method is operating within its statistical parameters. The validity of the model can be regularly assessed by using standards.

11.5.2 Data-pretreatment and regression analysis

NIR spectra are subject to sampling interferences such as light scattering effects, variations in optical path length, random noise (either sample- or instrument-related), and different crystalline forms within the sample matrix. Without correcting for these effects, any calibration model will not be robust. Thus, the goal of the data pretreatment methods is to reduce, filter out, or standardize the impact of these effects on spectra. Data pretreatments may be divided into two basic classes: (1) the normalization methods like standard normal variant[81,85,88] and multiplicative scatter correction[81,82,91]; and (2) the derivative methods like Savitzky–Golay, Taylor and orthogonal signal correction.[19,81,84,85,91] Typically, normalization methods are used for baseline-offset corrections caused by variation in path length or light scattering. Derivative methods can also be used to reduce baseline offsets as well as to improve the resolution of overlapping bands. A careful choice of wavelength range and data pretreatment method should be finalized before the regression analysis is conducted. Often, different combinations of data pretreatment methods need to be tested to assess which regime yields the optimum spectra.

The simplest regression approach is the well-known Beer–Lambert law which is so often used in other spectroscopic techniques. Unfortunately this approach does not lend itself to NIR spectra, which are multivariate in nature, and is therefore rarely used.[18] Multi-linear regression (MLR) seeks to establish a linear mathematical relationship between a small number of wavelengths on the one hand, and a property of the samples on the other. Each wavelength is evaluated by itself and correlated with the desired property. Wavelengths that succeed in approximating the property are then kept as part of the calibration wavelengths. A model is then generated between the calibration wavelength set and the reference values of sample property. This older method is no longer used as frequently.[81,91] Principle component regression (PCR) is a twofold regression method.[87] It first requires that the spectral data be analyzed by principal component analysis (PCA),[84,85,88] which is a variable-reduction method. Since NIR spectra are multivariate in nature, i.e. they consist of an abundance of correlated variables, often referred to as colinearity, it is extremely helpful to simplify the data by removing all irrelevant variables. PCA seeks to accomplish this by mathematically resolving the spectral data into eigenvectors (orthogonal components called principal components or PC) whose linear combinations approximate the original data. PCR then performs MLR upon the resultant PCs. Lastly, partial least squares regression (PLSR) compares the spectral data set with the reference value set and identifies the wavelengths which best describe the reference values. Linear combinations of these wavelengths, called PLS factors, then make up the PLSR model. PLSR is probably the most used regression model today.[19,81,82,84,85,89,91]

11.5.3 Disadvantages and limitations of NIRS

NIRS exhibits low sensitivity offering detection limits on the order of 0.1% (w/w). The cost for purchasing the instrument, developing and implementing the method is initially high but pays for itself later through rapid analyses. The development of calibration models requires highly trained personnel, accurate and robust calibration sets with sufficient variability and size which are not easy to come by, and continuous maintenance. Furthermore, the transferability of NIRS methods between different instruments remains complicated. Lastly, NIRS requires an accurate, primary reference method for the calibration data set.

References

1. Yoshioka, S., Stella, V. J., Eds. *Stability of Drugs and Dosage Forms;* Springer, 2000.
2. Warner, A. W.; Stafford, J. D.; Warburton, R. E. *Establishing a Microbial Control Strategy in Active Pharmaceutical Ingredients,* Vol. 9; American Laboratory, 2007, 15–18.
3. *Specification: Test Procedures and Acceptance criteria for New Substances and Drug Products: Chemical Substances (Q6A),* The International Conference on Harmonization of Technical Requirements for Registration of Pharmaceuticals for Human Use, 1994.
4. *Specification: Test Procedures and Acceptance criteria for New Substances and Drug Products: Chemical Substances (Q6A),* The International Conference on Harmonization of Technical Requirements for Registration of Pharmaceuticals for Human Use, 1999.
5. United States Pharmacopeia; 731; Loss on Drying.
6. United States Pharmacopeia; 921; Water Determination.
7. European Pharmacopeia, Supplements 4.1 to the 4th Ed. 2002.

8. Streim, H. G.; Boyce, E. A.; Smith, J. R. Determination of Water in 1,1-Dimethylhydrazine, Diethylenetriamine, and Mixtures. *Anal. Chem.* **1961,** *33,* 85–89.
9. Knight, H. S.; Weiss, F. T. Determination of Traces of Water in Hydrocarbons. A Calcium Carbide Gas–Liquid Chromatography Method. *Anal. Chem.* **1962,** *34,* 749–751.
10. Quiram, E. R. Wide-Diameter Open Tubular Columns in Gas Chromatography. *Anal. Chem.* **1963,** *35,* 593–595.
11. Hogan, J. M.; Engel, R. A.; Stevenson, H. F. Versatile Internal Standard Technique for the Gas Chromatographic Determination of Water in Liquids. *Anal. Chem.* **1970,** *42,* 249–252.
12. MacDonald, J. C.; Brady, C. A. Substitute Method for the Karl Fischer Titration. Gas Chromatographic Determination of Water in Ketonic Solvents by Use of the Method of Standard Addition. *Anal. Chem.* **1975,** *47,* 947–948.
13. Houston, T. E. Methods for the Determination of Water in Coatings. *Met. Finish.* **1997,** *95,* 36–38.
14. Nußbaum, R.; Lischke, D.; Paxmann, H.; Wolf, B. Quantitative GC Determination of Water in Small Samples. *Chromatographia* **2000,** *51,* 119–121.
15. Lee, C.; Chang, S. A GC-TCD Method for Measuring the Liquid Water Mass of Collected Aerosols. *Atmos. Environ.* **2002,** *36,* 1883–1894.
16. O'Keefe, W. K.; Ng, F. T. T.; Rempel, G. L. Validation of a Gas Chromatography/Thermal Conductivity Detection Method for the Determination of the Water Content of Oxygenated Solvents. *J. Chromatogr. A* **2008,** *1182,* 113–118.
17. Jayawardhana, D. A.; Woods, R. M.; Zhang, Y.; Wang, C.; Armstrong, D. W. Rapid, Efficient Quantification of Water in Solvents and Solvents in Water Using an Ionic Liquid-Based GC Column. *LC–GC N. Am.* **2012,** *30,* 142.
18. Vornheder, P. F.; Brabbs, W. J. Moisture Determination by Near-Infrared Spectrometry. *Anal. Chem.* **1970,** *42,* 1454–1456.
19. Zhou, X.; Hines, P. A.; White, K. C.; Borer, M. W. Gas Chromatography as a Reference Method for Moisture Determination by Near-Infrared Spectroscopy. *Anal. Chem.* **1998,** *70,* 390–394.
20. Jamin, E.; Guerin, R.; Retif, M.; Lees, M.; Martin, G. J. Improved Detection of Added Water in Orange Juice by Simultaneous Determination of the Oxygen-18/Oxygen-16 Isotope Ratios of Water and Ethanol Derived from Sugars. *J. Agric. Food Chem.* **2003,** *51,* 5202–5206.
21. Pinheiro, C.; Lima, J. C.; Parola, A. J. Using Hydrogen Bonding-Specific Interactions to Detect Water in Aprotic Solvents at Concentrations Below 50 ppm. *Sens. Actuators B* **2006,** *B114,* 978–983.
22. Sun, H.; Wang, B.; DiMagno, S. G. A Method for Detecting Water in Organic Solvents. *Org. Lett.* **2008,** *10,* 4413–4416.
23. Taylor, G. B.; Selvig, W. A. Method for Rapid Analysis of Graphite. *Bull.* **1920,** *112,* 43–45.
24. Ahuja, S., Scypinski, S., Eds., Vol. 3; Academic Press, 2011; p 566.
25. Field, C.. *Developing a Method for Suitable Loss on Drying by TGA,* 2003.
26. Lindquist, J. Determination of Water in Penicillins Using Fast Karl Fischer Reagents and Electronic End-Point Optimization. *J. Pharm. Biomed. Anal.* **1984,** *2,* 37–44.
27. Connors, K. A. The Karl Fischer Titration of Water. *Drug Dev. Ind. Pharm.* **1988,** *14,* 1891–1903.
28. Cachet, T.; Hoogmartens, J. The Determination of Water in Erythromycin by Karl Fischer Titration. *J. Pharm. Biomed. Anal.* **1988,** *6,* 461–472.
29. MacLeod, S. K. Moisture Determination Using Karl Fischer Titrations. *Anal. Chem.* **1991,** *63,* 557.
30. Zhou, L.; Socha, J. M.; Vogt, F. G.; Chen, S.; Kord, A. S. A systematic Method Development Strategy for Water Determinations in Drug Substance using Karl Fischer Titrations. *Am. Pharm. Rev.* **2010.**
31. Fischer, K. A New Method for the Analytical Determination of the Water Content of Liquids and Solids. *Angew. Chem.* **1935,** *48,* 394–396.
32. Smith, D. M.; Bryant, W. M. D.; Mitchell, J. Jr. Analytical Procedures Employing Karl Fischer's Reagent. I. Nature of the Reagent. *J. Am. Chem. Soc.* **1939,** *61,* 2407–2412.

33. Scholz, E. In*Karl Fischer Titration, Determination of Water; Karl Fischer Titration, Determination of Water;* Springer-Verlag: New York, 1984; p 138.
34. Eberius, E.; Kowalski, W. The Effect of Methanol on Water Determination with Karl-Fischer Solution. *Fresenius' Z. Anal. Chem.* **1956,** *150,* 13–20.
35. Cedergren, A. Reaction Rates between Water and the Karl Fischer Reagent. *Talanta* **1974,** *21,* 265–271.
36. Verhoef, J. C.; Barendrecht, E. Mechanism and Reaction Rate of the Karl-Fischer Titration Reaction. Part I. Potentiometric Measurements. *J. Electroanal. Chem. Interfacial Electrochem.* **1976,** *71,* 305–315.
37. Verhoef, J. C.; Barendrecht, E. Mechanism and Reaction Rate of the Karl-Fischer Titration Reaction. Part II. Rotating Ring-Disk Electrode Measurements. *J. Electroanal. Chem. Interfacial Electrochem.* **1977,** *75,* 705–717.
38. Wünsch, G.; Seubert, A. Stoichiometry and Kinetics of the Karl-Fischer Reaction in Methanol as Solvent. *Fresenius' Z. Anal. Chem.* **1989,** *334,* 16–21.
39. Scholz, E. Karl Fischer Reagents without Pyridine. 7. Two-Component Reagents Containing Imidazole. *Fresenius' Z. Anal. Chem.* **1982,** *312,* 462–464.
40. Scholz, E. Karl Fischer Titrations of Aldehydes and Ketones. *Anal. Chem.* **1985,** *57,* 2965–2971.
41. Schöffski, K. The Long Way to Nontoxic Karl-Fischer Titration. *GIT Labor-fachz.* **1998,** *42,* 681.
42. Kaagevall, I.; Aastroem, O.; Cedergren, A. Determination of Water by Flow-Injection Analysis with the Karl Fischer Reagent. *Anal. Chim. Acta* **1980,** *114,* 199–208.
43. Nordin-Andersson, I.; Aastroem, O.; Cedergren, A. Determination of Water by Flow Injection Analysis with the Karl Fischer Reagent. Minimization of Effects Caused by Differences in Physical Properties of the Samples. *Anal. Chim. Acta* **1984,** *162,* 9–18.
44. Nordin-Andersson, I.; Cedergren, A. Spectrophotometric Determination of Water by Flow Injection Analysis Using Conventional and Pyridine-Free Two-Component Karl Fischer Reagents. *Anal. Chem.* **1985,** *57,* 2571–2575.
45. Cedergren, A.; Lundstroem, M. Electrochemical Determination of Water in Environmental Hydraulic Fluids Using the Karl Fischer Reaction. *Anal. Chem.* **1997,** *69,* 4051–4055.
46. Cedergren, A.; Luan, L. Potentiometric Determination of Water Using Spent Imidazole-Buffered Karl Fischer Reagents. *Anal. Chem.* **1998,** *70,* 2174–2180.
47. www.emdmillipore.com.
48. Johansson, A. Determination of Water by Titration (a Modified Karl Fischer Method). *Sven. Papperstidn.* **1947,** *50,* 124–126.
49. Meyer, A. S., Jr.; Boyd, C. M. Determination of Water by Titration with Coulometrically Generated Karl Fischer Reagent. *Anal. Chem.* **1959,** *31,* 215–219.
50. Isengard, H. Rapid Water Determination of Foodstuffs. *Trends Food Sci. Technol.* **1995,** *6,* 155–162.
51. Bruttel, P.; Schlink, R. Water Determination by Karl Fischer Titration. *Metrohm Monograph* **2006,** 8.026.5013-2006-02.
52. Isengard, H.-D.; Kerwin, H. Proposal of a New Reference Method for Determining Water Content in Butter Oil. *Food Chem.* **2003,** *82,* 117–119.
53. Kurath, P.; Jones, P. H.; Egan, R. S.; Perun, T. *J. Experientia.* **1971,** *27,* 362.
54. Sherman, F.; Kuselman, I. Water Determination in Drugs Containing Ascorbic Acid. *Drug Dev. Ind. Pharm.* **1999,** *25,* 1115–1119.
55. Sherman, F. Isr.Pat.Appl. Patent 114,938, 1995.
56. Sherman, F. **1996**, *43,* 1035.
57. Sherman, F.; Kuselman, I.; Shenhar, A. Determination of Water Samples and Ene-Diols or Thiols in Samples Inaccessible for Direct K. Fischer Titration. *Talanta* **1996,** *43,* 1035–1042.
58. Sherman, F. U.S.Patent 5,755,044, 1998.
59. Hoffmann, H.; Felgner, A. Water Determination in Pharmaceutical Compounds. *Analytix*, Article 4, 9.

60. Isengard, H. D.; Striffler, U. Karl Fischer Titration in Boiling Methanol. *Fresenius J. Anal. Chem.* **1992,** *342,* 287–291.
61. Scholz, E. New Developments in Karl Fischer Titration. *LaborPraxis* **1984,** *8,* 630–631.
62. Oguchi, R.; Yamaguchi, K.; Shibamoto, T. Determination of Water Content in Common Organic Solvents by a Gas Chromatograph Equipped with a Megabore Fused-Silica Column and a Thermal Conductivity Detector. *J. Chromatogr. Sci.* **1988,** *26,* 588–590.
63. Bennett, F. Water Determination by Gas Chromatography. *Anal. Chem.* **1964,** *36,* 684.
64. Andrawes, F. F. Analysis of Liquid Samples by Capillary Gas Chromatography and Helium Ionization Detection. *J. Chromatogr.* **1984,** *290,* 65–74.
65. Huang, K.; Han, X.; Zhang, X.; Armstrong, D. W. PEG-Linked Geminal Dicationic Ionic Liquids as Selective, High-Stability Gas Chromatographic Stationary Phases. *Anal. Bioanal. Chem.* **2007,** *389,* 2265–2275.
66. Sun, P.; Armstrong, D. W. *Anal. Chim. Acta* **2010,** *661,* 1–16.
67. Han, X.; Armstrong, D. W. Ionic Liquids in Separations. *Acc. Chem. Res.* **2007,** *40,* 1079–1086.
68. Anderson, J. L.; Armstrong, D. W. Immobilized Ionic Liquids as High-Selectivity/High-Temperature/High-Stability Gas Chromatography Stationary Phases. *Anal. Chem.* **2005,** *77,* 6453–6462.
69. Armstrong, D. W.; Payagala, T.; Sidisky, L. M. The Advent and Potential Impact of Ionic Liquid Stationary Phases in GC and GCxGC. *LC–GC N. Am.* **2009,** *27*, 596, 598, 600–602, 604-605.
70. Breitbach, Z. S.; Armstrong, D. W. Characterization of Phosphonium Ionic Liquids through a Linear Solvation Energy Relationship and Their Use as GLC Stationary Phases. *Anal. Bioanal. Chem.* **2008,** *390,* 1605–1617.
71. Payagala, T.; Zhang, Y.; Wanigasekara, E.; Huang, K.; Breitbach, Z. S.; Sharma, P. S.; Sidisky, L. M.; Armstrong, D. W. Trigonal Tricationic Ionic Liquids: A Generation of Gas Chromatographic Stationary Phases. *Anal. Chem.* **2009,** *81,* 160–173.
72. Armstrong, D. W.; He, L.; Liu, Y. Examination of Ionic Liquids and Their Interaction with Molecules, When Used as Stationary Phases in Gas Chromatography. *Anal. Chem.* **1999,** *71,* 3873–3876.
73. Soukup-Hein, R. J.; Warnke, M. M.; Armstrong, D. W. Ionic Liquids in Analytical Chemistry. *Annu. Rev. Anal. Chem.* **2009,** *2,* 145–168.
74. Anderson, J. L.; Armstrong, D. W.; Wei, G. Ionic Liquids in Analytical Chemistry. *Anal. Chem.* **2006,** *78,* 2893–2902.
75. Jayawardhana, D. A.; Woods, R. M.; Zhang, Y.; Wang, C.; Armstrong, D. W. Rapid, Efficient Quantification of Water in Solvents and Solvents in Water Using an Ionic Liquid-Based GC Column. *LC–GC Eur.* **2011,** *24,* 516–529.
76. Center for Drug Evaluation and Research (CDER), US Food and Drug Administration (FDA), 1994, pp 1–30.
77. Center for Drug Evaluation and Research (CDER), US Food and Drug Administration (FDA), 1996, pp 1–10.
78. Reich, G. Near-Infrared Spectroscopy and Imaging: Basic Principles and Pharmaceutical Applications. *Adv. Drug Deliv. Rev.* **2005,** *57,* 1109–1143.
79. Roggo, Y.; Chalus, P.; Maurer, L.; Lema-Martinez, C.; Edmond, A.; Jent, N. A Review of Near Infrared Spectroscopy and Chemometrics in Pharmaceutical Technologies. *J. Pharm. Biomed. Anal.* **2007,** *44,* 683–700.
80. Jamrogiewicz, M. Application of the Near-Infrared Spectroscopy in the Pharmaceutical Technology. *J. Pharm. Biomed. Anal.* **2012,** *66,* 1–10.
81. Bruells, M.; Folestad, S.; Sparen, A.; Rasmuson, A.; Salomonsson, J. Applying Spectral Peak Area Analysis in Near-Infrared Spectroscopy Moisture Assays. *J. Pharm. Biomed. Anal.* **2007,** *44,* 127–136.
82. Berntsson, O.; Zackrisson, G.; Ostling, G. Determination of Moisture in Hard Gelatin Capsules Using Near-Infrared Spectroscopy: Applications to At-Line Process Control of Pharmaceutics. *J. Pharm. Biomed. Anal.* **1997,** *15,* 895–900.
83. Findlay, W. P.; Peck, G. R.; Morris, K. R. Determination of Fluidized Bed Granulation End Point Using Near-Infrared Spectroscopy and Phenomenological Analysis. *J. Pharm. Sci.* **2005,** *94,* 604–612.

84. Zhou, G. X.; Ge, Z.; Dorwart, J.; Izzo, B.; Kukura, J.; Bicker, G.; Wyvratt, J. Determination and Differentiation of Surface and Bound Water in Drug Substances by Near Infrared Spectroscopy. *J. Pharm. Sci.* **2003,** *92,* 1058–1065.
85. Grohganz, H.; Fonteyne, M.; Skibsted, E.; Falck, T.; Palmqvist, B.; Rantanen, J. Role of Excipients in the Quantification of Water in Lyophilized Mixtures Using NIR Spectroscopy. *J. Pharm. Biomed. Anal.* **2009,** *49,* 901–907.
86. Tran, C. D.; De, P. L.; Silvia, H.; Oliveira, D. Absorption of Water by Room-Temperature Ionic Liquids: effect of Anions on Concentration and State of Water. *Appl. Spectrosc.* **2003,** *57,* 152–157.
87. Buice, R. G., Jr.; Gold, T. B.; Lodder, R. A.; Digenis, G. A. Determination of Moisture in Intact Gelatin Capsules by Near-Infrared Spectrophotometry. *Pharm. Res.* **1995,** *12,* 161–163.
88. De Beer, T. R. M.; Alleso, M.; Goethals, F.; Coppens, A.; Vander Heyden, Y.; Lopez De Diego, H.; Rantanen, J.; Verpoort, F.; Vervaet, C.; Remon, J. P.; Baeyens, W. R. G. Implementation of a Process Analytical Technology System in a Freeze-Drying Process Using Raman Spectroscopy for In-Line Process Monitoring. *Anal. Chem.* **2007,** *79,* 7992–8003.
89. Derksen, M. W. J.; van de Oetelaar, P. J.; Maris, F. A. The Use of Near-Infrared Spectroscopy in the Efficient Prediction of a Specification for the Residual Moisture Content of a Freeze-Dried Product. *J. Pharm. Biomed. Anal.* **1998,** *17,* 473–480.
90. Dziki, W.; Bauer, J. F.; Szpylman, J. J.; Quick, J. E.; Nichols, B. C. The Use of Near-Infrared Spectroscopy to Monitor the Mobility of Water within the Sarafloxacin Crystal Lattice. *J. Pharm. Biomed. Anal.* **2000,** *22,* 829–848.
91. Zhou, X.; Hines, P.; Borer, M. W. Moisture Determination in Hygroscopic Drug Substances by Near Infrared Spectroscopy. *J. Pharm. Biomed. Anal.* **1998,** *17,* 219–225.

PART

Specific Tests: Drug Product

4

CHAPTER

Dissolution

12

Vivian A. Gray*, Thomas W. Rosanske†

* *V. A. Gray Consulting, Inc., Hockessin, DE, USA,* † *T.W. Rosanske Consulting, Overland Park, KS, USA*

CHAPTER OUTLINE

Specification of Drug Substances and Products. http://dx.doi.org/10.1016/B978-0-08-098350-9.00012-6

12.1 INTRODUCTION

The dissolution test is a required test for almost all pharmaceutical products that are not true solutions. Dissolution testing monitors the rate at which a solid or semisolid pharmaceutical dosage form releases the active ingredient(s) into a liquid medium under standardized conditions of liquid/solid interface, temperature, and media composition. Dissolution, or in vitro release, of the drug substance from the product into a typically aqueous-based medium, is linked to the release of the drug into the body, making it available for absorption, and then efficacy or clinical outcome. Scientists have been conducting dissolution studies for many years. However, it was not until 1970 that dissolution testing was officially recognized as a product quality indicator when it was incorporated into 12 monographs in the United States Pharmacopeia/National Formulary (USP/NF).[1] In the current USP 35, nearly all solid dosage form monographs include a dissolution test.[2] Dissolution tests are defined as Category III by the USP, i.e. "Analytical method for the determination of performance characteristics…".[3] Dissolution testing is primarily used in the pharmaceutical industry as a quality control tool to monitor the formulation and manufacturing processes of the dosage form. Dissolution is considered by most regulatory agencies as a highly critical quality characteristic for most solid dosage forms.

The regulatory agencies use the dissolution test to provide a quality connection from a pivotal biobatch to the commercialized product. For this reason, the dissolution test development and validation are critical factors in insuring that the test is robust and clinically relevant. Clinical relevance comes from developing a test that provides understanding of the product release mechanism(s) and, in the highest form, an in vivo–in vitro correlation. The specifications attributed to a clinically relevant test are most important and useful.

It is not the primary intent of this chapter to discuss the test development aspects of the dissolution test, but rather to explain in detail the validation aspects. There are many sources for method development in the literature.[4–8] There is also a comprehensive chapter in the USP, General Chapter <1092>, The Dissolution Procedure: Development and Validation, that is an excellent resource for both method development and validation.[9] The International Conference on Harmonization (ICH) Harmonized Tripartite Guideline Q2 (R1): Validation of Analytical Procedures: Text and Procedures, is also a useful resource.[10]

An important development has been the FDA Guidance "Waiver of in vivo bioavailability (BA) and bioequivalent (BE) studies for immediate release solid oral dosage forms based on a biopharmaceutics classification system (BCS)".[11] This guidance is of extreme significance, stating that under some circumstances dissolution testing can be used instead of BA and/or BE studies. This guidance applies only to immediate release products. The four classes of the BCS are defined and the methods for determining the three aspects—solubility, permeation, and dissolution—are described.

In order to validate a dissolution method, it is important to have a good understanding of the theory of dissolution and the roles of the key parameters of a dissolution test. A complete dissolution validation package would consider at a minimum the dissolution apparatus used, equipment qualification requirements, and any appropriate governmental or regulatory guidelines. Therefore, it is important to address these issues here in the context of dissolution test validation.

12.2 THE DISSOLUTION TEST

In order for a dissolution test to demonstrate the unique dissolution characteristics of the dosage form, the dissolution procedure should be based on the physical and chemical properties of the drug substance as well as the dosage form characteristics. Some of the physicochemical properties of the drug substance which influence the dissolution characteristics are:

- solubility in water and other appropriate solvents
- ionization constants
- solution stability
- particle size and surface area
- crystal form
- common ion effects
- ionic strength
- buffer effects
- octanol/water partition coefficients
- effect of temperature on solubility

Once the drug substance properties have been determined, the actual dosage form needs to be considered. The analyst developing the dissolution test needs to know, for example, whether the dosage form is a tablet, capsule, semisolid (ointment or cream), or transdermal patch, and whether it is designed for immediate release or controlled release of the drug product. Of key importance is the potency of the dosage form or the amount of drug to be delivered and the rate at which the drug is to be delivered. This is related directly to the mathematical expression of dissolution rate, which is defined by the Noyes–Whitney equation:

$$\mathrm{d}W/\mathrm{dt} = \mathrm{k}_1 S(C_{\mathrm{sat}} - C_{\mathrm{sol}}) \tag{12.1}$$

where:

$\mathrm{d}W/\mathrm{dt}$ = the dissolution rate
k_1 = a dissolution constant
C_{sat} = the concentration of a saturated solution
C_{sol} = the concentration of the solution at any given time
S = the surface area of the solid

In vivo, the gastrointestinal tract acts as a natural sink, i.e. the drug is diluted and absorbed as it dissolves. In vitro, sink conditions are simulated by using either a large volume of dissolution medium or by replenishing the medium with fresh solvent at a specific rate. By keeping the volume of dissolution medium at least three times greater than the saturation volume, sink conditions are approximated.[12] When sink conditions are achieved, $C_{\mathrm{sat}} \gg C_{\mathrm{sol}}$, and Equation (12.1) simplifies to:

$$\mathrm{d}W/\mathrm{dt} = \mathrm{k}_2 S \tag{12.2}$$

In this case, the dissolution rate is characteristic of the release of active ingredient from the dosage form rather than its solubility in the dissolution medium. Therefore, sink conditions are one of the main experimental parameters to be controlled in dissolution testing. Sink conditions can be achieved by the appropriate selection of the dissolution apparatus and dissolution medium. The selection of the appropriate dissolution system should be based on the drug substance and dosage form characteristics.

12.2.1 Apparatus

For a dissolution test to be used universally to control the consistency of a pharmaceutical dosage form, some controls must be placed on the type of apparatus used. USP General Chapter <711>, "Dissolution",[13] describes several apparatus types used in dissolution testing. The USP Apparatus 1 (basket) and Apparatus 2 (paddle) are by far the most frequently used for immediate release and most sustained release dosage forms. Apparatus 1 and Apparatus 2 are described in some detail below. USP Apparatus 3–7 are less frequently used but are coming into increased use with novel dosage forms.[14] Compendia other than the USP, e.g. the Japanese Pharmacopoeia (JP),[15] the British Pharmacopoeia (BP),[16] and the European Pharmacopoeia (EP)[17] all contain the Apparatus 1 and 2 dissolution equipment as described in the USP.

12.2.1.1 USP Apparatus 1 (basket)

This apparatus, shown schematically in Fig. 12.1, consists of a covered vessel of specified shape and dimensions with a capacity of 1000 mL (smaller volume vessels are used in certain instances), a metallic shaft one end of which attaches to a motor, and a cylindrical metallic mesh basket that attaches to the opposite end of the shaft. The dosage form is placed inside the basket and the basket assembly is immersed in the dissolution vessel containing dissolution medium and rotated at a specified speed.

12.2.1.2 USP Apparatus 2 (paddle)

The paddle is currently the most frequently used apparatus for solid dosage forms. The dissolution vessel used with this apparatus is the same as for the USP Apparatus 1. However, the basket assembly is replaced by a paddle of specified dimensions as shown schematically in Fig. 12.2. With this apparatus, the dosage form is dropped directly into the vessel containing the dissolution medium and allowed to sink into the bottom; the paddle is then rotated at a specified speed. The paddle can be immersed in the vessel prior to dropping the dosage form, but paddle rotation should not begin until the dosage form has been dropped; this is the standard industry practice. The USP specifies placing the dosage form in the apparatus and immediately operating at a specified rotational speed. The USP does not state whether the paddle can or cannot be immersed prior to addition of the dosage form, only that the rotation be started after the dosage form has been added.

12.3 VALIDATION AND METHOD DEVELOPMENT ASPECTS

Validation of a dissolution test method consists of two parts. The first part, and the part that will be given the most emphasis in this chapter, is the validation of the dissolution method, that is, the actual dissolution run and taking of the sample aliquot. The second part is the analytical determinative step, in

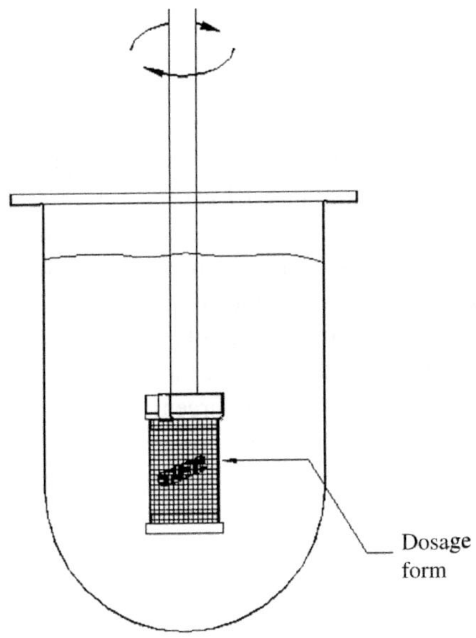

FIGURE 12.1

USP Apparatus 1.

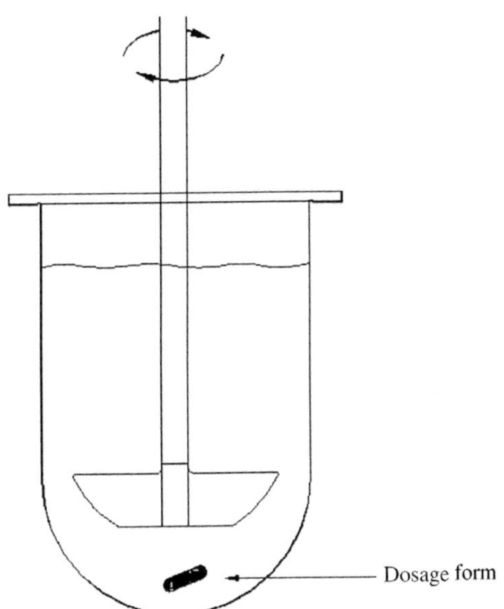

FIGURE 12.2

USP Apparatus 2.

which the samples and standards are analyzed, typically by a high performance liquid chromatography (HPLC) or an ultraviolet (UV) method.

There are numerous aspects of the dissolution procedure that require validation and there are also different levels of validation. In the early product development stages, filtration, deaeration, linearity, precision, solution stability, selectivity and accuracy/recovery should be considered as part of the validation. During later development, i.e. phase III and beyond, full validation is necessary, and the aspects of intermediate precision, automation and robustness are added. This chapter will focus on validation of methods used in later stages of development, where full validation is necessary.

It is presumed that full validation will be conducted on the final dissolution test method established for a registration formulation. The final method should exhibit a moderate to low variability in the dissolution results, the solutions should be stable in the dissolution media, and there should be an established dissolution profile, using a minimum of three time points. The dissolution profile should be gradual with at least two points at or below 85% to satisfy the rules for f_2 analysis.[11] The f_2, or similarity factor, is a critical tool for demonstrating bioequivalence, though it is not needed for BCS Class 1 drugs, where the product is rapidly dissolving (e.g. 85% in 15 min) in the pH range 2–7.[6] Ideally, the dissolution test method should have some ability to detect change in the product critical quality attributes, as these attributes influence the release of the drug and can be characteristics of the drug substance, the formulation, the manufacturing process, and/or stability.

12.3.1 Sampling, filtration, and analytical instrumentation

A single dissolution test generally requires the sampling and analysis of several samples (usually six per sampling interval). As such, it is necessary to have efficient sampling systems and rapid analytical systems in place. Sampling devices can be either a syringe and cannula combination with a filter at the tip of the needle where the sample is drawn through the filter first, or a syringe and cannula without an in-line filter of any kind. Typically, the sample is drawn into the syringe and quickly passed through a filter prior to analytical determination.

Sampling can be performed manually or by using automated sampling systems. Manual sampling can be quite labor intensive and tedious, however. The manual sampling from six vessels for a single time point in a dissolution test can take several minutes, making establishment of dissolution profiles for rapidly dissolving dosage forms quite difficult at early time points. Many automated sampling systems are commercially available in today's market. The automated sampling systems are typically microprocessor or computer controlled, allowing for precisely timed sampling at frequent intervals, if necessary. Some automated sampling systems transfer the samples to collection tubes for manual transfer to the analytical instrument, while other sampling systems transfer the samples directly into the analytical instrument for analysis.

Carry-over between samples should be determined, especially when common sampling pathways are used. Errors associated with automated sampling systems are usually related to partial or complete blockage of the sampling lines. Therefore, it is recommended that the appropriate flow rates be determined and used prior to each use of an automated sampling system.

The filtering step must be evaluated to determine if the appropriate amount of solution has been passed through the filter to saturate the filter material with drug substance or excipients so that the final filtered solution is not biased due to drug loss through filter adsorption.

The filtration step is a critical yet underestimated source of problems in the dissolution test and needs to be adequately evaluated with both the standard solution and the sample solution. Filtration is usually necessary to prevent undissolved particles from entering the analytical sample and further dissolving. Care should be taken that the filter pore size is not larger than the drug particle size; this is especially important with micronized or nanoparticles. Filtration also removes insoluble excipients that may otherwise cause a high background.

Validation of the filter generally includes preparation of a suitable standard solution (lowest and highest profile concentrations are recommended) and a completely dissolved sample solution. For the standard solutions, results of the filtered solution (appropriate discard volumes should be determined beforehand) are compared to those of an unfiltered standard. For the filtered sample solution, results should be compared to a fully dissolved and centrifuged sample solution. The acceptable range for standard and sample filtration efficiency is generally between 98% and 102% of the unfiltered standards solutions and unfiltered but centrifuged sample solution.

Because of the large number of samples normally generated in a dissolution test, the analytical system should be relatively rapid, allowing for a high throughput of samples. The most common analytical instruments used for dissolution testing are UV–visible spectrophotometers and HPLC with UV detection (HPLC-UV or Ultra-HPLC-UV).

UV–visible spectroscopy allows for rapid analysis of samples. Diode array-based instruments with sophisticated computer-enhanced data analysis have improved the quality and speed of single-component as well as multicomponent analysis by UV–visible spectroscopy. When this mode of analysis is used, however, the accuracy should be confirmed by a more discriminating and selective mode of analysis such as HPLC. When HPLC analysis is required, a rapid chromatographic method is desirable for the analysis. It is important to note that the analytical instrumentation should be checked for wavelength accuracy and repeatability as well as photometric accuracy and repeatability.

12.3.2 Single-point test versus dissolution profile

The sampling and analysis procedures used for a dissolution test may to a large extent depend on the type of dosage form and purpose of the test. Ideally, the in vitro dissolution rate for a given formulation and dosage form will be reflective in some way of in vivo availability of the drug, thus allowing for establishment of a correlation between the in vitro dissolution behavior and one or more pharmacokinetic parameters (C_{max}, t_{max}, Area Under the Curve (AUC), etc.). A well-established correlation will allow for a reasonable prediction of the in vivo behavior of formulations without performing additional bioavailability studies. Finding the appropriate correlation has been the focus of numerous studies.[18–20] The establishment of in vitro–in vivo correlations has been the subject of much discussion recently in the arena of quality by design and setting clinically relevant acceptance criteria.[21]

It is generally more difficult to generate useful or predictive correlations between measured in vitro dissolution rate and bioavailability for immediate release dosage forms. Since dissolution rates for such products are by design relatively rapid, it is often found that dissolution of the drug may not be the rate-limiting factor for in vivo activity. For immediate release dosage forms, usually only single-point dissolution acceptance criteria are required. However, for method development, BE studies, and scale up and post approval changes[22], dissolution profiles are established.

Extended or controlled release formulations are designed to release drug from the product matrix over an extended period of time, generally 12–24 h or longer. This implies that the rate at which a drug dissolves from the formulation matrix is a controlling factor in the bioavailability of the drug. For such products, it is critical to establish a correlation at multiple time points in order to ensure batch to batch product consistency over the entire release profile of the product. Thus, it is obvious that single point dissolution tests are inadequate for control of extended or controlled release products, and dissolution profiles are necessary as part of the routine testing.

Approaches to in vitro–in vivo correlations for extended release products are discussed in detail in the FDA guidance "Extended release solid oral dosage forms: development, evaluation, application of in vitro/in vivo correlations",[23] and USP General Chapter <1088> "In Vitro and In Vivo Evaluation of Dosage Forms".[12]

12.4 ANALYTICAL VALIDATION

Once the appropriate dissolution conditions have been established, the analytical method should be suitably validated. The validation parameters may vary depending on the intended use, but will typically include at a minimum linearity and range, accuracy, precision, specificity, solution stability and robustness. Each of these analytical parameters is discussed in detail elsewhere in this book in other contexts, but the principles are the same for dissolution. This section will discuss these parameters only in relation to issues unique to dissolution testing. All dissolution testing must be performed on a qualified dissolution apparatus meeting the specified mechanical and performance standards and with qualified analytical instrumentation.

12.4.1 Linearity and range

Linearity and range will show the ability of the dissolution method to obtain test results over the range of expected concentrations that are directly (linearly) proportional to the concentration of the analyte in the sample. Detector linearity should be checked over the entire range of concentrations expected during the procedure. For immediate release formulations, a concentration range of at least 50% of the lowest concentration expected in the dissolution vessel to 120% of the highest expected concentration is sufficient. For controlled release products, the concentration range should extend from approximately 10–120% of that expected from dissolution of the entire dose. If a controlled release product is formulated in multiple strengths, the detector linearity should be confirmed from 10% of the lowest concentration of the lowest strength to 120% of the highest concentration of the highest strength. Typically, the concentration range is divided into five evenly spaced concentrations. Linearity testing of the dosage form should cover the entire specification range of the product. All samples should be heated to 37 °C (or the specified dissolution temperature) prior to analysis; this is especially important if the samples are close to sink conditions.

The drug substance itself is used in this experiment and may be prepared with placebo to show linearity and range of the sample solution. All solutions should be made from a common stock if possible, using the dissolution medium as the diluting solvent. Organic solvents may be used (if necessary), to enhance the solubility of the drug, following the restriction of not more than 5% organic in the final solution. The diluted solutions are read in triplicate (i.e. three test tubes of the same solution) with spectrophotometric analysis and two injections of the same solution from different vials for HPLC analysis. For the highest concentration, the absorbance values must not exceed the linearity limits of the instrument. Linearity is generally considered acceptable if the correlation coefficient is ≥ 0.997, with the y-intercept not significantly different from zero at the 95% confidence limit.

12.4.2 Accuracy

The accuracy or recovery of the dissolution method is the closeness of the test results obtained by the method to the true value. In general, accuracy is determined by preparing multiple samples containing the drug and other constituents present in the dosage form (e.g. excipients, coating materials, etc.) ranging in concentration from at or below the expected concentration of the lowest profile point to above the highest concentration upon full release. Capsule shells, coating blends, inks, and sinkers, are also to be added where appropriate. The samples can be prepared either in situ (in the vessel) or on the benchtop in a flask with media heated to 37 °C. Note: the addition order of the drug substance to the media either before or after the excipients may be critical. For example, some drugs need to be in the media without the excipients to wet properly or vice versa, where the excipients need to mix first before the drug substance is introduced, this being a case by case determination. The mixing, especially with poorly soluble drugs may take some time before the drug is fully dissolved. The drug must be fully dissolved before the accuracy aliquot is taken. In some cases, the simple drug powder cannot be used directly for this determination, and a solution is appropriate. This would be the case, for example, when the drug concentration is too small to accurately weigh out or the drug is electrostatic, making accurate weighing difficult. The generally acceptable measured recovery is 95–105% of the amount added. Also, low recoveries may be due to excipient interference[24] or incomplete dissolution.

12.4.3 Infinity point

The final time point selected for a dissolution test often does not, and is not intended to, indicate complete dissolution of the dosage form. It is, however, important to establish that, at some point, or under some conditions, all of the drug material can be accounted for. Performing an infinity point, or fast stir, test in the early product development phases or routinely on samples in testing can give an indication of the recovery/accuracy, assuming that the drug is formulated at 100%, of the label claim. To obtain an infinity point, during the normal test after the last time point is pulled, without stopping the test, the paddle or basket speed is increased to at least 150 rpm for 30–60 min, after which time a further sample is taken. Although there is no requirement for 100% dissolution in the profile, the infinity point can provide supportive data when compared to the content uniformity data. The infinity point data for all six vessels can be compared to the content uniformity values in terms of the mean and variability. This may give useful information as to any artifacts from the dissolution method and/or the dosage form interacting with the dissolution medium.

12.4.4 Intermediate precision and reproducibility

Intermediate precision evaluates the effects of random events or variations in the laboratory on the precision of the dissolution method. It is advisable to change the laboratory and environmental conditions as much as possible, as this study can be an early read to method transfer to other laboratories. Many times method transfer studies are not successful because not enough emphasis is put on the intermediate precision study to really challenge the method to variations. Intermediate precision should be evaluated at all profile points and the drug product should be used in the evaluation.

The dissolution profiles on the same drug product should be determined by at least two different analysts, with the analysts each preparing their own standard solutions and media (with different lots of reagents if available). The analysts should use different dissolution baths, HPLC equipment (including columns) or UV detectors, and autosamplers (if not performing the test manually), and the tests should be performed on different days. Each analyst should run $N = 12$ samples. The difference in the mean value between each profile point of the dissolution results of each analyst is compared. Results are generally considered acceptable if the absolute mean difference does not exceed 10% at time points with less than 85% dissolved and does not exceed 5% for time points equal to or above 85%.[9] The 10% acceptance criteria may appear generous, but high variability of some products, especially at the earlier time points, is well known. However, if the data for the product are usually tight and a 10% difference in means obtained, it may be prudent to look for causes of the bias.

Reproducibility, another form of precision, follows the general concepts of intermediate precision but is performed across multiple laboratories.

12.4.5 Repeatability/method precision

A third form of precision, repeatability, or method precision, is determined by replicate measurements of standard and/or sample solution prepared by a single analyst on a single instrument. The drug substance is used for this evaluation and may be prepared with placebo to show precision of the sample solution. It is important that the drug be completely in solution for this evaluation. As such, preparation of a standard solution can use a small amount of a solvent in the case of a poorly soluble drug. This avoids incomplete dissolution of the drug in media where the dissolving step may take time, even with the aid of sonication. The solvent content, however, should not exceed 5% of the final solution, as amounts in excess of 5% may shift the chromophore maximum. General acceptance criteria for repeatability for HPLC procedures are $\leq$1% relative standard deviation (RSD) and for UV analysis $\leq$2% RSD.

12.4.6 Specificity

The dissolution analysis method must be specific for the drug substance in the presence of the placebo. The analytical method used for the accuracy testing can also be used to establish specificity. To establish appropriate specificity, accuracy solutions should be monitored for degradation. Simply monitoring the UV spectra of the solutions is insufficient to determine degradation since many degradation products will have the same or very similar UV spectra as the parent compound. The dissolution test is not necessarily a stability indicating assay but the analyst should know if there is interference or bias from an excipient or impurities at the UV wavelength used for the test. Degradation products, unless greater than 2–3% at the detection wavelength, are usually ignored. It is good practice, however, to run an HPLC test if a UV test is proposed just to see if any dramatic stability issues may be present in the dissolution medium. Therefore, specificity testing should be confirmed by analyzing accuracy samples with a selective analytical methodology such as HPLC. If there is capsule shell interference with detection of the drug, the USP allows for a correction for the capsule shell interference. However, corrections greater than 25% of labeled content are unacceptable.[13]

12.4.7 Solution stability

Solution stability is the determination of the stability of sample and standard solutions over time under normal laboratory conditions. The standard solution is stored under conditions typical in the laboratory over a period of time (at least 8 days is recommended) and results are compared with those obtained on freshly prepared standard solutions. The standard solution should be at expected profile point concentrations not exceeding 100% dissolved. Sample solutions should also be compared in the same manner to a freshly prepared sample; a placebo plus drug solution could be used instead of an actual sample. The acceptable range for standard and sample stability samples is between 98% and 102% of the freshly prepared solutions. A certain time frame (sample preparation to sample analysis) should be specified in the method to reflect the stability data.

12.4.8 Robustness

Robustness testing should be conducted to determine the critical test parameters for a particular dissolution method. Robustness is typically evaluated by subjecting each dissolution parameter to slight variations. Evaluation of the data will establish the necessary control required for the test parameters in order to maintain consistency in the method. This will facilitate any method transfers and minimize the need for troubleshooting. A robustness study should include evaluation of the effect of varying media pH, media volume, flow rate, rotation speed, sample position in the apparatus, sinkers (if applicable), media deaeration, media surfactant level, temperature, and filters. If the analysis is performed by HPLC, the effect of columns and mobile phase conditions should also be addressed.

12.4.9 Automation

Automated systems have become much more widely used in recent years. If possible, when conducting a validation of an automated method, there should be a comparison to the manual sampling method. All profile points should be evaluated. This validation can be done in one of two ways: (1) when the drug dissolution results are not highly variable, and understanding the effect of an in-residence probe is desired, two concurrent runs (same sampling intervals, $n = 6$) using manual and automated sampling methods are compared using the criteria established for intermediate precision, or (2) if the dissolution results are highly variable (i.e. the RSD is above 20% in time points at 10 min or earlier and 10% RSD or above in later time points), the analysis can be performed by pulling samples from the vessel simultaneously by manual and automated sampling methods for each time point. Note that the correction for the volume withdrawn from the medium is doubled in the latter case.

12.4.10 Sinkers

Dosage forms that tend to float or move around during the course of a dissolution test can create issues with variability and biased results. Sinkers are often used to hold such dosage forms in place during a dissolution test. There are many types of sinkers used in dissolution testing, some homemade, some commercially available, and the type of sinker used can affect results. In the case where sinkers are necessary, detailed sinker descriptions and an explanation of why a sinker is used must be stated in the method and any regulatory submission. When comparing different sinkers (or sinkers versus no sinker), tests must be run concurrently with each sinker. Each sinker type must be evaluated based on its ability to maintain the dosage at the bottom of the vessel without inhibiting drug release. A comparison of the different sinker types (or no sinker) is performed using the same criteria as for intermediate precision. All profile points should be evaluated.

Sinkers can significantly influence the dissolution profile of a drug. The use of sinkers, therefore, must be part of case-by-case dissolution validation. The sinker design must be stated clearly in the method. When transferring the method, the sinkers must be duplicated as closely as possible in any subsequent testing facility.

12.4.11 Other validation parameters

Other aspects of validation may include carryover of residual drug, effect of an in-residence probe (simultaneous sampling as mentioned in Section 12.4.10 may not be suitable in this case), adsorption of drug, and cleaning and/or rinse cycles. These parameters should be evaluated on a case by case basis when the circumstances warrant.

12.5 SOURCES OF ERROR IN DISSOLUTION TESTING

There are many factors that can contribute to errors or bias in a dissolution test, and it is wise for the analyst to be aware of these as the test is under development or being conducted on a routine basis. Some of the more prevalent sources of error are described below.

12.5.1 Drug substance properties

Knowledge of drug substance properties, especially solubility as a function of pH or in solutions containing surfactants, is essential. One needs to anticipate precipitation of the drug as the pH changes in solution, or as the amount of drug in solution increases. The analyst also needs to be aware that complete dissolution of the drug in the standard solution may be more difficult than expected. It is customary to use a small amount of alcohol to dissolve the standard completely, provided the additive does not affect the results. A history of the typical absorptivity range of the standard can also be very useful in determining if the standard has been prepared properly.

12.5.2 Drug product properties

Provided the drug product is manufactured with consistency, highly variable dissolution results may indicate that the method is not robust, and this can cause difficulty in identifying trends and effects of formulation changes. There are two major causal factors that influence dissolution variability: mechanical and formulation. Mechanical contributions to variability can arise from the dissolution method conditions chosen, e.g. lack of deaeration, paddle speed, etc. Careful observation of the vessels during the dissolution process can often provide the analyst with an indication of whether mechanical conditions, such as rotational speed need to be altered.

Contributions from the formulation can be several. The formulation may have poor content uniformity, and reactions and/or degradation may be occurring in situ. Film coating may cause sticking of the dosage form to the vessel walls, thereby changing the hydrodynamics of the test. Upon aging, capsule shells are known for pellicle formation. Tablets may become harder or softer, depending upon the excipients and drug interaction with moisture, and this may affect the disintegration and hence the dissolution rate. Careful visual observation of the dissolution process may help identify some of the causes of high variability.

12.5.3 Equipment

The major components of dissolution equipment are the tester, water bath, paddles, baskets and shafts, vessels, samplers, and analyzers. Mechanical aspects of the equipment, such as medium temperature, paddle or basket speed, shaft centering and wobble, and vibration can all have a significant impact on the dissolution of the product.

12.5.4 Deaeration

Dissolved atmospheric gases in the dissolution medium may affect dissolution, and the deaeration of the medium has become a standard practice. There are numerous methods for deaeration of medium, some manual and some automated. The method described in USP 35 uses heat, filtration, and vacuum. Helium sparging is also a typical method for deaeration. Dissolved oxygen and other gases can result in the presence of bubbles, which are commonly observed in non-deaerated medium. USP General Chapter <711> on dissolution states that bubbles can interfere with dissolution test results and should be avoided.[13] Bubbles adhering to either a tablet surface or basket screens create a barrier to the medium contact and can slow down dissolution. Particles can cling to bubbles on the glass surface of the vessel or shafts. The dissolution test should be performed immediately after deaeration of the medium. In some laboratories it is a common practice to rotate the paddle to help bring the temperature to equilibrium (the paddle rotation is stopped before adding the dosage form), but it is best not to have the paddle rotating before adding the dosage form, as paddle movement aerates the medium.

12.5.5 Standard solutions

Preparation of standards is highly important in dissolution testing. Care should be taken when preparing standard solutions, especially if the standard must be dried prior to the solution preparation. Care should also be taken to ensure that the drug powder is completely dissolved. In the case of USP Prednisone Reference Standard, for example, the powder becomes very hard upon drying, making it more difficult to dissolve. Dissolving the powder first in a small amount of alcohol often helps to eliminate this problem.

12.5.6 Vibration

A common problem with dissolution equipment is vibration.[25–27] Vibration can be minimized by ensuring that the top plate and lids for Apparatus 1 and 2 are properly leveled. Also, careful maintenance of the equipment is critical. For example, within the spindle assembly of Apparatus 1 and 2, the bearings can become worn and cause vibration and wobble of the shaft. In addition, the drive belts should be checked for wear and dirt and the tension adjustments for the belt should be optimized for smooth operation. Surging of spindles, though difficult to detect without closely scrutinizing the tester operation, can also cause spurious results. The dissolution vessels need to be locked in place so that they do not move with the flow of water in the bath.

External vibration sources might include other equipment on benchtops, such as shakers, centrifuges, or sonicators. Local construction in the area or within the building is a common, though often overlooked, source of vibration. The testers should not be near refrigerators, hoods or significant air flow sources. Heavy foot traffic and door slamming also should be avoided.

The water bath itself is rarely a source of vibration because water bath designs have evolved to eliminate noisy circulators near the bath. Monitoring of the temperature of the media in all the vessels used in a test (rather than just one) can assure the temperature uniformity. The bath water level should always be maintained at the top of the vessels to ensure uniform heating of the medium.

12.5.7 Apparatus

USP Apparatus 1 and 2 can be sources of error if not closely inspected before use. Obviously, dimensions should be as specified. In the cases of both baskets and paddles, shafts must be straight and true. The paddles are sometimes partially coated with Teflon. This coating can peel and partially shed from the paddle, causing flow disturbance of hydrodynamics within the vessel. Paddles can rust and become nicked or dented; this can adversely affect dissolution hydrodynamics and also be a source of contamination. Thorough cleaning of the paddles is important to preclude carryover of drug or medium.

The baskets need special care and examination. They can become frayed, misshapen, or warped with use. Screen mesh size may change over time, especially when used with acidic medium. There are different designs for attaching baskets to shafts. The attachment can be with clips or with O-rings. These attachment variations can affect dissolution results, depending upon the product; therefore this factor should be taken into consideration when evaluating the robustness of a dissolution method.[28] Baskets are especially prone to gelatin or excipient build up if not thoroughly cleaned immediately after use.

12.5.8 Vessels

Problems caused by vessel irregularities are often overlooked. Vessels are manufactured from large glass tubing, from which the vessel bottom is molded. Depending upon techniques used in the molding process, irregular surfaces can occur and the uniformity of vessel bottom roundness can vary. Cheaply made vessels are notorious for this problem. Close examination of vessels when newly purchased is very important, as surface irregularity can cause dissolution results to differ significantly.[29–31] Another common problem with vessels is residue build up either from oily products or sticky excipients. Insoluble product, if not rinsed well from previous testing, can cause contamination. Vessels that become scratched and etched after repeated washing should be discarded. Lids always need to be in place during a dissolution run to prevent evaporation. As mentioned before, vessels should be locked down to avoid vibration.

12.5.9 Method considerations

The best way to avoid errors and data "surprises" is to put a great deal of effort into selecting and validating methods. Some areas of testing are especially troublesome. Sample introduction can be tricky and, unfortunately at times, uncontrollable. Products can have a dissolution rate that is "position dependent". For example, if the tablet is off-center, the dissolution rate may be higher due to shear forces, or, if it is in the center, coning may occur and the dissolution rate will decrease. Film-coated tablets can be sticky and pose problems related to tablet position. Little can be done in this case except to use a basket (provided there is no gelatinous or excipient build up) or a sinker.

Suspensions can be introduced in a variety of ways: manually, using syringes or pipettes, pouring from a tared beaker, or automated delivery using calibrated pipettes. Each method has its own set of limitations, although automated methods may show less variability. Mixing of the suspension sample will generate air bubbles; therefore the mixing time of suspension samples must be strictly uniform to reduce erroneous or biased results.

The medium is a critical component of the test that can cause problems. One cause of inaccurate results may be that the volume of medium withdrawn through multiple sampling without replacement is too large, therefore adversely influencing sink conditions.

Surfactants can present a significant cleaning problem, especially if the concentration is high, e.g. over 0.5%. In the sampling lines, surfactants such as sodium lauryl sulfate (SLS) may require several rinsings to assure complete removal. The same is true with carboys and other large containers. Some surfactants, and SLS in particular, have other limitations, as quality can vary, depending upon grade and age, and the dissolving effect can consequently change, depending upon the surface-active impurities and electrolytes.[32] The foaming nature of surfactants can make it very difficult to effectively deaerate the medium as well. Some pumps used in automated equipment are simply not adapted to successful use with surfactants. One caution when lowering a basket into surfactant medium is that surface bubbles can adhere to the bottom of the basket and decrease the dissolution rate substantially. When performing HPLC analysis using surfactants in the medium, several sources of error may be encountered. The auto-injectors may need repeated needle washing to be adequately cleaned. Surfactants, especially at high concentrations, may be too viscous for accurate delivery. Surfactants can also affect HPLC column packing to a great degree, resulting in extraneous peaks or poor chromatography. Finally, basic media, e.g. above pH 8, may cause issues with the determinative step, e.g. HPLC column degradation.

Sinkers are defined in the USP as "not more than a few turns of a wire helix..." Other sinkers may be used, but the analyst should be aware of the effect different types of sinkers may have on mixing.[33] Sinkers can be barriers to dissolution when the wire is wound too tightly around the dosage unit.

Filters are used in almost all analyses; many types or different materials are used in automated and manual sampling. Validation of the pre-wetting or discard volume is critical for both the sample and standard solutions. Plugging of filters is a common problem, especially with automated devices.

Manual sampling techniques can introduce error by virtue of variations in strength and size of the human hand from analyst to analyst. Therefore, the pulling velocity through the filter may vary considerably. Too rapid a movement of liquid through the filter can compromise the filtration process itself.

12.5.10 Observations

One of the most useful tools for identifying sources of error is close visual observation of the test. A trained analyst can pinpoint many problems because he or she has developed a knowledge and understanding of the cause and effect relationships of certain observations. Accurate, meaningful dissolution occurs when the product dissolves without disturbance from barriers to dissolution, or disturbance of vessel hydrodynamics from any source. The particle disintegration pattern must show freely dispersed particles. Anomalous dissolution usually involves one or more of the following observations: floating chunks of tablet, spinning, coning, mounding, gumming, swelling, capping, "clam shell" erosion, off-center positioning, sticking, particles adhering to apparatus or vessel walls, sacs, swollen/rubbery mass, or clear pellicles. Along with good documentation, familiarity with the dissolution behavior of a product is essential in quickly identifying changes in stability or changes associated with a modification of the formulation. One may notice a change in the size of the dissolving particles, excipients floating upward, or a slower erosion pattern. Changes in the formulation or an increase in strength may produce previously unobserved basket screen clogging. If contents of the basket immediately fall out and settle to the bottom of the vessel, a spindle assembly surge might be indicated. If the medium has not been properly deaerated, the analyst may see particles clinging to the vessel walls. The presence of bubbles always indicates that deaeration is necessary.

Lastly, the water bath should contain clean water so visual observations of the dissolution test can be performed clearly and easily.

12.5.11 Automation

While automation of dissolution sampling is very convenient and laborsaving, errors often occur with automation devices because the analysts tend to overlook problem areas. Sample lines are often a source of error for a variety of reasons: unequal lengths, crimping, wear beyond limits, disconnection, carryover, mix-ups or cross-connections, and inadequate cleaning. The volume dispensed, purged, recycled, or discarded should be routinely checked. Pumping tubes can wear out through normal use or repeated organic solvent rinsings and may need to be replaced.

The use of flow cells may generate variability in absorbance readings. Air bubbles can become caught in the cell, either introduced via a water source containing bubbles or by air entering inadvertently into poorly secured sample lines. Flow rate and dwell time should be evaluated, so absorbance readings can be determined to have reached a steady plateau. Detector cells need to be cleaned frequently to avoid build up of drug, excipient, surfactant, or buffer salts from the dissolution medium.

12.5.12 Cleaning

The analyst should take special care to examine this aspect when validating the method. In many laboratories, where different products are tested on the same equipment, cleaning is a critical issue that, if inadequately monitored, may be a cause of inspection failures and erroneous results.

12.5.13 Method transfer

Problems occurring during transfer of methods can often be traced to not having used exactly the same type of equipment, such as baskets/shafts, sinkers, dispensing apparatus, or sampling method. A precise description of medium and standard preparation, including grade of reagents, in the method is useful. The sampling technique (manual versus automated), and sample introduction, should be uniform.

12.6 PERFORMANCE VERIFICATION OF DISSOLUTION EQUIPMENT

In order for a dissolution method to be considered valid, the dissolution apparatus must be set up, qualified, and operated in compliance with appropriate compendia, as applicable. USP 35 General Chapter <711> on dissolution lists apparatus specifications, the apparatus suitability test (now called performance verification testing (PVT)), the dissolution medium requirements, as well as specific procedural requirements for USP Apparatus 1–4.[13] The acceptance criteria of the PVT for Apparatus 1 and Apparatus 2 were changed in 2010. The new acceptance criteria include a geometric mean and standard deviation. There is a "single-stage" test consisting of two consecutive runs of six (or eight, depending on the apparatus's configuration) and a "two-stage" test in which one run is evaluated and, if it does not pass the criteria, another run is performed. A detailed explanation of these new criteria was published in 2009.[34] A calculation tool for evaluating whether a PVT passes or fails is available at the USP Web site.[35]

Valid use of the PVT requires choosing either the one-stage or two-stage test before testing begins. In fact, choosing which test to use after examining the data invalidates the use of this compendial tool/worksheet. The PVT is valuable because it tells the analyst whether the equipment is operating properly, and it is always preceded by mechanical calibration. The toolkit at the USP Web site provides comprehensive information on mechanical calibration.

As discussed previously, the major sources of dissolution variability remain vibration, vessel design, and deaeration. A detailed look at the PVT and industry trends on the use of mechanical tests and PVT was published in a special edition of *Dissolution Technologies* in May 2010.[36–39] There is a new FDA guidance: "Use of mechanical calibration of dissolution apparatus 1 and 2 – current good manufacturing practice (cGMP)".[40] This guidance has created quite a bit of controversy as it states that an *enhanced* mechanical calibration (MC) can be used as an alternative to the current Apparatus Suitability procedure (PVT) for Apparatus 1 and 2 described in the USP Dissolution General Chapter <711>. Both procedures executed according to a written protocol will satisfy the cGMP requirements for calibration of laboratory apparatus and mechanical equipment for manufacturing, as set forth in Code of Federal Regulations (CFR) Sections 211.160(b)[4] and 211.68, respectively. An FDA protocol for MC can be found on the Internet.[41]

12.7 REGULATORY GUIDELINES

The regulatory agencies for the various global regions generally address dissolution guidelines in terms of the particular testing necessary to demonstrate the appropriate or intended release from a dosage form. These guidelines relate more to the development of an appropriate dissolution method than the actual validation procedure. The specifics of the analytical validation for dissolution procedures are not as a rule separated from the discussions of general method validation as most of the critical validation analysis parameters do not differ between dissolution methods, and, for example, assay methods. As the regional regulatory and ICH Guidelines are discussed in detail elsewhere in this book, the reader is referred to those chapters.

Acknowledgment

The authors appreciate the drawings for Figs 12.1 and 12.2 provided by Hanson Research Corporation.

References

1. *USP 18–NF 13;* USP: Rockville, MD, USA, 1970.
2. *USP 35–NF 30, Second Supplement;* USP: Rockville, MD, USA, 2012.
3. *USP 35–NF 30, Second Supplement, Validation of Compendial Procedures 1225;* USP: Rockville, MD, USA, 2012.

4. Wang, Q.; Fotaki, N.; Mao, Y. Biorelevant Dissolution: Methodology and Application in Drug Development. *Dissolution Technol.* **2009,** *16* (3), 6–12.
5. Brown, C. K. In *Dissolution Method Development: An Industry Perspective, Pharmaceutical Dissolution Testing;* Dressman, J., Krämer, J., Eds.; Taylor and Francis: Boca Raton, FL, 2005 (chapter 12).
6. Hanson, R.; Gray, V. *Solving Practical Problems, Method Development, and Method Validation, Handbook of Dissolution Testing,* 3rd ed.; Dissolution Technologies, Inc: Hockessin, DE, 2004 (chapter 7).
7. Martin, G.; Gray, V. General Considerations for Dissolution Methods: Development, Validation, and Transfer. *J. Valid. Technol.* **2011,** *17* (1), 8–11.
8. Rohrs, B. R. Dissolution Method Development for Poorly Soluble Compounds. *Dissolution Technol.* **2001,** *8* (3), 6–12.
9. *USP 35–NF 30, Second Supplement, The Dissolution Procedure: Development and Validation 1092;* USP: Rockville, MD, USA, 2012.
10. http://www.ich.org, 2013 (verified on 30 January 2013).
11. *Waiver of In Vivo Bioavailability and Bioequivalence Studies for Immediate-Release Solid Oral Dosage Forms Based on a Biopharmaceutics Classification System: Guidance for Industry.* U.S. Department of Health and Human Services, Food and Drug Administration, Center for Drug Evaluation and Research (CDER); U.S Government Printing Office: Rockville, MD, USA, 1999.
12. *USP 35–NF 30, Second Supplement, In Vitro and In Vivo Evaluation of Dosage Forms 1088;* USP: Rockville, MD, USA, 2012.
13. *USP 35–NF 30, Second Supplement, Dissolution 711;* USP: Rockville, MD, USA, 2012.
14. Brown, C.; Friedel, H.; Barker, A.; Buhse, L.; Keital, S., et al. FIP/AAPS Joint Workshop Report: Dissolution/In Vitro Release Testing of Novel/Special Dosage Forms. *AAPS Pharm. Sci. Tech* **2011,** *12* (2).
15. *Japanese Pharmacopeia.* Division of Pharmacopoeia and Standards for Drugs, Office of Standards and Guidelines Development, 16th ed.; Pharmaceuticals and Medical Devices Agency (PMDA): Tokyo, Japan, 2011.
16. *The British Pharmacopoeia;* The Stationary Office: London, UK, 2013.
17. *European Pharmacopoeia,* 7th ed.; European Directorate for the Quality of Medicines and HealthCare, Council of Europe: Strasbourg, France, 2013.
18. Eddington, N. D.; Marroum, P.; Uppoor, R.; Hussain, J.; Augsburger, L. Development and Internal Validation of an In vitro–In vivo Correlation for a Hydrophilic Metoprolol Tartrate. *Pharm. Res.* **1998,** *15* (3), 466–473.
19. Rohrs, B. R. Dissolution Assay Development for In vitro–In Vivo Correlations. *Am. Pharm. Rev.* **2003,** *6* (1), 8–12.
20. Retting, H.; Mysicka, J. IVIVC: Methods and Applications in Modified-Release Product Development. *Dissolution Technol.* **2008,** *15* (1), 6–9.
21. Marroum, P. Clinically Relevant Dissolution Methods and Specifications. *Am. Pharm. Rev.* **2012,** *15* (1), 36–41.
22. *Immediate Release Solid Oral Dosage Forms, Scale-up and Postapproval Changes: Chemistry, Manufacturing, and Controls; In Vitro Dissolution Testing and In Vivo Bioequivalence Documentation: Guidance for Industry.* U.S. Department of Health and Human Services, Food and Drug Administration, Center for Drug Evaluation and Research (CDER); U.S Government Printing Office: Rockville, MD, USA, 1995.
23. *Extended Release Oral Dosage Forms: Development, Evaluation, and Application of in Vitro/in Vivo Correlations: Guidance for Industry.* U.S. Department of Health and Human Services, Food and Drug Administration, Center for Drug Evaluation and Research (CDER); U.S Government Printing Office: Rockville, MD, USA, 1997.
24. Rohrs, B.; Thamann, T.; Gao, P.; Stelzer, D.; Bergren, M.; Chao, R. Tablet Dissolution Affected by a Moisture Mediated Solid-State Interaction between Drug and Disintegrant. *Pharm. Res.* **1999,** *16,* 1850–1856.

25. Vangani, S.; Flick, T.; Tamayo, G.; Chiu, R.; Cauchon, N. Vibration Measurements on Dissolution Systems and Effects on Dissolution Prednisone Tablets RS. *Dissolution Technol.* **2007,** *14* (1), 6–14.
26. Gao, Z.; Moore, T. W.; Doub, W. H. Vibration Effects on Dissolution Tests with USP Apparatus 1 and 2. *J. Pharm. Sci.* **2008,** *97* (8), 3335–3343.
27. Collins, C. C. Vibration: What Is It and How Might It Effect Dissolution Testing. *Dissolution Technol.* **1998,** *5* (4), 16–18.
28. Gray, V.; Beggy, M.; Brockson, R.; Corrigan, N.; Mullen, J. A Comparison of Dissolution Results Using O-ring versus Clipped Basket Shafts. *Dissolution Technol.* **2001,** *8* (4), 8–11.
29. Tanaka, M.; Fujiwara, H.; Fujiwara, M. Effect of the Irregular Inner Shape of a Glass Vessel on Prednisone Dissolution Results. *Dissolution Technol.* **2005,** *12* (4), 6–14.
30. Scott, P. Geometric Irregularities Common to the Dissolution Vessel. *Dissolution Technol.* **2005,** *12* (1), 18–21.
31. Eaton, J.; Deng, G.; Hauck, W.; Brown, W.; Manning, R.; Wahab, S. Perturbation Study of Dissolution Apparatus Variables–A Design of Experiment Approach. *Dissolution Technol.* **2007,** *14* (1), 20–26.
32. Crison, J. R.; Weiner, N. D.; Amidon, G. L. Dissolution Media for In vitro Testing of Water-Soluble Drugs, Effect of Surfactant Purity and Electrolyte on In vitro Dissolution of Carbamazepine in Aqueous Solutions of Sodium Lauryl Sulfate. *J. Pharm. Sci.* **1997,** *86* (3), 384–388.
33. Soltero, R. A.; Hoover, J. M.; Jones, T. F.; Standish, M. Effects of Sinker Shapes on Dissolution Profiles. *J. Pharm. Sci.* **1989,** *78* (1), 35–39.
34. Hauck, W. W.; Manning, R. G.; Cecil, T. L.; Brown, W. E.; Williams, R. L. Proposed Change to Acceptance Criteria for Dissolution Performance Verification Testing. *Pharm. Forum* **2007,** *33* (3), 574–579.
35. http://www.usp.org/usp-nf/compendial-tools, 2013 (verified 30 January 2013).
36. Martin, G. P.; Gray, V. A. Overview of Dissolution Instrument Qualification, Including Common Pitfalls. *Dissolution Technol.* **2011,** *18* (2), 6–10.
37. Krämer, J.; Schwan, R. Practical Aspects of Dissolution Instrument Qualification–A European Perspective. *Dissolution Technol.* **2011,** *18* (2), 11–15.
38. Yan, B.; Lu, X.; Lozano, R. Feasibility Study on Qualification of USP Dissolution Apparatus 1 and 2 Using the Enhanced Mechanical Calibration Procedure. *Dissolution Technol.* **2011,** *18* (2), 17–23.
39. Salt, A.; Glennon, J. Enhanced Mechanical Calibration of Dissolution Test Equipment. *Dissolution Technol.* **2011,** *18* (2), 25–29.
40. *The Use of Mechanical Calibration of Dissolution Apparatus 1 and 2 – Current Good Manufacturing Practice (CGMP): Guidance for Industry.* U.S. Department of Health and Human Services, Food and Drug Administration, Center for Drug Evaluation and Research (CDER); U.S Government Printing Office: Rockville, MD, USA, 2010.
41. www.fda.gov/downloads/AboutFDA/CentersOffices/OfficeofMedicalProductsandTobacco/CDER/UCM142492.pdf, 2013 (verified 4 February 2013).

CHAPTER

Extractables and leachables

13

Kurt L. Moyer, James Scull
NSF Pharmalytica, Bristol, CT, USA

CHAPTER OUTLINE

Specification of Drug Substances and Products. http://dx.doi.org/10.1016/B978-0-08-098350-9.00013-8

13.1 INTRODUCTION

Leachables are compounds that migrate into a drug product from the sample container closure (SCC) system under normal storage conditions. Both the primary SCC in direct contact with the drug product (metered dose inhaler, prefilled syringe, eye dropper, IV bag, HDPE bottle, LDPE ampoule, etc.) and the secondary SCC, which does not contact the drug product (printed label, cardboard box, foil pouch, environmental exposure, etc.), can be sources of leachables. Leachables present a potential risk to the patient both from the toxicity of the leachable and from the possible negative impact upon stability and efficacy of the drug product. Examples of common leachables are shown in Table 13.1.

Although many types of materials can be used in a primary SCC system, the three most common are glass, polymers, and elastomers. One may expect the manufacturer of the component of the SCC to be able to provide a complete list of the formulation and process used to manufacture the component; however this may not always be the case. The two main reasons manufacturers may not provide this information are that the manufacturer may consider the information to be proprietary or the manufacturer may not have the information. The absence of manufacturer information is particularly common for polymers. The main reason for this among the manufacturers of polymer SCCs is that

Table 13.1 Examples of Common Leachables

Class	Specific Example	Structure of Specific Examples
Lubricants	Oleamide	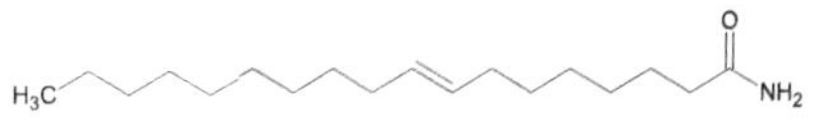
Plasticizers	Bis(2-ethylhexyl) phthalate	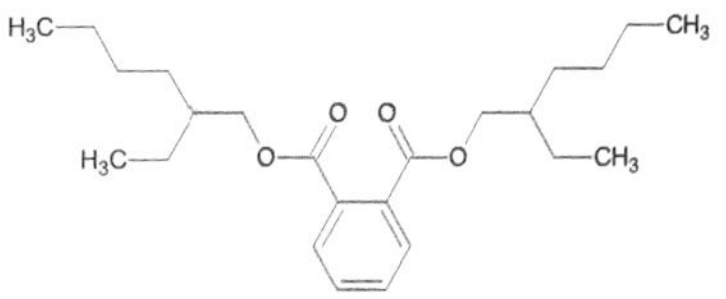
Monomers	Bisphenol A	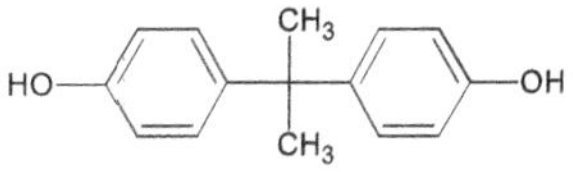
"Small" antioxidants	Butylhydroxytoluene	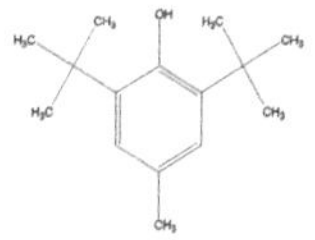
"Large" antioxidants	Irganox 1010	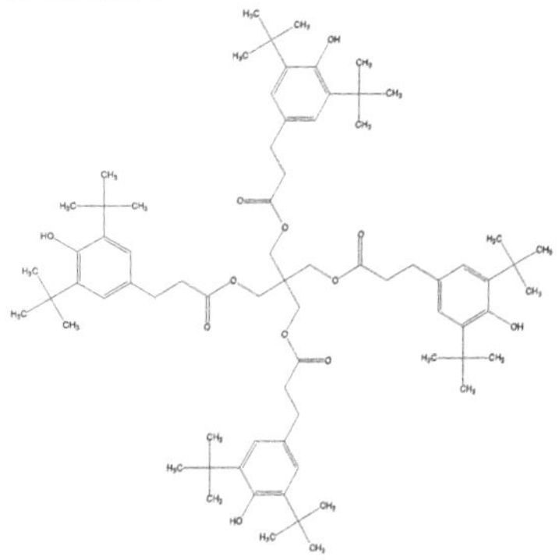
Organic impurities (alkanes, alcohols and aldehydes)	Butanol	

Table 13.2 Risk of Leachables Based on the Route of Administration and Interaction of Drug Products with SCC for Common Drug Product Types

Risk Associated with the Route of Administration	Risk of Drug Product Interaction with SCC		
	High	**Medium**	**Low**
Highest	• Inhalation aerosols and solutions • Injectable solutions and suspensions	• Sterile powders and powders for injection • Inhalation powders	
High	• Ophthalmic solutions and suspensions • Transdermal ointments and patches • Nasal aerosols and sprays		
Low	• Topical solutions and suspensions • Topical and lingual aerosols • Oral solutions and suspensions	• Topical powders • Oral powders	• Oral tablets and capsules

their upstream suppliers do not need to place strict control over their processes. For example, a resin manufacturer will set specifications for their product on its physical characteristics only, and then sell the same resin to a manufacturer of a pharmaceutical SCC and a manufacturer of lawn furniture. In this example, the resin manufacturer may not have needed to keep accurate records on the amounts and type of antioxidants used as long as the resin met the manufacturer's specifications, but these antioxidants do have the potential to leach into a drug product.

Leachables can enter any type of drug product, including solid dosage forms. Generally, orally inhaled and nasal drug products (OINDPs) and parenteral and ophthalmic drug products (PODPs) are the most common drug products at a high risk for leachables. Table 13.2 summarizes the risk for most common drug products. Low risk is not the same as no risk, as evidenced by several high profile recalls of solid dosage forms due to leachables. An assessment of the risk of leachables into a given drug product needs to be done when considering a testing strategy for leachables.

The toxicity of a leachable is dependent upon the route of entry into the body. For example, levels of a compound that can be safely ingested orally can have a toxic effect when the same level is inhaled. As a result, the potential route of administration of a leachable must be considered when assessing the risk of a leachable.

Leachables present unique analytical challenges. Since leachables are not related to the drug product, the analytical methods used to detect impurities in the drug product may not be able to detect the leachables. Even when leachables could be detected by drug product impurity methods, the leachables are often at levels which are orders of magnitude lower than drug degradation products or related substances, and thus below the sensitivity of the method. Thus, separate analytical methods are usually needed for the analysis of leachables in the drug product.

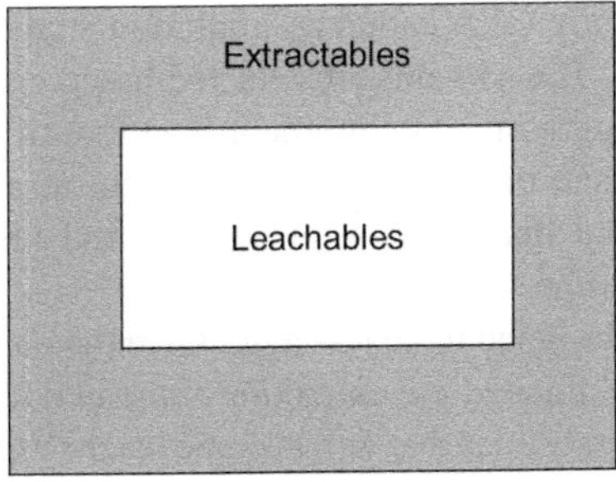

FIGURE 13.1

In an ideal system, leachables are a subset of extractables.

Potential leachables need to be identified before an analytical method for leachables can be developed. This is done by performing an extraction study on the SCC under exaggerated conditions with the goal of identifying the observed extractables. Extractables are the compounds that can be extracted from the SCCs that "might" become leachables. Figure 13.1 illustrates the ideal relationship between extractables and leachables.

The conditions of an extraction study are selected based upon the drug product and are designed to mimic a "worst-case" scenario for the intended drug product. Care must be taken in the selection process so that conditions are aggressive enough to ensure that the extractables include all leachables while not being too aggressive, thus generating an impractically large number of extractables that are not leachables. The extraction study should not lead to a complete deformulation of the material.

At the time that this chapter was written, the only guidance available on extractable and leachable testing had been written by the Product Quality Research Institute (PQRI) and titled "Safety Thresholds and Best Practices for Extractables and Leachables in Orally Inhaled and Nasal Drug Products".[1] As stated in the title of this guidance, it only pertains to OINDPs. The PQRI is currently working on a guidance on PODPs, however this guidance has not been issued at the time of this writing.

13.2 OVERVIEW OF THE STUDY DESIGN

13.2.1 Overview of extraction studies

The first step toward evaluating leachables is to perform extraction studies. There are two types of extraction studies: Controlled Extractions (CEs) and Simulated Use Extractions. These two extractions can be done in series or in parallel. In some cases, just one of the extraction studies may be sufficient.

A CE study (also called materials characterization study) involves extracting the SCC in two or three solvents of varying polarities. The solvents are selected based upon the drug product, with one of the solvents representing a "worst-case scenario". The extraction conditions used are aggressive, typically reflux or Soxhlet extraction. The combination of the "worst-case scenario" solvent with the aggressive extraction conditions is intended to yield a high number of extractables. The end result of this approach is that all potential leachables (except those that react or have a unique affinity for the drug product) will be identified.

A simulated use extraction study (also called a simulation study) involves extracting the SCC in two solvents of varying polarities. The solvents are selected based upon the drug product with the goal of representing a slightly more aggressive environment than that of the drug product. The extraction conditions are usually static storage of the SCC in the solvent at a temperature above the intended storage condition of the packaged final drug product. The end result of this approach is that the observed extractables are likely to be leachables.

A simulated use extraction is designed to be less aggressive than a CE study; thus fewer extractables are expected to be identified in a simulated use extraction compared to a CE. The simulated use study is more likely to identify only the extractables that will become leachables compared to the CE study which will potentially identify many extractables that will not become leachables. However, a simulated use study is more likely to "miss" a potential leachable than a CE study. Both studies reveal useful information on the potential leachables from a given material, but the project team must be aware of the strengths and weaknesses of each study. In some cases, for example, if the CE study results in a low number of extractables, the project team may decide that only one type of extraction study is necessary.

Regardless of the type of extraction study performed, once completed, the sample extracts are analyzed using a minimum of three methods, including gas chromatography coupled to mass spectrometry (GC–MS), liquid chromatography coupled to mass spectrometry (LC–MS), and inductively coupled plasma with mass spectrometry (ICP–MS). The goal of these analyses is to identify as many extractables as possible and to semiquantitatively determine the level of each extractable. Based on the material, additional analysis may be required for specific extractables known to be highly toxic. Since the methods are designed to detect unknowns, these methods cannot be validated. Results from these analyses are reported as the amount of the extractable (usually in micrograms) per weight (usually in grams) or surface area (usually in square centimeters) of the SCC component.

13.2.2 Selection of target leachables for analytical methods

Following the completion of the extraction studies, a list of extractables is generated. From this list, the target leachables for the analytical methods are selected. Since the extractables from the simulated use extraction are likely to be leachables, all the extractables from the simulated use extraction should be selected as target analytes for the analytical method. Any additional highly toxic extractable observed in the CE study may be selected as a target leachable if it is deemed to have the potential to migrate into the drug product. An extractable can be eliminated from selection if it is present at a low enough level not to present a toxicity concern. If a high number of extractables is present, selecting representative compounds for a group of target leachables based on structure may be necessary.

After the target leachables for the methods have been selected, the analytical evaluation threshold (AET) for the leachable in drug product is calculated (discussed in more detail in Section 13.4.5). The AET represents the level above which leachables must be reported, and is similar to a specification for a drug product impurity.

13.2.3 Development and validation of analytical methods for leachables

The goal of the analytical methods is to have sufficient sensitivity so that the limit of quantitation (LOQ) of the method is at or below the AET. Extensive sample preparation may be necessary to ensure sufficient sensitivity.

The analytical methods are then validated. Since the challenge of these methods is to be able to detect very low levels of leachables in often complex drug product matrices, some allowances may need to be made in other aspects of method performance to allow sufficient sensitivity. These allowances may be seen in higher acceptance criteria than in drug product impurity methods.

13.2.4 Analysis of leachables in drug product

The analytical methods are then used to analyze drug product stored in the SCC under the intended storage conditions. Ideally, this testing can be done as a part of the product stability program, but it can also be done in a separate migration study. Results from an analysis are reported as the concentration of the leachable in the drug product. If a leachable is observed at a level above the AET, additional experimentation may be necessary to confirm the identification of the leachable. Additional toxicological evaluation may also be needed to assess the risk associated with the particular leachable.

13.3 EXTRACTION STUDY

13.3.1 Sample

13.3.1.1 Sample selection

All components of the SCC that directly contact the drug product either during storage or during the administration of the drug product are considered to be primary components of the SCC. All the components of the SCC that do not contact the drug but do have the potential to interact with the primary SCC are considered to be secondary components of the SCC. A secondary SCC component will either contact the primary SCC or contact another secondary SCC component that does directly contact the primary SCC. Table 13.3 shows some common examples of primary and secondary components of SCCs.

All primary SCC components should be included in the extraction. If in the final SCC the components are to be pretreated in any way (e.g. sterilized) before being filled with the drug product, the samples to be used in the extraction study should be pretreated in a similar manner to ensure that the extraction profile correctly models the SCC exposed to the drug product.

Table 13.3 Examples of Primary and Secondary SCCs

Examples of Primary SCC	Examples of Secondary SCC
• All components in a metered dose inhaler	• Printed labels (attached either to the primary SCC or another component of the SCC)
• The barrel, plunger and tip cap in a prefilled syringe	• Boxes
• The film and the ports on an IV bag	• Aluminum pouches
• The aluminum tube and cap of a topical cream	• Skids and overwrap during storage and shipment
• The web and foil of a blister pack	• Environment exposure

Selection of secondary SCC components for inclusion in the extraction study is based upon a risk assessment. In this risk assessment the likelihood of the secondary SCC component giving rise to leachables and the likelihood of these leachables being able to contact and penetrate the primary SCC are considered. One secondary SCC component that will usually need to be included in the extraction study is a printed label if it is to directly contact a part of the primary SCC. This is due to the observation that printing inks are prone to contain many potential leachables.[2]

13.3.1.2 Sample size

For a CE study, the sample size is based upon the desired limit of detection (LOD) of extractables in the component. The sample weight can be calculated as follows:

$$\text{Sample weight (g)} = \frac{\text{Extraction solvent volume (mL)} \times \text{Analytical method LOD} \left(\frac{\mu g}{mL}\right)}{\text{Target extractable limit of detection in component} \left(\frac{\mu g}{g}\right)} \qquad (13.1)$$

For example, a SCC component has a target detection limit for extractables in the component at 1.0 ppm. If this component is extracted in 50 mL of solvent with the sample extracts analyzed by an analytical method with an LOD of 0.1 μg/mL, the sample weight of the component extracted would have to be 5.0 g.

For a simulated use extraction, the sample surface area of the SCC component exposed to the extraction solvent should represent a slightly exaggerated condition from the intended final product. For example, a SCC component will have 5 cm^2 of its surface exposed to 10 mL of drug product. In the simulated use extraction the sample would be exposed to solvent at a ratio <2 mL of solvent per square centimeter of the surface of the SCC component.

13.3.1.3 Sample preparation

No cleaning or pretreatment of the samples is done unless those steps will be done on the final SCC before being filled with the drug product. Samples can be cut as needed to obtain the desired size and shape for the extraction, but attempts should be made to minimize the change in surface area.

13.3.2 Controlled extraction

13.3.2.1 Definition

A CE study, which can also be called a material characterization study, is a study designed to generate a complete list of potential leachables. In a CE study the components of the SCC are exposed to two or three solvents of varying polarities. The solvents are selected based upon the drug product, with the first solvent approximating the polarity of the drug product, and the other solvents selected to be increasingly less polar. The extraction conditions used are aggressive, including possibly exposing the component of the SCC to the solvent at the boiling point of the solvent or under exhaustive extraction conditions. The combination of the increasingly less polar solvents with the aggressive extraction conditions is intended to yield a high number of extractables and a "worst-case scenario" of potential leachables. The strength of this approach is that all potential leachables (except those that result from a reaction with the drug product or are unstable under the extraction conditions) will likely be found. The weakness of this approach is that a potentially large number of extractables may be observed without a reliable method to identify which ones will be leachables, thus complicating the evaluation process.

Table 13.4 Example Extraction Solvents

Polar	Intermediate Polarity	NonPolar
• Water	• Ethanol	• Hexane
• 50 mM Buffer at pH of drug product	• Water/Isopropanol, 50/50	• Methylene chloride
• 0.9% Saline solution		

13.3.2.2 Recommended use

A CE study using three solvents is required for drug products at the highest risk based upon the route of administration in Table 13.2. A CE study using two solvents is recommended for drug products in the other two risk categories for the route of administration if the risk of packaging component–dosage form interaction is high (Table 13.2).

13.3.2.3 Solvent selection

Solvents are selected based upon the drug product. The most polar solvent should be similar in polarity to the drug product, and the second and third (when applicable) solvents should be increasingly less polar. Table 13.4 shows example extraction solvents for an aqueous drug product.

13.3.2.4 Extraction types

13.3.2.4.1 Reflux

In a reflux extraction, a sample is placed into a flask with a set volume of extraction solvent and boiling chips. The flask is attached to a cooling condenser and then heated to the boiling point of the solvent. See Fig. 13.2 for an example of a reflux extraction apparatus. Reflux extraction can be done with neat solvents, solvent mixtures, or aqueous buffers.

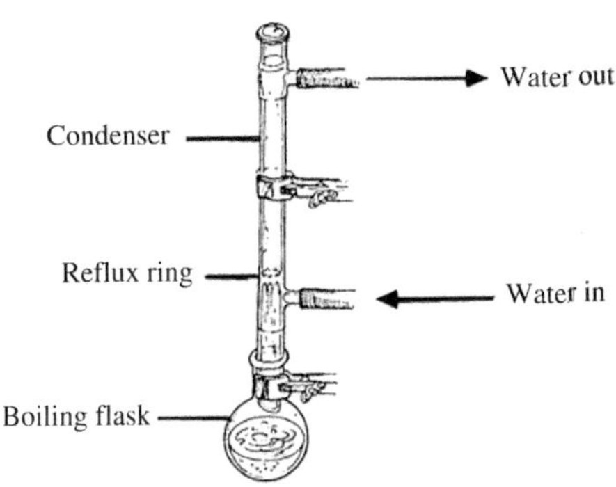

FIGURE 13.2

Diagram of a reflux apparatus.

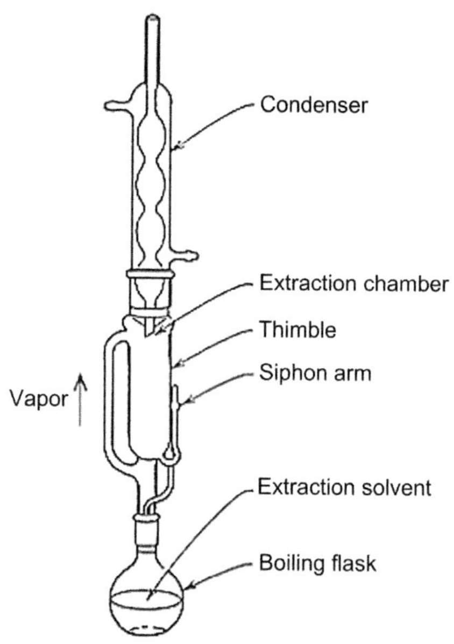

FIGURE 13.3

Diagram of a Soxhlet extractor.

13.3.2.4.2 Soxhlet

In a Soxhlet extraction, a sample is placed into a "thimble" of the Soxhlet extractor. A flask with a set volume of extraction solvent and boiling chips is attached to the bottom of the Soxhlet extractor and a cooling condenser is attached to the top. The solvent is then heated to boiling. See Fig. 13.3 for an example of a Soxhlet extraction apparatus. Since the sample is continuously exposed to the pure condensed solvent and not limited by extractable solubility, Soxhlet is an exhaustive extraction and can only be done using neat solvents.

13.3.2.4.3 Oven

When analyzing for all extractable types by oven extraction, samples are placed in a sealed vessel with solvent. The sealed vessel is then placed in an oven at an elevated temperature that is below the boiling point of the solvent. The vessel is agitated and incubated for a set time, usually 24–48 h. For analysis of volatile extractables, samples are placed in sealed vials neat without the extraction solvent. The sealed vials are then heated for about 1 h at an elevated temperature below the melting point of the material, and then the headspace is analyzed directly using GC–MS.

13.3.2.4.4 Other techniques

Other extraction techniques, such as sonication, may be used if one of the above techniques is not appropriate for a given material. Care must be taken in the design of the experiment to ensure that the conditions are aggressive enough to yield a complete extractable profile.

13.3.2.5 Extraction blanks

Regardless of the extraction technique, a representative blank of the extraction solvent will greatly facilitate the identification of extractables. One method to get a representative extraction solvent blank is to take an aliquot of the solvent that has been exposed to the extraction apparatus. For example, if a sample is to be extracted by reflux in 50 mL of solvent, first add 60 mL of solvent. Start the reflux for a short time interval, and after the solvent has cooled, remove a 10-mL aliquot to use as the extraction blank in the analyses. The sample is then extracted in the remaining 50 mL of solvent.

13.3.2.6 Extraction time

The completeness of the extraction is demonstrated when the total concentration of extractables has become constant. This can either be determined by one of the chromatographic methods described below or by measuring the total organic carbon in an aqueous extract. The completeness of extraction needs to be balanced against extractable instability at elevated temperatures, so for some materials it may not be practical to completely extract the material.

13.3.3 Simulated use extraction

13.3.3.1 Definition

A simulated use extraction (SU) study, which can also be called a simulation study, is a study designed to generate a list of the most likely potential leachables. In a SU study the components of the SCC are exposed to two solvents of different polarities. The solvents are selected based upon the drug product, with the first solvent approximating the polarity of the drug product, and the other solvent selected to be slightly less polar. The extraction conditions used are intended to exaggerate the storage conditions of the drug product in the SCC. The end result of this approach is that the observed extractables are likely to be leachables, thus facilitating the selection of the target analytes for the analytical methods.

13.3.3.2 Recommended use

A SU study is recommended when a large number of extractables are expected from a CE study, or when the drug product will have a large dose volume (and therefore an extremely low AET).

13.3.3.3 Solvent selection

Solvents are selected based upon the drug product. The most polar solvent should be of a similar polarity to the drug product, and the second solvent should be slightly less polar. For example, for an aqueous drug product, the first solvent would be an aqueous buffer and the second would be 80/20 water/ethanol.

13.3.3.4 Extraction conditions

The components of the SCC can be exposed to the solvents either independently or in their intended final configuration. When possible, the exposure should be set so that more surface area of the component will be exposed per unit of volume of the solvent compared to the ratio in the drug product. The extraction temperature should be higher than the intended storage temperature. The extraction time should be sufficient to ensure that all potential leachables are observed. Typically 60 days is sufficient.

13.3.3.5 Extraction blanks

An aliquot of each extraction solvent should be stored in an inert container under the same conditions as the extraction samples.

13.4 EXTRACTABLE ANALYSIS

13.4.1 Overview

In an extractable analysis, since all the extractables cannot be accurately predicted, the analytical methods cannot be validated before the analysis. Instead, appropriate analytical methods are selected that should detect most volatile, semivolatile and nonvolatile organic extractables and inorganic extractables. The most common methods include GC–MS, LC–UV/MS, and ICP–MS. Since the methods for organic extractables will detect extractables of differing volatilities, the methods are not expected to detect the same extractables although overlap is sometimes observed. Extractables are identified from the observed mass spectra and confirmed by comparison to authentic substances when possible.

13.4.2 MS screening methods

13.4.2.1 Volatile organic extractables

Headspace GC–MS is the most commonly used technique for volatile organic extractables. Some representative starting conditions are given in Table 13.5. Analyzing the sample neat directly from the

Table 13.5 Example Headspace GC–MS Instrument Conditions for Analysis of Volatile Extractables

Column	**Capillary Column, 30 m x 0.25 mm, 0.25 μm or Equivalent**
Headspace conditions	
Vial size	20 mL
Sample volume	2 mL
Headspace oven temperature	90 °C
Loop temperature	100 °C
Transfer line temperature	110 °C
Vial heat time	60 min
Temperature gradient	
Initial temperature	40 °C for 1 min
Ramp	To 300 °C at 10 °C/min
Final temperature	300 °C for 10 min
MS detector	
Ionization mode	EI
Scan range (*m/z*)	40–650

Table 13.6 Example Direct Inject GC–MS Instrument Conditions for Analysis of Semivolatile Extractables

Column	**Capillary Column, 30 m x 0.25 mm, 0.25 μm or Equivalent**
Injection volume	1 μL
Temperature gradient	
Initial temperature	40 °C for 1 min
Ramp	To 300 °C at 10 °C/min
Final temperature	300 °C for 10 min
MS detector	
Ionization mode	EI
Scan range (*m/z*)	40–650

headspace vial following an oven incubation as part of the instrument method will usually yield a higher level of extractables than analyzing sample extracts.

13.4.2.2 Semivolatile organic extractables

Direct injection GC–MS is the most commonly used method for semivolatile organic extractables. Some representative starting conditions are given in Table 13.6. Care must be taken when working with aqueous sample extracts because most columns are not compatible with water. A liquid:liquid extraction solves this problem with the added benefit of concentrating the extracts if a smaller volume of extraction solvent is used than the volume of sample extract. Adjusting the pH of the sample before the liquid:liquid extraction may be helpful when a specific acidic or basic extractable is expected, but is generally not useful since most acidic or basic extractables are poor analytes for GC–MS and are more likely to be detected by LC–UV/MS.

13.4.2.3 Nonvolatile organic extractables

For nonvolatile organic extractables, LC–UV/MS is the most commonly used technique. Some representative starting conditions are given in Table 13.7. Care must be taken when working with the nonpolar solvents in that the solvent strength can significantly alter peak shape, which leads to loss of sensitivity. Since the target extractables in this analysis are nonvolatile, a simple fix to this problem is to evaporate the nonpolar sample extracts to dryness and reconstitute in mobile phase.

13.4.2.4 Inorganic extractables

For inorganic extractables, ICP–MS is the most commonly used technique. Some representative starting conditions are given in Table 13.8. The method analyzes for the most common inorganic extractables in their elemental form.

Table 13.7 Example LC–UV/MS Instrument Conditions for Analysis of Nonvolatile Extractables

Equipment	HPLC or UPLC with photodiode array and MS detectors
Column	C8 or C18 reverse phase column
Photodiode array detector	200–400 nm
Mobile phases	Water (A) and acetonitrile (B)
Postcolumn infusion	50 mM ammonium acetate at 10 μL/min
Gradient	50% B to 100% B in 6 min (UPLC) or 35 min (HPLC)
MS detector	
Ionization source	APCI
Polarity	Positive and negative
Source temperature	150 °C
Probe temperature	550 °C
Corona (kV)	±3
Scan range	200–2000 amu

Table 13.8 Example ICP–MS Instrument Conditions for Analysis of Inorganic Extractables

Equipment	ICP mass spectrometer
Sample introduction	Peristaltic pump
Detector	Dual stage discrete dynode electron multiplier (off axis)

13.4.2.5 Special cases

Depending on the material of the SCC, additional methods may be needed for specific extractables of concern. One common example is that when analyzing for extractables from elastomers, analytical methods that can detect low levels of nitrosamines and 2-mercaptobenzothiazole need to be used. Another common example is that if the SCC contains carbon black, analytical methods that can detect low levels of polycyclic aromatic hydrocarbons need to be used.

13.4.3 Standard selection

Standards selected for organic extractable analyses should represent the anticipated extractables from the material. Some recommended standards for organic extractables from plastics are listed in Table 13.9. For LC–UV/MS, standards need to be selected so that at least one standard is detected in each detection mode (positive ionization, negative ionization, or UV). Since the MS standards would be expected to have different responses, the concentrations of the standards would be expected to be different.

Standards for inorganic extractables are usually the target elements. Table 13.10 is an example of the standards used for inorganic extractables.

13.4.4 Initial identifications of extractables

13.4.4.1 Identification of chromatographic peaks as extractables

Only peaks that are not observed in the extraction blank are labeled as extractables. If a peak observed in the blank is also observed at a significantly higher response level in the sample extract (e.g. peak area in the sample extract is three times higher than in the blank), the peak should also be labeled as an extractable and background corrected for the area in the blank during the semiquantitation step described later.

13.4.4.2 Volatile and semivolatile organic extractables

For volatile and semivolatile organic extractables detected by GC–MS methods, the extractables can be identified from standard GC–MS libraries. The use of libraries is possible because GC–MS uses

Table 13.9 Example Standards for Analysis of Organic Leachables from Plastics

Analytical Method	Standard
Headspace GC–MS	Butylated hydroxytoluene Hexane Octamethylcyclotetrasiloxane Butanol
Direct inject GC–MS	Butylated hydroxytoluene Dodecane Hexadecamethylcyclooctasiloxane 4-Methyl-2-pentanone Methyl stearate
LC–UV/MS	Irgafos 168 (positive ionization, UV) Bis(2-ethylhexyl) phthalate stearic acid (negative ionization, UV) (positive ionization, UV) Oleamide (positive ionization) Cyanox 1790 (positive and negative ionization, UV) Irganox 1010 (negative ionization, UV) Stearic acid (negative ionization) Bisphenol A (UV)

Table 13.10 Standards Used in ICP–MS for Inorganic Extractables

Lithium	Selenium	Samarium
Beryllium	Rubidium	Europium
Boron	Strontium	Gadolinium
Sodium	Yttrium	Terbium
Magnesium	Zirconium	Dysprosium
Aluminum	Niobium	Holmium
Silicon	Molybdenum	Erbium
Phosphorus	Ruthenium	Thulium
Potassium	Rhodium	Ytterbium
Calcium	Palladium	Lutetium
Scandium	Silver	Hafnium
Titanium	Cadmium	Tantalum
Vanadium	Indium	Tungsten
Chromium	Tin	Rhenium
Manganese	Antimony	Iridium
Iron	Tellurium	Platinum
Cobalt	Cesium	Gold
Nickel	Barium	Mercury
Copper	Lanthanum	Thallium
Zinc	Cerium	Lead
Gallium	Praseodymium	Bismuth
Germanium	Neodymium	Uranium
Arsenic	Promethium	Thorium

electron impact ionization, which is a high energy ionization that gives reproducible spectra independent of instrument type or condition. The National Institute of Standards and Technology (NIST) is a commonly used GC–MS library that contains many observed extractables, but other libraries are available that are also useful.[3]

Identification of specific alkanes and alkenes is often difficult because of the similarities observed in the mass spectra of these compounds. This is due to the observation that once the chain lengths exceed eight carbons, the resulting mass spectra look almost identical regardless of size or branching. Size can be at least estimated by comparison to retention times of standards of known carbon-chain length.

13.4.4.3 Nonvolatile organic extractables

If a high-resolution mass detector is used in the LC–MS analysis, extractables can be identified from library matching of the exact mass of the observed molecular ion. The NIST library is commonly used, but other libraries are also available.[3] If a low-resolution mass detector is used in the LC–MS analysis, the exact mass of the observed molecular ion will not have sufficient resolution to allow for library matching. Instead, the observed molecular weight is used to identify the extractable based on the user generating a list of expected extractables from the literature and experience. Confirmation of extractables identified in this manner is strongly recommended and is described in a following section.

13.4.4.4 Inorganic extractables

Inorganic extractables are identified only in their elemental form. These identifications are done by matching the observed mass to the anticipated elemental exact mass.

13.4.4.5 Confirmation of identifications

When possible for organic extractables, identifications should be confirmed by analyzing authentic materials. If the retention time and observed mass spectrum of the authentic material match the extractable, the identification of the extractable is confirmed. If an authentic material is not available, identification of an organic extractable can still be confirmed if the following are observed:

1. Mass spectrum matches to a library
2. The molecular weight is confirmed or the elemental composition is determined
3. The mass spectral fragmentation pattern is consistent with the structure

If an identification of an organic extractable cannot be confirmed, it can still be reported as a confident identification if the mass spectrum matches with the one found in a library. If a library match to an observed mass spectrum is not found, the organic extractable is reported as an unknown. However, reporting as much structural information as can be learned from the mass spectrum (e.g. unknown aromatic compound) is strongly recommended. All inorganic extractables identified by ICP–MS are considered confirmed.

13.4.4.6 Semiquantitation

The same results that were used to identify the extractables can be used to quantify them if standards were used that were detected in the same mode, and if appropriate system suitability criteria were met.

For extractables where the identification was confirmed with an authentic material that was used as a standard during the analysis, obviously the authentic material standard would be used to reliably quantitate the extractable. For extractables where the identification was confirmed with an authentic material that was not used as a standard, a response factor between the authentic material and the standards from the analysis can be calculated. If the response factor is correctly calculated, the amount of the extractable can be reliably calculated.

For extractables where an authentic material is not available, a model compound that is structurally similar to the identified extractable should be used as described above for the authentic material. A similar approach is used for unknown extractables where some structural information is available. If the extractable is an unknown with no structural information available, the model compound selected is the standard with the closest retention time to the unknown extractable. In all these cases, the reliability of the results are dependent upon how close the model compound is to the structure of the extractable and are less reliable than if an authentic material had been used. This weakness in the results is compensated by the uncertainty factor used in the AET discussed next.

13.4.5 Analytical evaluation threshold

Following the completion of the extraction studies, a list of extractables is generated. The challenge at this point is to select which extractables are the most likely leachables, which are the highest toxicological risk, and what levels are acceptable in the drug product.

The most likely leachables are those that were observed in the most polar extraction solvents that modeled the intended drug product. This will apply to extractables from both controlled extraction and simulated use studies. Extractables observed in intermediate nonpolar solvents can also be selected as target leachables if the intermediate nonpolar solvent is deemed to represent a realistic model for the drug product. Extractables observed only in the worst-case scenario solvents are only selected as target leachables when the extractable is highly toxic.

To evaluate the toxicity of each observed extractable, the safety concern threshold (SCT) is used. The SCT is the absolute highest acceptable exposure of a patient to a leachable in the drug product, and is usually expressed in terms of micrograms of leachable per day. If an SCT is not known, the PQRI recommends an SCT of 0.15 μg of each individual leachable per day. The PQRI selected this SCT as representing a threshold below which leachables would have negligible safety concerns from carcinogenic and noncarcinogenic toxic effects. This SCT cannot be used with the previously discussed special case extractables discussed in Section 13.4.2.5. Compounds listed in this section must use an SCT determined from the available literature.

To apply the SCT to a given drug product, an AET is calculated based on the SCT of an individual leachable (usually in micrograms per day), the number of doses of the drug product administered per day, the number of doses contained in the SCC, and the volume of drug product in the SCC. The AET is defined as follows:

$$\text{AET} = \frac{\text{SCT}\left(\frac{\mu\text{g}}{\text{day}}\right)}{\text{\#of doses/day}} \times \frac{\text{\#of doses/SCC}}{\text{weight of component of SCC (g)}} \times \text{uncertainty factor} \tag{13.2}$$

The AET will have units of micrograms per gram unless other units were used in the calculation. Surface area of the component of the SCC instead of the weight may be applicable in some situations. The uncertainty factor is an adjustment for the confidence in the identification and quantitation of the extractables. For extractables where the identification was confirmed by authentic material, the uncertainty factor can be 1.0. For extractables where the confirmation was only confident or the extractable remained an unknown, the uncertainty factor should be ≤ 0.5.

All extractables present at levels above their AET should be selected as target leachables.

An example set of extractable results are shown Table 13.11 for a fictional SCC with an AET of 10.0 ppm for each extractable. The results listed in italics are well below the AET and would not be selected as target leachables. The results that have been bolded are significantly above the AET and would definitely need to be selected as target leachables. The results that are underlined represent results that would require additional consideration since the results are close to the AET. The SCT and

Table 13.11 Example Set of Extractables from a Fictional SCC with an AET of 10.0 ppm

Extractable	ppm
Cyclohexane	*5.02*
BHT	**16.03**
2,6-Bis(1,1-dimethylethyl)phenol	*1.84*
Oleamide	*3.75*
Palmitic acid	10.02
Bis(2-ethylhexyl)phthalate	9.06
Irganox 1010	9.90
Phosphate of Irgafos 168	**102.22**
Irgafos 168	**57.75**
Irganox 1076	**20.67**

the uncertainty factor should be reevaluated before selecting or dismissing those extractables as leachables. A conservative decision to include an extractable that is just below the AET as a target leachable is an acceptable and common practice.

13.5 LEACHABLE STUDY

13.5.1 Overview

After the completion of the extraction studies, the next three steps are

1. Develop analytical methods with sufficient sensitivity to detect all the target leachables.
2. Validate the analytical methods.
3. Analyze for leachables in the drug product from samples stored in the SCC under the intended storage conditions over the intended shelf life.

13.5.2 Analytical method requirements

The first step in selecting analytical methods for analysis of target leachables identified in the extraction study is to determine the required sensitivity of the method. This is done by converting the AET of each target leachable to an AET in the drug product (AET_{dp}) as follows:

$$AET_{dp} = AET\left(\frac{\mu g}{g}\right) \times \frac{\text{Weight of component of SCC (g)}}{\text{Total volume of drug product in SCC (mL)}} \tag{13.3}$$

The AET_{dp} will have units of micrograms per milliliter unless other units are entered into the equation (i.e. micrograms per gram). The most important criteria for the analytical methods will be to have an LOD at or below the AET_{dp}.

The analytical methods selected do not need to be the same as those used in the extraction studies. If a wide range of target leachables have been identified, multiple analytical methods may be needed. The use of more universal and rugged techniques like GC with flame ionization detection and high-performance liquid chromatography (HPLC) with UV or charged aerosol detection is recommended when those methods can achieve the needed sensitivity.

A significant amount of effort is usually required to develop methods that can reach the required sensitivity. The case where the AET_{dp} is several orders of magnitude lower than the impurity specification for the drug product is common. In addition, the matrices for OINDP and PODP are often complicated with the drug itself being a major interference for detection of the leachables. As a result, extensive sample preparations using various extraction techniques are often needed.

13.5.3 Method validation

Leachable methods are validated as impurities following ICH Q2 (R1)[4] when possible. The following parameters are typically sufficient for the validation of leachables methods:

- LOD/LOQ
- Specificity
- Accuracy and precision
- Linearity

Ruggedness may also be included based on the complexity of the method and the intended number of times the method will be used. Due to the unique challenges of leachable methods, some allowances in the acceptance criteria as described below may be required. System suitability requirements are set from the validation results.

13.5.3.1 LOD/LOQ

The LOD and LOQ of the method need to be established for each target leachable in the drug product. The LOD must have a *S/N* ratio of $\geq$3:1 in six replicates and the LOQ must have a *S/N* ratio of $\geq$10:1 in six replicates.

13.5.3.2 Specificity

The specificity of the method is established by demonstrating that there are no interferences from the drug product that would negatively impact the detection of the target leachables. A clear distinction needs to be made between interference and the leachable being present in the drug product. Due to the ubiquitous nature of some of the common leachables, detecting these leachables in the drug product or placebo before being exposed to the SCC is common. The specificity acceptance criteria should then be set based on the method's ability to detect a small change in the target leachable based on the LOQ and the level of leachable already present. If the target leachables are not present in the drug product or placebo, the acceptance criteria should be no interference in the drug product or placebo >30% of the LOQ.

13.5.3.3 Accuracy and precision

Accuracy and precision of the method are established by spiking replicate samples ($N = 6$) of drug product with the target leachables at a level of approximately three times the LOQ if the LOQ is equal to the AET_{dp}. If the LOQ is significantly less than the AET_{dp}, include a second level at the AET_{dp}. Unspiked replicates ($n = 3$) of each drug product are also analyzed and can be used for background area correction. Ideally, the acceptance criteria should be 30.0% for both accuracy and precision. However, wider acceptance criteria may be needed if a complex sample preparation and analysis are needed to reach the necessary LOQ.

13.5.3.4 Range

The range of the method is established by analysis of five levels between the LOQ and up to 10 times the LOQ of each target leachable. If the analytical method is expected to be linear (e.g. GC with flame ionization detector), the acceptance criteria for a linear fit will have an $R^2 \geq 0.995$. If the analytical method is not expected to be linear (e.g. GC–MS), a quadratic fit can be used with appropriate acceptance criteria. The analytical method will need to include an appropriate number of standards if a quadratic fit is used.

13.5.3.5 Ruggedness

The level of ruggedness testing involved in validation is dependent upon the intended use of the method in the leachable study. If the method is intended to be used on less than three time points by the same laboratory, limited or no ruggedness testing is required. If the method is intended to be used on more than three time points or by multiple laboratories, appropriate ruggedness testing should be performed.

13.5.3.6 Setting system suitability

At a minimum, system suitability criteria should be set to ensure the precision and sensitivity observed in validation is maintained. If not all of the target leachables are included in the standards, an additional chromatographic parameter, such as resolution or relative retention time, should be added to system suitability. If an analytical method has multiple detection modes, system suitability needs to be established in each detection mode.

13.5.4 Leachable study design

13.5.4.1 Storage conditions and study duration

Leachables need to be evaluated in the drug product stored in the SCC under the intended storage conditions for the entire shelf life. The simplest and best approach to accomplish this is to include the testing for leachables in the stability study to evaluate the shelf life of the drug product under the intended storage conditions. Since migration rates are typically slow, leachable analysis does not need to be done at the early time points in the stability study. Evaluating leachables at the midpoint and the end of a stability study can be sufficient if the levels of leachables remain low.

Leachable analysis can be done under accelerated stability conditions but care needs to be taken in evaluation of the results. The concentration of leachables present in a drug product is controlled by the rate of diffusion of the leachable through the SCC and the partitioning between the SCC and the drug product (which can decrease as temperature increases), not a chemical reaction. Therefore the Arrhenius equation does not apply to leachables.

Generally samples are stored in their intended container orientation for leachables analysis. In certain cases, the orientation may be changed to maximize the contact of the drug product with the component of the SCC deemed to be the highest risk. For example, a parenteral aqueous drug product to be stored in a glass vial with a rubber stopper could be stored inverted for the leachables study, which would give a more confident assessment of the risk from leachables.

13.5.4.2 Control samples

Inclusion of control samples can greatly simplify the analysis of the leachable results. One recommended control is to store the drug product in a different SCC under the same storage conditions for the same length of time. Ideally the different SCC would be expected to yield significantly less and/or different leachables (e.g. a glass vial with a Teflon coated lid). This control is used to distinguish the degradation products of the matrix from leachables.

13.5.4.3 Placebo samples

The inclusion of placebo samples stored in the SCC as part of the leachable study is strongly recommended. These samples can help confirm the presence of leachables observed in the active drug product, and leachables that might have been missed in the active drug product may be observed in the placebo samples. However, one cannot assume a peak that is observed in the placebo, but not in the active drug product, is not a leachable. The active drug product may facilitate the migration of the leachable or could react with the leachable.

13.5.5 Leachable sample analysis

Samples from the leachable study are analyzed at each time point using the validated analytical methods. All leachables above the AET_{dp} are reported.

13.5.5.1 Leachables with confirmed identifications

If the identification of the leachable was confirmed in the extraction study and the authentic material was used to validate the method, no further action is required of the analytical methods. If the identification of the leachable was confirmed in the extraction study but the authentic material was not used to validate the method, the authentic material should be used to perform additional validation experiments to demonstrate the method is appropriate for the leachable. The additional validation experiments could lead to the calculation of a response factor which would change the result. In this case, only the result calculated with the response factor is reported.

13.5.5.2 Leachables with confident identifications

If the identification of the leachable was only confident in the extraction study, additional experiments need to be done at this point to confirm the identification. Once the identification has been confirmed, if authentic material is available, it should be used to perform additional validation experiments to demonstrate the method is appropriate for the leachable. If authentic material is not available, a structurally similar model compound can be used to perform the additional validation experiments. The additional validation experiments could lead to the calculation of a response factor which would change the result. In this case, only the result calculated with the response factor is reported.

13.5.5.3 *Unknown leachables*

Additional experiments need to be performed to confirm the identity of all unknown leachables. The experimental approach, described below, is based on whether or not the leachable was observed in the extraction study. Once the identification is confirmed, the additional method validation experiments as described above are followed.

13.5.5.3.1 Leachables observed in the extraction study

If an unknown leachable was observed as an extractable, the extraction conditions where the leachable was observed should be used as the starting point of generating sufficient amounts of the leachable to allow for structural elucidation. Sample extraction followed by chromatographic fractionation may be required to collect a sufficient quantity of material at an acceptable purity. All relevant analytical techniques should be employed, including but not limited to Fourier transform infrared spectroscopy (FTIR), MS, and nuclear magnetic resonance spectroscopy (NMR).

13.5.5.3.2 Leachables observed only in the leachable study

If an unknown leachable was not observed as an extractable, consider the possible reasons why the leachable was not observed as an extractable when designing experiments to identify the unknown leachable. The first possible reason would be that in the extraction study a large number of extractables were generated and this led to the leachable not being observed. In this case, the extraction results should be reexamined to confirm that the leachable was not observed. The second possibility would be that the leachable is actually a reaction product of the drug (or some other drug product component) and one of the observed extractables. The data on the unknown leachable should be evaluated for this possibility, and experiments should be done to prove the hypothesis that the leachable is in fact a reaction product. The third possible reason would be that the leachable was not detected by the analytical methods used in the extraction studies. In this case, additional sample extracts may be analyzed by methods similar to the leachable methods to determine if the unknown leachable could be detected as an extractable. A fourth possibility would be that the drug product was not well modeled by the extraction solvents and thus the observed leachable would not have been an extractable. This unfortunate scenario, although rare, is most likely to happen in biologics where a protein may have a specific affinity for a leachable. In this case, additional extraction experiments may yield the leachable as an extractable.

13.5.5.4 *Unidentifiable unknown leachables*

If an unknown leachable detected above the AET_{dp} cannot be identified, the project team may have no choice but to assume the default SCT is correct.

13.5.6 Toxicological review

At the end of the leachable study, the observed leachables should be submitted for toxicological review. At this point, the SCT for each observed leachable should be determined by a toxicologist and the default SCT no longer applied. If sufficient data are not available in the literature to evaluate the SCT for a given leachable, studies to assesses the toxicity of the leachable are strongly encouraged. Once the SCT values have been determined for each leachable, AET_{dp} values specific to each leachable are calculated.

13.6 IMPACT OF LEACHABLES IN THE DRUG PRODUCT ABOVE THE AET

If a leachable is observed in the drug product above the AET_{dp} calculated for that specific leachable, the project team must take actions to prevent patients from being exposed to this level of the leachable. The options are shortening the shelf life to insure the levels of leachable do not exceed the AET_{dp}, or selection of a different SCC. If a different SCC is selected, the entire extractable and leachable testing must be repeated.

If no leachables are observed in the drug product above the AET_{dp}, the project team can set the shelf life and storage conditions of the drug product in the SCC based solely on drug product stability. Thankfully, this is generally the case!

13.6.1 Material qualification

If a leachable is of particular concern to a project team, methods should be developed to evaluate each lot of SCC before being filled. These methods will need to include an extraction followed by an analytical method. Depending upon the material and the leachable of concern, the analytical method can be as simple as FTIR or total organic carbon or the methods can be as complex as a sample preparation involving sample concentration followed by a highly sensitive chromatographic method.

The extraction will need to be validated in addition to the analytical method. At a minimum, the precision of the extraction ($n = 6$) for three different lots of material should be evaluated. If a representative material is available with a known quantity of the leachable, this representative material should be used to evaluate if the extraction is exhaustive. Representative materials with known quantities of leachables are rare.

Simple analytical methods need to be validated as appropriate for the technique. Chromatographic techniques should be validated in the extraction solvent as was previously discussed for the leachable methods in drug product.

13.7 COMBINATION MEDICAL DEVICES

For many therapies, the distinction between a medical device and an SCC can become blurred (e.g. an insulin pump with prefilled cartridges). If a component of a combination medical device in contact with the drug is the primary storage vessel for the drug product and if the component is not intended to have direct tissue contact, this component should be evaluated as an SCC. A detailed discussion of extractables and leachables from other types of combination medical devices is beyond the scope of this chapter; however, some general considerations when evaluating combination medical devices are based on the following two questions:

Does the component of the combination medical device have direct tissue contact with the patient?

Does the component of the combination medical device have direct contact with the drug product?

If the answer to the first question is yes, the extractables and leachables evaluation studies should be based on the experiments described in ISO 10,993-12.[5] If the answer to the second question is yes, the extractable and leachable evaluation studies should be based on the experiments described previously for an SCC. If the answer to both questions is yes, a risk assessment needs to be made to determine which route is the most likely for a leachable to enter the patient, and the extractable and leachable evaluation studies designed accordingly. If the answer to both questions is no, extractable and leachable studies are probably not necessary unless the component directly contacts a second component for which the answer to one of the questions is yes. In this case, a risk assessment should be made to determine if this component should be included in the extractable and leachable studies.

13.8 CONCLUSIONS

Leachables present a unique challenge in ensuring drug product safety and efficacy. The experimental approach discussed in this chapter represents a rational experimental approach to evaluate this risk. When a project team designs experiments based on this approach, the more information the team gathers on the material composing the SCC and on the drug product, the more effective and efficient the experimental strategy will be.

Acknowledgments

The authors would like to thank Michael Ruberto of Material Needs Consulting for his contributions to this chapter. The authors would also like to thank Christopher Jensen and Cassandra Tellarini of NSF Pharmalytica for their assistance.

References

1. Safety Thresholds and Best Practices for Extractables and Leachables in Orally Inhaled and Nasal Drug Products. PQRI Leachables and Extractables Working Group, September 9, 2006. Available at http://www.pqri.org/pdfs/LE-Recommendations-to-FDA-09-29-06.pdf.
2. *Guidance for Industry: Inhalation Drug Products Packaged in Semipermeable Container Closure Systems;* Food and Drug Administration Center for Drug Evaluation and Research, July 2002.
3. Other Common MS Libraries Include but are Not Limited to Chemspider (Available at http://www.chemspider.com) and the Wiley Registry (Available at http://onlinelibrary.wiley.com).
4. International Conference on Harmonisation of Technical Requirements for Registration of Pharmaceuticals for Human Use, Ich Harmonised Tripartite Guideline, Validation of Analytical Procedures: Text and Methodology, Q2(R1), November 2005.
5. Biological Evaluation of Medical Devices—Part 12: Sample Preparation and Reference Materials. ISO 10993–12, Reference Number ISO 10993-12:2007(E), Corrected Version 2008-02-15.

PART

5

Pharmacopeial Methods

CHAPTER

Pharmacopeial methods and tests 14

Eric B. Sheinin
Sheinin & Associates LLC, North Potomac, MD, USA

CHAPTER OUTLINE

Specification of Drug Substances and Products. http://dx.doi.org/10.1016/B978-0-08-098350-9.00015-1

14.1 INTRODUCTION

The International Conference on Harmonization of Technical Requirements for Registration of Pharmaceuticals for Human Use (ICH) Q6A guideline includes a discussion of pharmacopeial tests and acceptance criteria in chapter 2.8.[1] The importance of these tests and acceptance criteria is indicated by the statement, "Wherever they are appropriate, pharmacopeial procedures should be utilized." While ICH quality guidelines are intended for new drug substances and new drug products not previously registered in one of the ICH regions (Europe, Japan, and United States), the pharmacopeial general chapter requirements generally are extended to generic drugs, where appropriate. The three compendia in these regions are the European Pharmacopoeia (EP), the Japanese Pharmacopoeia (JP), and the United States Pharmacopeia (USP). These compendia are considered the three major pharmacopeias of the world by most of the pharmaceutical community.

While emphasis in this chapter will be placed on the USP, the EP and the JP will be discussed where applicable. Among the topics to be discussed are the legal status of the pharmacopeias, requirements for inclusion of a monograph, the process for inclusion, types of tests included in monographs, compendial reference standards, and pharmacopeial harmonization.

14.2 LEGAL STATUS

14.2.1 United States Pharmacopeia

Following a meeting of 11 physicians on January 1, 1820, to establish a pharmacopeia for the United States, the first edition of the *Pharmacopoeia of the United States of America* was published on December 15, 1820, as a compendium of recipes of the best and most fully established medicines used in the United States at that time.[2] Thus, the USP predates the U.S. Food and Drug Administration (FDA). USP's standards were recognized in the 1848 Drug Import Act enacted to stop the dumping of inferior drugs from Europe. Initially, the USP was published at 10-year intervals until 1942, when publication was changed to five-year intervals. In 2002, the schedule was changed to annual editions. The American Pharmaceutical Association published the first National Formulary (NF) as *The National Formulary of Unofficial Preparations.*[2] The United States Pharmacopeial Convention, Inc. (USPC) acquired the NF in 1975 and began publishing the two compendia in a single volume. Today, the USP–NF is published annually in a three-volume set, with two annual supplements, by the USPC. The USPC was incorporated in the District of Columbia in July, 1900.[2]

The United States Pharmacopeial Convention is a private, nongovernmental, nonprofit corporation that establishes standards, included in the USP–NF, for pharmaceuticals and other articles legally marketed in the United States. For example, the USP has monographs for purified cotton and purified rayon as well as monographs for dietary supplements. In addition, the USPC publishes the USP Dietary Supplements Compendium as a two-volume set. Its standards are enforced by the FDA. The legal standing of the USP and the NF arises through the Federal Food, Drug, and Cosmetic Act (The Act) of 1906 and its amendments. "*The term "drug" means (A) articles recognized in the official United States Pharmacopoeia, official Homeopathic Pharmacopoeia of the United States, or official National Formulary, or any supplement to any of them,....or (D) articles intended for use as a component of any article specified in (A), (B), or (C).*"[3] Thus, drug substances, drug products, and excipients fall within the definition of a "drug". "*The term* "official compendium" *means the official United States Pharmacopoeia, official Homeopathic Pharmacopoeia of the United States, or official National Formulary or any supplement to any of them.*"[4] The Act also states that a "*drug or device shall be deemed to be adulterated if it purports to be or is represented in an official compendium, and its strength differs from, or its quality or purity falls below, the standard set forth in such compendium. Such determination as to strength, quality, or purity shall be made in accordance with the tests or methods of assay set forth in such compendium.*"[5] According to The Act, a drug defined in the USP or NF must meet the monograph criteria when tested according to monograph procedures. The applicable USP or NF standard applies to any article marketed in the United States that: (1) is recognized in the compendium, and (2) is intended or labeled for use as a drug or as an ingredient in a drug. The applicable standard applies to such articles whether or not the added designation "USP" or "NF" is used.[6]

14.2.2 European Pharmacopoeia

The EP was established by the Convention for the Elaboration of an EP in 1964. The first edition was published over a period of about 13 years in a series of volumes and supplements, the last of which appeared in 1977.[7] According to the EP section of the Council of Europe "*36 member states and the European Union are signatories to the Commission. Eight European countries, 16 non-E. countries, and the World Health Organization are observers to the Convention. Member states are able to participate in sessions of the EP Commission and to vote on technical matters. Observers are able to participate, but not vote, in these sessions.*" The European Directorate for the Quality of Medicines (EDQM) was created in 1996[8] and now is responsible for the preparation of the EP. The EP is updated every three years with eight supplements between new editions—two in the first year and three in each of the next two years. The 8th edition will be implemented January 1, 2014.

14.2.3 Japanese Pharmacopoeia

The JP was first published in June, 1886, and was implemented in July, 1887. The 16th edition of the JP was implemented on April 1, 2011, pursuant to Paragraph 1, Article 41-1 of the Pharmaceutical Affairs Act (PAL) (Law No. 145, 1960).[9] "*The intent of this article is to standardize and control the properties and quality of drugs, the Minister shall establish and publish the JP, after hearing the opinion of the Pharmaceutical Affairs and Food Sanitation Council (PAFSC).*"[10] The JP is published at least every 10 years according to Article 41-2 of the PAL.[9]Currently the JP is prepared by the JP Secretariat in the Pharmaceuticals and Medical Devices Agency of the Ministry of Health, Labor, and Welfare (MHLW) and is subject to regular revision every five years with partial revision as necessary.

14.3 REQUIREMENTS FOR INCLUSION

14.3.1 United States Pharmacopeia

In order for the U. S. Centers for Medicare and Medicaid Services to provide reimbursement for certain drugs and biologics provided in a physician's office, these drugs and biologics must be included or approved for inclusion in the USP or NF, or be approved by a hospital's pharmacy and drug therapeutics committee.[i,11] In order to be deemed "approved for inclusion" the manufacturer must submit information, including proposed standards and analytical procedures, sufficient to initiate development of a monograph. The USP provides guidance for the information that should be included in this submission in its *Guideline for Submitting Requests for Revision to the* USP–NF.[12] The manufacturer also must agree in writing to address concerns raised by either USP scientific staff, or the appropriate expert committee assigned to evaluate the submission, within six months after the date the proposed monograph was approved for inclusion. The manufacturer also must agree in writing to provide reference standard material no later than the time the monograph is ready for publication in the Pharmacopeial Forum (PF) so that the necessary reference standard(s) is (are) available at the time the monograph becomes official.

PF is USP's journal of standards development and official compendia revision. It is published bimonthly and provides interested parties an opportunity to review and comment on the new or revised standards of the USP–NF. PF is available as a free, online publication. Subscribing information is available on the USP website, www.USP.org.

Monographs are included in the USP–NF only for articles legally marketed in the United States. However, the pending monographs may be developed and published on the USP website for articles that have not been approved by the FDA, provided the monograph sponsor is seeking, or plans to seek, FDA approval. The submission guideline[12] includes templates to assist sponsors in assembling the necessary documentation to support the proposed monograph. All analytical procedures must either be fully validated if they are not compendial procedures, or verified if they are compendial procedures. General Chapters <1225> *Validation of Compendial Procedures* and <1226> *Verification of Compendial Procedures* provide guidance in this regard.

14.3.2 European Pharmacopoeia

The EDQM has a guide for authors of monographs.[14] The guide also serves as a means of communicating the principles for the elaboration of monographs to the users of the EP. In addition, it may serve as a guideline in the elaboration of specifications intended for inclusion in the licensing applications. The EP monographs are mandatory standards, and must be applicable in licensing the procedures in all member states of the Convention on the Elaboration of an EP.

[i]Under the Social Security Act (Act), the Medicare program provides reimbursement for certain medical and other health services, including drugs and biologicals that cannot be self-administered and are provided in a physician's office as part of a physician's professional service. The Act defines the term "drugs and biologicals" to include those drugs and biologicals that are included or approved for inclusion in the United States Pharmacopeia (USP), the National Formulary, or those approved by a hospital's pharmacy and drug therapeutics committee.[13]

The methods to be included in an EP monograph should be methods already included in the EP. These methods may be included in the general methods as well as in published monographs for similar materials, provided the methods are adequate for their specific purposes. For example, if the monograph includes a test for the pH, the method to be used should be described in General Method 2.2.3 *Potentiometric Determination of pH*. Monograph methods must be validated as described in the section on analytical validation.[15] The test procedures should be verified in two or more laboratories.

14.3.3 Japanese Pharmacopoeia

No similar guide was located with regard to requirements for inclusion in the JP. However, the preface to JP 16 indicates that JP articles should cover drugs that are important from the viewpoint of healthcare and medical treatment, clinical results and frequency of use, as soon as possible after they reach the market.[16]

14.4 PROCESS FOR REVISION

The three pharmacopeias follow similar processes for revision and updating their contents, and all three rely on input from experts and the industry.

14.4.1 United States Pharmacopeia

The USP publishes all proposed revisions for public comment in its bimonthly journal PF. Revisions are proposals for either new or revised standards of existing content. As shown in Fig. 14.1, the process begins with the submission of a proposal for revision. Guidance for sponsors is found in the USP *Guideline for Submitting Request for Revision to the USP–NF.*[12] The guideline includes general information for all submissions as well as specific information related to small-molecule drug substances and drug products, excipients, biologics and biotechnology drug substances and drug products, vaccines, and blood, plasma, and cellular blood components. Templates are provided to assist sponsors with compilation of submissions. Proposals can be submitted by any interested party, e.g. companies or individuals with an interest in having public standards for articles legally marketed in the United States.

The revision request is assigned to one of the USP's scientific liaisons for review. The liaison will contact the sponsor and request additional information or clarification if the information provided is incomplete or the liaison has any questions. Once the request is complete, the liaison forwards the request for publication in the PF. Interested parties have 90 days to comment on the proposed revision. Any comments received by USP are collated by the scientific liaison, and forwarded to one of USP's expert committees. The USP expert committees are responsible for developing and revising standards for medicines and foods that appear in the USP and NF. They develop and review monographs, general chapters, and general test methods. Revisions are adopted by majority vote of the expert committee members.[17] Expert committee members are elected based on their experience and background, and come from the pharmaceutical industry, academia, and governments (both domestic and foreign). Members are volunteers and do not represent their companies, universities, or governments.

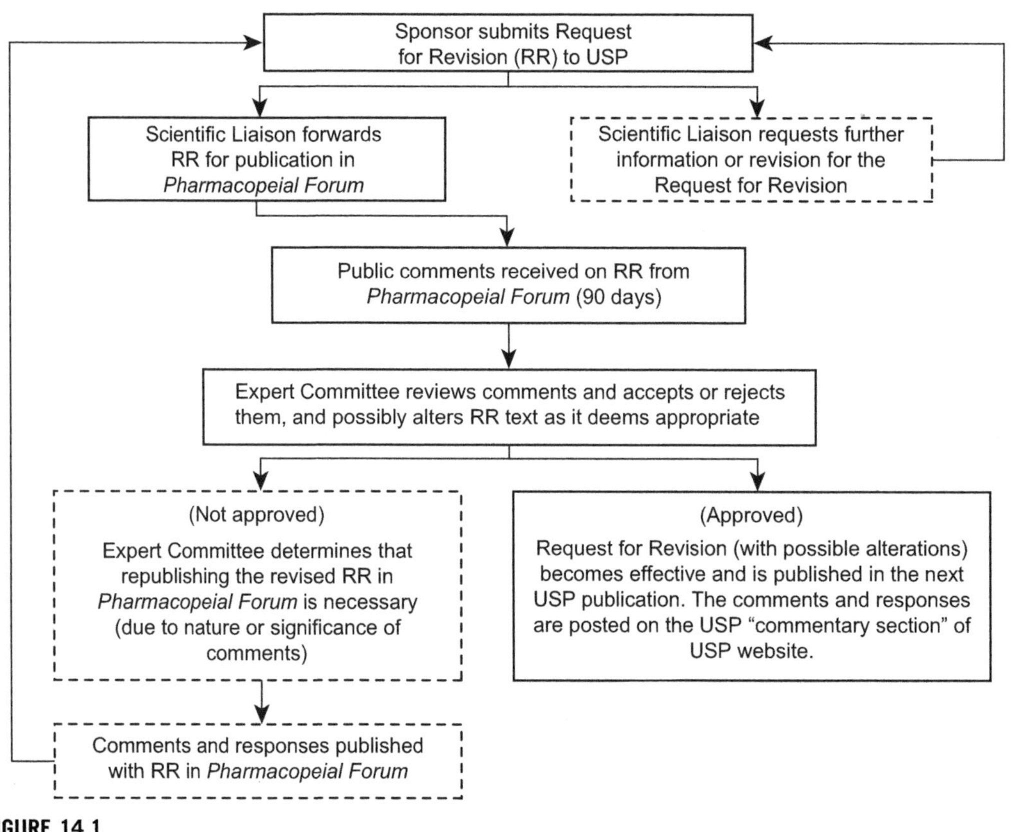

FIGURE 14.1

USP request for revision process.

Source: Copyright 2012 United States pharmacopeial convention. Used by permission.

The assigned expert committee reviews and evaluates the comments that were received. If no comments are received or the comments received are deemed minor, the expert committee members vote to approve the revision (with possible alterations) for official status. Approval may occur through publication in the next edition of the USP–NF or one of the two annual supplements, through an *Interim Revision Announcement* (*IRA*) in the PF, or through a revision bulletin that is posted on the USP website. *Interim Revision Announcements* are published in PF for the 90-day comment period. If there are no significant comments, the *IRA* becomes official in the "Official Text" section of the USP website, with the official date indicated.[18] Revision bulletins are used when circumstances require rapid publication of the official text. *Revision Bulletins* are posted on the USP website with the official date indicated.[18] Regardless of the route, the responses and comments are always posted on the USP website. If the expert committee determines that the comments are significant, the request for revision is republished for public comment in the PF. The comments and responses are then included with the revision proposal in the PF.

14.4.2 European Pharmacopoeia

The work program of monographs in the EP for active substances, excipients, and certain classes of medicinal products that are approved for use in member states, is decided by the European Pharmacopoeial Commission. Each monograph or general chapter is allocated to a group of experts or working party. The proposal is circulated by the Secretariat to the public via industry associations, manufacturers' pharmacopeial liaisons, the EDQM website, and *Pharmeuropa. Pharmeuropa* is the EDQM's mechanism for informing users about proposed revisions to the EP. The Secretariat contacts manufacturers/suppliers of the substance, asking them to supply current production batches and small amounts of the known impurities, in-house specifications, analytical procedures, method validation data, and stability data. If possible, a batch that can be used as a reference substance is desired.[19]

After receipt of any samples and documentation, the Secretariat sends the material to the coordinator, who agrees to a target date for completion of laboratory work to be performed by the regulatory authorities, preferably in six months or less, and initiates the work. After the laboratory work is completed, comments are sent to the coordinator who keeps the Secretariat updated on the progress. The first draft is produced by the coordinator, ideally within three months of completion of the labwork. The draft is presented to the Secretariat. The draft monograph and a report of the work completed are presented to the assigned group of experts. The experts are selected by their individual member states.[19]

If the group of experts feels additional work is required, it is performed by the coordinator of the EP laboratory. In general, the draft is approved in no more than two meetings. Once the group of experts has approved the draft, the monograph is published in *Pharmeuropa* for consultation. *Pharmeuropa* is similar to PF. Comments are due within three months and are provided to the coordinator who reviews them, tries to resolve any difficulties, and carries out any necessary labwork. Comments are considered by the group of experts. The group of experts submits a text for adoption by the Commission while proposing further work on unresolved matters. The Secretariat prepares the document for the Commission and submits it for adoption by the next session.[19]

14.4.3 Japanese Pharmacopoeia

The process for revising or updating the JP begins with the Pharmaceutical Affairs and Food Sanitation Council of MHLW who establishes the basic policy for the update, and provides a list to PMDA for consideration. The PMDA seeks public comment through publication on its website. The PMDA reports back to the MHLW council, and the candidate drug list is finalized and sent back to PMDA. The PMDA seeks draft monographs from the industry and prepares draft monographs based on companies' input as well as PMDA's drafts. The list is sent to the appropriate JP draft committees who seek input from the industry. The committees then prepare the final drafts. Public comment is again requested from the industry. If necessary, the final monographs are revised and sent to the PAFSC for preparation of an announcement of the next edition of JP. The Council then prepares the English edition of JP.[20]

14.5 TYPES OF TESTS

The ICH guideline on specifications (Q6A) provides guidance on tests that should be included in applications submitted to the three ICH regions' regulatory authorities.[21] These same tests should be included in the compendial monographs where appropriate. As indicated in the Introduction, use of pharmacopeial tests and procedures are recommended where appropriate. Q6A divides these tests into Universal and Specific, with the former essentially applying to all drug substances and drug products, and the use of specific tests depending on the intended use of the drug substance and the type of dosage form (see Chapter 2 for more details).

14.5.1 United States Pharmacopeia

14.5.1.1 General information

Monographs include information on packaging and storage as well any labeling requirements. An article purported to be a drug, the name of which is recognized in an official compendium, may be considered adulterated and misbranded unless it is packaged and labeled as prescribed therein.[22] Articles shall be stored and distributed as indicated in an individual monograph unless a different storage temperature is appropriate based on the stability studies of that particular formulation.[23] In most instances, the monograph is in agreement with the approved New Drug Applications (NDA)/ Abbreviated New Drug Applications (ANDA) specification. The FDA comments on PF proposals are considered along with all other comments when decisions are made by the expert committees. The FDA cannot provide information in their comments that is confidential; i.e. specific information is not communicated to the USP. However, if a PF proposal for a new drug substance monograph includes an impurity procedure that is not included in any approved application, neither NDA nor ANDA, the FDA could comment that the proposed impurity procedure is not used in any approved application without revealing any confidential information.

Legally, all information submitted to the FDA in support of a marketing application is highly confidential, and cannot be divulged outside of the Agency. These applications include Investigational New Drug Applications (IND), NDAs, ANDAs, DMFs, and any amendments or supplements to any of these. DMFs provide a mechanism for suppliers and contractors to an applicant to provide confidential information to the FDA without revealing this information to the applicant. For example, the applicant may purchase the drug substance from an outside source and the source then provides all of the chemistry, manufacturing, and controls information for that drug substance to a DMF.

14.5.1.2 Universal tests

14.5.1.2.1 Description

The description is not included in the monograph itself but appears in a reference table—Description and Solubility. This table is not intended to replace, nor should it be interpreted as replacing, the definitive requirements stated in the individual monographs. Table 14.1 presents the USP definition of solubility terms.[24]

14.5.1.2.2 Identification

Generally two identification tests are included in drug substance monographs, especially where the assay is a chromatographic procedure, most often high-performance liquid chromatography (HPLC). Identification should be specific and unequivocal; hence, a single chromatographic procedure may not be sufficient.[25] Infrared spectroscopy is the most commonly included identification test, combined with a comparison of the chromatographic retention times for the sample and standard solutions prepared for the assay. UV is used less frequently and spectra for samples and standards must be obtained concomitantly.[26]

Table 14.1 USP Solubility Definitions

Descriptive Term	Parts of Solvent Required for 1 Part of Solute
Very soluble	<1
Freely soluble	From 1 to 10
Soluble	From 10 to 30
Sparingly soluble	From 30 to 100
Slightly soluble	From 100 to 1000
Very slightly soluble	From 1000 to 10,000
Practically insoluble or insoluble	>10,000

14.5.1.2.3 Assay

Both ICH Q6A[27] and the USP monograph submission guideline[12] indicate that the assay should be specific and stability indicating. Both allow for scenarios where a nonspecific assay is justified; supporting analytical procedures should be provided demonstrating overall specificity. Current experience indicates that, while FDA was a signatory to the ICH quality guidelines, it is not likely to accept a nonspecific assay for either the drug substance or the drug product. In actual practice, the vast majority of new and revised monograph assay procedures utilize HPLC.

14.5.1.2.4 Impurities

USP has been moving to the Q6A[28] nomenclature of inorganic and organic impurities as well as the ICH Q3A[29] nomenclature of specified and unspecified impurities in the drug substance and Q3B[30] nomenclature of specified degradation products and unspecified degradation products in the drug product (see Chapter 6). The general chapter on residual solvents[31] is based on ICH Q3C[32] (see Chapter 7). Testing for residual solvents is not included in most USP and NF monographs unless the procedure or the acceptance criteria differ from those in the general chapter. All drug substances and products are subject to relevant control of solvents likely to be present in a drug substance or product.[31] Further, the general notices state, "*All USP and NF articles are subject to relevant control of residual solvents, even when no test is specified in the individual monograph.*"[33]

Older USP monographs generally do not provide the names of organic impurities. For example, the monograph for *Ibuprofen* includes a test for the limit of ibuprofen compound C but the chemical name is not included.[34] Newer monographs as well as monographs that have been revised in a redesigned style generally include a table of impurities with their common names as well as footnotes with their chemical names, e.g. allopurinol, which lists 6 impurities along with their chemical names.[35]

Inorganic impurities have traditionally been monitored using the General Chapter <231>, Heavy Metals.[36] Based on comments from the pharmaceutical community, USP has created two new general chapters, *Elemental Impurities—Limits* <232>[37] and *Elemental Impurities—Procedures* <233>[38] (see Chapter 8). These new general chapters were scheduled to become official on December 1, 2012, in the 2nd *Supplement* to *USP 35—NF30*. However, USP has posted a *Revision Bulletin* on its website announcing the postponement of the official date for these new chapters. "*This is to allow adequate time for the Executive Committee of the Council of Experts to adjudicate and render a decision on three appeals related to the two general chapters.*"[39]

14.5.1.2.5 Flexible monographs

Drug substances used in generic drug products tend to be manufactured by synthetic routes that may differ from the routes used for drug substances developed by the innovator. This often creates a situation where the test for organic impurities in a USP drug substance monograph is not suitable for monitoring and controlling the organic impurities that arise from these new synthetic routes. In order to prevent the exclusion of any approved drug products from the U.S. market due to noncompliance with the associated USP monograph, the USP created the flexible monograph approach. This approach has been used for many years for drug release testing of extended release drug products and is best exemplified by the monograph for *Theophylline Extended-Release Capsules*, which includes 10 different dissolution tests each with its own conditions and acceptance criteria.[40] This approach was necessary to accommodate differing release mechanisms for approved drug products.

Applying this concept to differing synthetic routes to a drug substance, each with its own impurity profile, results in more than one organic impurities test in a given monograph. If Test 1 is used, no addition to the drug substance labeling is required. However, if a test other than Test 1 is followed, this must be indicated in the labeling. For example, the drug substance monograph for *Loratadine* contains two different tests, Test 1 and Test 2. There is a note under *Related Compounds* that reads "*On the basis of the synthetic route, perform either Test 1 or Test 2. Test 2 is recommended if 4,8-dichloro-6,11-dihydro-5H-benzo[5,6]cyclohepta[1,2-b]pyridin-11-one is a potential related compound.*"[41] Other similar examples include monographs for *Meloxicam*,[42] *Paroxetine*[43] and *Propofol*[44] among others.

14.5.1.2.6 Uniformity of dosage units

While uniformity of dosage units is not included in the list of universal tests, it is included in most drug product monographs. For all but a small minority of drug products, this test is included only as a release test and is not included in stability protocols.

14.5.1.3 Specific tests

Individual tests and acceptance criteria should be included in the specification when the tests have an impact on the quality of the drug substance and drug product for batch control.[45] Specific tests are included in the USP monographs to describe and control a drug substance or drug product when they cannot be described adequately using the four universal tests described in the Guideline.[12] The use of optional tests will require strong rationale, adequate procedures, and full validation, as described in General Chapter <1225>, *Validation of Compendial Procedures*. Examples of drug substance optional tests are loss on drying, water determination, pH, optical rotation, refractive index, polymorphic considerations, particle size, melting range, and microbial limits.[12] Examples of drug product specific tests include dissolution/drug release, pH, water determination, microbial limits, antimicrobial preservative content, antioxidant preservative content, extractables and leachables, reconstitution time, sterility, endotoxins/pyrogens, particulate matter, and osmolarity, as well as others depending on the nature of the drug product.[45]

14.5.2 European Pharmacopoeia

With some exceptions, such as radiopharmaceutical drug products, the EP includes monographs for drug substances only.

14.5.2.1 Universal tests

Monographs in the EP include a test for "Characters", which includes the description. For example, the characters test in the EP monograph for *chloramphenicol* reads, "*A white grayish-white or yellowish-white, fine, crystalline powder or fine crystals, needles, or elongated plates, slightly soluble in alcohol and in propylene glycol, slightly soluble in ether. A solution in ethanol is dextrorotatory and a solution in ethyl acetate is laevorotatory.*"[46]

14.5.2.2 Specific tests

Specific tests in EP monographs are similar to those found in the USP.

14.5.3 Japanese Pharmacopoeia

14.5.3.1 Universal tests

Monographs in the JP include the description in the body of the monograph. Universal and specific tests are similar to those found in the USP and the EP. For example, the description test in the JP monograph for *Cloxazolam* reads "*Cloxazolam occurs as white crystals or crystalline powder. It is odorless and tasteless. It is freely soluble in acetic acid (100), sparingly soluble in dichloromethane, slightly soluble in ethanol (99.5), and in diethyl ether, very slightly soluble in ethanol (95), and practically insoluble in water. It dissolves in dilute hydrochloric acid. It is gradually colored by light. Melting point about 200 °C (with decomposition).*"[47] Table 14.2 presents the JP definition of solubility terms.[48] Section G1 of the General Information section of the JP includes a discussion of residual solvents[49] and references the *Guideline for Residual Solvents in Pharmaceuticals*,[50] which contains the acceptance criteria. While this document could not be located, it appears to be the original, unrevised ICH Q3C guideline.[32]

Table 14.2 JP Solubility Definitions

Descriptive Term	Volume of Solvent Required for Dissolving 1 g or 1 mL of Solute
Very soluble	Less than 1 mL
Freely soluble	From 1 mL to less than 10 mL
Soluble	From 10 mL to less than 30 mL
Sparingly soluble	From 30 mL to less than 100 mL
Slightly soluble	From 100 mL to less than 1000 mL
Very slightly soluble	From 1000 mL to less than 10,000 mL
Practically insoluble or insoluble	10,000 mL and over

14.5.3.2 Specific tests

Specific tests in JP monographs are similar to those found in the USP and the EP.

14.6 COMPENDIAL REFERENCE STANDARDS

Compendial reference standards are highly characterized materials, which are used for comparison to compendial articles in identification tests, for assay and impurity tests, and for use in performing other compendial tests. Candidate materials for compendial reference standards have been voluntarily donated by industry as part of the monograph development process, as well as following the time when the respective monographs became official. Replacement lots of these reference standards are needed as current lots are depleted or expire.

14.6.1 United States Pharmacopeia

The USP has created a guideline for parties interested in donating candidate reference standard materials.[51] In instances where suitable donors of candidate materials cannot be located, USPC has turned to contract manufacturers as a source of these materials. Regardless of the source, all candidate materials are subjected to a collaborative study. Participants in the collaborative study include USP laboratories, donor companies, FDA and other regulatory authority laboratories, and other industry and contract laboratories as appropriate. Not all laboratories participate in every study. Results of the studies are evaluated by USP reference standard scientists and by the appropriate expert committee.

For certain compendial articles, chemical reference standards are not useful. For example, Graftskin, which is a living, bilayered skin substitute derived from human fetal foreskins.[52]

Authentic Visual References (AVRs) are USPC Reference Standards, but unlike chemical reference materials, AVRs are not used in chemical analyses. Instead, AVRs are visual images used by analysts to compare certain test articles to ensure that they meet compendial requirements (See Chapter 5). An AVR is incorporated by reference into the monograph[53].

The USP also provides certified reference materials (CRMs), which are Reference Standards that provide certified property values with associated uncertainties and metrological traceability, in accordance with International Organization for Standards (ISO) Guides 30–35. Correct use of these CRMs support traceability of results to SI units and comparability of procedures.[52]

14.6.2 European Pharmacopoeia

In addition to chemical reference substances, the EP provides reference spectra for use in tests and assays described in the compendium. Candidate materials are characterized in the EDQM laboratory using methods found in the EP and elsewhere. Certain reference materials may be subjected to an international collaborative study.[54] Participants include experts of the EP, academia, industry, and Official Medicine Control Laboratories. The EDQM also provides CRMs characterized by validated procedures for one or more specified properties accompanied, by a certificate that states the value of the specified property, and a statement of metrological traceability.[55]

14.6.3 Japanese Pharmacopoeia

Reference standards in the JP are produced and distributed by the Pharmaceutical and Medical Device Regulatory Science Society of Japan (PMRJ). A list of frequently asked questions can be found on the PMRJ website. PMRJ also distributes the USPC reference standards to domestic users through a distribution agreement with the USPC.[56]

14.7 HARMONIZATION

14.7.1 Pharmacopeial discussion group

The Pharmacopeial Discussion Group (PDG) was formed as an informal body in 1989 in response to industry proposals to harmonize certain monographs and general chapters so as to reduce the testing burden on multinational companies seeking to market drugs in the European Union, Japan, and the United States. The PDG was formed with representatives of the EDQM, the JP, and the USPC. In May 2001, the PDG welcomed the World Health Organization (WHO) as an observer to the PDG. The group generally meets twice a year. The PDG work plan includes excipient monographs and general chapters. The PDG definition of a harmonized monograph reads, "*A pharmacopeial general chapter or other pharmacopeial document is harmonized when a pharmacopeial substance or product tested by the document's harmonized procedure yields the same results and the same accept/reject decision is reached.*"[57]

When using a fully harmonized pharmacopeial monograph or general chapter, an analyst will perform the same procedures and reach the same accept/reject decisions irrespective of which PDG pharmacopeia is referenced. This approach is called interchangeability, and each pharmacopeia will identify, in an appropriate manner, such a monograph or general chapter.

When full harmonization of a pharmacopeial monograph or general chapter is not possible, the PDG works to harmonize it using an approach termed harmonization by attribute. In this approach, some elements of a monograph or general chapter may be harmonized but others may not. When a monograph is harmonized by attribute, a combination of approaches is needed. For nonharmonized elements, reliance on the individual PDG pharmacopeia is necessary.[57]

Harmonization of pharmacopeial documents in the PDG is based upon decisions of the expert bodies of each pharmacopeia. The PDG works transparently in many ways, principally through public notice and comment procedures of each pharmacopeia. The PDG working procedure is described in the USP General Chapter <1196>, Pharmacopeial Harmonization.[57] Each PDG pharmacopeia presents the status of harmonization in their respective compendia.

14.7.2 ICH Q4 expert working group

Consensus by the regulatory authorities in the three PDG regions was considered necessary in order for PDG-harmonized documents to be considered interchangeable. As the PDG traditionally met in conjunction with the ICH Expert Working Group (EWG) meetings and reported on progress to the ICH steering committee, the PDG requested the ICH Steering Committee to form an EWG to consider interchangeability of pharmacopeial documents. The Q4B EWG on pharmacopeial harmonization was established by the ICH Steering Committee in November 2003. The Q4B Guideline "Evaluation and Recommendation of Pharmacopoeial Texts for Use in the ICH Regions" achieved the ICH Steering Committee approval November 1, 2007.[58] It should be noted that the Q4B EWG did not consider interchangeability of any of the PDG harmonized excipient monographs.

Table 14.3 presents the list of PDG harmonized general chapters that have been found to be interchangeable by the Q4B Expert Working Group.[59]

Table 14.3 Interchangeable PDG-Harmonized General Chapters

PDG Harmonized General Chapter	ICH Q4B Guideline Number	Date Finalized by Q4B Expert Working Group
Residue on ignition/ sulphated ash	Q4B Annex 1R1	November 2007
Test for extractable volume of parenteral preparations	Q4B Annex 2R1	September 2010
Test for particulate contamination: sub-visible particles	Q4B Annex 3R1	September 2010
Microbiological examination of non-sterile products: microbial enumeration test	Q4B Annex 4AR1	September 2010
Microbiological examination of nonsterile products: test for specified micro-organisms	Q4B Annex 4BR1	September 2010
Microbiological examination of nonsterile products: acceptance criteria for pharmaceutical preparations and substances for pharmaceutical use	Q4B Annex 4CR1	September 2010
Disintegration test	Q4B Annex 5R1	September 2010
Uniformity of dosage units	Q4B Annex 6R1	September 2010
Dissolution	Q4B Annex 7R2	November 2010
Sterility test	Q4B Annex 8R1	September 2010
Tablet friability	Q4B Annex 9R1	September 2010
Polyacrylamide gel electrophoresis	Q4B Annex 10R1	September 2010
Capillary electrophoresis	Q4B Annex 11	June 2010
Analytical sieving	Q4B Annex 12	June 2010
Bulk density and tapped density of powders	Q4B Annex 13	June 2012
Bacterial endotoxins test	Q4B Annex 14	June 2010

References

1. *Specifications: Test Procedures and Acceptance Criteria for New Drug Substances and Drug Products: Chemical Substances (Q6A)*, The International Conference on Harmonization of Technical Requirements for Registration of Pharmaceuticals for Human Use, 1999; Chapter 2.8.
2. *USP 35-NF 30, Mission and Preface;* United States Pharmacopeial Convention, Inc, 2012.
3. Federal Food, Drug, and Cosmetic Act, section 201(g)(1) (§321).
4. Federal Food, Drug, and Cosmetic Act, section 201(j) (§321).
5. Federal Food, Drug, and Cosmetic Act, section 501(b) (§501).
6. USP35-NF 30, 1st Supplement, General Notices, 3.10.10, Applicability of Standards to Drug Products, Drug Substances, and Excipients.
7. EP 6.1, Preface, European Pharmacopoeial Convention
8. European Directorate for the Quality of Medicines, EDQM website, http://www.edqm.eu/en/EDQM-history-93.html.
9. *JP 16, Introductory Statement;* Pharmaceuticals and Medical Devices Agency: Japan, April 1, 2011.
10. Pharmaceutical Administration and Regulations in Japan, March 2010, JPMA website, http://www.jpma.or.jp/english/parj/1003.html.
11. *USP Guideline on Drugs Approved for Inclusion*, USP Website http://www.usp.org/usp-nf/development-process/policies-guidelines/usp-guideline-drugs-approved-inclusion.
12. *USP Guideline for Submitting Requests for Revision to the* USP-NF, May 27, 2011, USP website, http://www.usp.org/usp-nf/development-process/submit-new-monographs/submission-guidelines.
13. 42 United States Code Sections 1395k, 1395x.
14. EDQM *Technical Guide for the Elaboration of Monographs.* Section 1, Introduction, 6th ed., 2011.
14. EDQM *Technical Guide for the Elaboration of Monographs.* Section 3, Analytical Validation, 6th ed., 2011.
16. *JP 16, Preface;* Pharmaceuticals and Medical Devices Agency: Japan, April 1, 2011.
17. USP Website, http://www.usp.org/usp-nf/development-process.
18. USP *Guideline on Use of Accelerated Processes for Revisions to the* USP-NF, Version 2.1, August 1, 2011.
19. *Guide for the Work of the European Pharmacopoeia*, European Directorate for the Quality Medicine, PA/PH/SG (11) 54 DEF.
20. *The Pharmaceuticals and Medical Devices Agency Annual Report* FY 2009, 56.
21. *Specifications: Test Procedures and Acceptance Criteria for New Drug Substances and Drug Products: Chemical Substances (Q6A)*, The International Conference on Harmonization of Technical Requirements for Registration of Pharmaceuticals for Human Use, 1999; Chapters 3.2 and 3.3.
22. Federal Food, Drug, and Cosmetic Act, section 502(g) (§352).
23. *USP 35-NF 20*, General Notices, 10.30, Storage Temperature and Humidity, United States Pharmacopeial Convention, Inc. 2012.
24. *USP 35-NF 30, Description and Solubility Table*, United States Pharmacopeial Convention, Inc., 2012.
25. *Specifications: Test Procedures and Acceptance Criteria for New Drug Substances and Drug Products: Chemical Substances (Q6A)*, The International Conference on Harmonization of Technical Requirements for Registration of Pharmaceuticals for Human Use, 1999; Chapter 3.2.1.b.
26. *USP 35–NF30, General Chapter Spectrophotometric Identification Tests* <197>, United States Pharmacopeial Convention, Inc. 2012.
27. *Specifications: Test Procedures and Acceptance Criteria for New Drug Substances and Drug Products: Chemical Substances (Q6A)*, The International Conference on Harmonization of Technical Requirements for Registration of Pharmaceuticals for Human Use, 1999; Chapter 3.2.1.c and 3.2.2.c.

28. *Specifications: Test Procedures and Acceptance Criteria for New Drug Substances and Drug Products: Chemical Substances (Q6A)*, The International Conference on Harmonization of Technical Requirements for Registration of Pharmaceuticals for Human Use, 1999; Chapter 3.2.1.d and 3.2.2.d.
29. *Impurities in New Drug Substances* (Q3A(R2)), The International Conference on Harmonization of Technical Requirements for Registration of Pharmaceuticals for Human Use, 2006; Chapter 6.
30. *Impurities in New Drug Substances* (Q3B(R2)), The International Conference on Harmonization of Technical Requirements for Registration of Pharmaceuticals for Human Use, 2006; Chapter 5.
31. *USP 35-NF 30, General Chapter Residual Solvents* <467>, United States Pharmacopeial Convention, Inc., 2012.
32. *Impurities: Guideline for Residual Solvents* (Q3C(R5)), The International Conference on Harmonization of Technical Requirements for Registration of Pharmaceuticals for Human Use, 2011.
33. *USP 35-NF 30, General Notices 5.6.20 Residual Solvents in* USP *and* NF *Articles*, United States Pharmacopeial Convention, Inc., 2012.
34. *USP 35- NF 30, Ibuprofen monograph*, United States Pharmacopeial Convention, Inc., 2012.
35. *USP 35-NF 30, Allopurinol monograph*, United States Pharmacopeial Convention, Inc., 2012.
36. *USP 35-NF 30, General Chapter Heavy Metals* <231>, United States Pharmacopeial Convention, Inc., 2012
37. *USP 35-NF 30, General Chapter Elemental Impurities – Limits* <232>, United States Pharmacopeial Convention, Inc., 2012.
38. *USP 35-NF 30, General Chapter Elemental Impurities – Procedures* <233>, United States Pharmacopeial Convention, Inc., 2012.
39. USP website, http://www.usp.org/usp-nf/official-text/revision-bulletins/elemental-impurities-limits-and-elemental-impurities-procedures.
40. *USP 35-NF 30, Theophylline Extended-Release Capsules monograph*, United States Pharmacopeial Convention, Inc., 2012.
41. *USP 35-NF 30, Loratadine monograph*, United States Pharmacopeial Convention, Inc. 2012.
42. *USP 35-NF 30, Meloxicam monograph*, United States Pharmacopeial Convention, Inc., 2012.
43. *USP 35-NF 30, Paroxetine monograph*, United States Pharmacopeial Convention, Inc., 2012.
44. *USP 35-NF 30, Propofol monograph*, United States Pharmacopeial Convention, Inc., 2012.
45. *Specifications: Test Procedures and Acceptance Criteria for New Drug Substances and Drug Products: Chemical Substances (Q6A)*, The International Conference on Harmonization of Technical Requirements for Registration of Pharmaceuticals for Human Use, 1999; Chapter 3.3.
46. *EP 4th Edition Chloramphenacol monograph*, European Directorate for the Quality of Medicine, 2002.
47. *JP 16. Cloxazolam monograph*, Pharmaceuticals and Medical Devices Agency, Japan, 2011.
48. *JP 16, General Information section G1 Guideline for Residual Solvents and Models for Residual Solvents Test*, Pharmaceuticals and Medical Devices Agency, Japan, 2011.
49. *JP 16, General Notices 29*, Pharmaceuticals and Medical Devices Agency, Japan, 2011.
50. *Guideline for Residual Solvents in Pharmaceuticals*, PAB/ELD Notification No. 307, March 30, 1998.
51. *USP Guideline for Donors of USP Reference Standard Candidate Materials*, version 2.1, USP Website, August 23, 2012.
52. *USP 35-NF 30, Graftskin monograph*, United States Pharmacopeial Convention, Inc., 2012.
53. *USP 35-NF 30, General Chapter USP Reference Standards* <11>, United States Pharmacopeial Convention, Inc., 2012.
54. EDQM Reference Standards, EDQM Website, http://www.edqm.eu/en/EDQM_Reference_Standards-649.html%3faMotsCles%3da%3A1%3A%7Bs%3A0%3A%22%22%3Ba%3A4%3A%7Bi%3A0%3Bs%3A6%3A%22pharm.%22%3Bi%3A1%3Bs%3A4%3A%22eur.%22%3Bi%3A2%3Bs%3A9%3A%22reference%22%3Bi%3A3%3Bs%3A9%3A%22standards%22%3B%7D%7D.
55. John, H. McB. Miller, WHO QCL Training Seminar, Tanzania, December 5-7, 2007.

56. PMRJ website, Reference Standards, http://www.pmrj.jp/hyojun/html/frm010e.php.
57. *USP 35, General Chapter Pharmacopeial Harmonization* <1196>.
58. *Evaluation and Recommendation of Pharmacopoeial Texts for Use in the ICH Regions*, The International Conference on Harmonization of Technical Requirements for Registration of Pharmaceuticals for Human Use, 2007.
59. The International Conference on Harmonization of Technical Requirements for Registration of Pharmaceuticals for Human Use Website.

PART

Microbial Methods

CHAPTER

15 Microbial methods*

Beth Ann Brescia

BioMonitoring NA, EMD Millipore Corporation, Billerica, MA, USA

CHAPTER OUTLINE

15.1 INTRODUCTION

Compendial microbiological methods, including United States Pharmacopeia General Chapter <61> "Microbiological Examination of Non-sterile Products: Microbial Enumeration Tests",[1] United States Pharmacopeia General Chapter <62> "Microbiological Examination of Non-sterile Products: Tests for Specified Microorganisms",[2] United States Pharmacopeia General Chapter <71> "Sterility Tests"[3] and their respective European Pharmacopoeia (EP) chapters have been in existence for many decades. Only minor changes to these tests have been incorporated over the years. However, more recently new rapid technologies have emerged and some have received the United States Food and Drug Administration's (FDA) and European Medicines Agency's (EMA) approval as alternatives to the traditional compendial tests.

This chapter will provide guidance on a phased approach for microbial testing during the drug development process, review of alternative rapid microbiological methods (RMMs), and validation of the alternative methods.

15.2 MICROBIAL TESTING (PHASED APPROACH)

Several guidance documents exist which provide direction for setting microbiological acceptance criteria for drug substances and drug products. For example, United States Pharmacopeia General Chapter

*The views expressed in this chapter are those of the author and may not represent those of EMD Millipore Corporation.

Specification of Drug Substances and Products. http://dx.doi.org/10.1016/B978-0-08-098350-9.00015-3

<1111> "Microbiological Examination of Nonsterile Products: Acceptance Criteria for Pharmaceutical Preparations and Substances for Pharmaceutical Use"[4] provides guidelines for setting acceptance criteria based upon the different dosage forms and the drug substances used in the dosage forms. The International Conference on Harmonization of Technical Requirements for Registration of Pharmaceuticals for Human Use (ICH) guideline Q6A "Specifications: Test Procedures and Acceptance Criteria for New Drug Substances and New Drug Products: Chemical Substances"[5] has been finalized since 1999 and provides guidance on the applicable microbial tests and factors to be considered when setting specifications for new drug substances or drug products at the time of submission for marketing approval. However, these documents do not provide instructions for testing during the drug development life cycle.

During early development phases of a drug product, many different formulations may be evaluated. Since it may not be feasible to confirm the suitability of the microbial method for each formulation, a more practical approach may be considered. Instead of testing everything from the drug substance to the finished product, a risk assessment should be performed in order to determine if reduced or no testing is acceptable. As several types of dosage forms exist, with each one warranting a slightly different approach, the following phased-approach discussion will be based on solid oral dosage forms.

The first step is to evaluate the raw materials. Those materials included in the USP/National Formulary (NF) have microbial limit test specifications that need to be tested. However, those raw materials that are not included in the USP/NF and are used in the formulation may not require testing. Several factors should be assessed, including the nature of the raw material (chemical origin or natural origin) and the pH of the material (does it fall in a range that may not be conducive to supporting microbial growth or viability?). Based upon the assessment, it is possible that little, if any, inherent bioburden exists in the raw material.

A similar approach can be taken for the drug substance. Is it chemically synthesized? If the drug substance is chemically synthesized then the likelihood of any microorganisms surviving the chemical process is extremely low. If water is used in the manufacturing process, what is the quality of the water used and in which steps of the process? What is the amount of drug substance in a single tablet?

The drug product manufacturing process needs to be reviewed. Are there controls in place such as laminar flow hoods or portable high-efficiency particulate air (HEPA) rooms to reduce environmental contamination? Is the equipment used to formulate the product cleaned and sanitized? Typically at early stages in the development process, tablets are uncoated. This reduces the potential of additional bioburden as tablet coatings are often aqueous based. What is the water activity of the tablet? Reduced water activity aids in the prevention of microbial proliferation. The water activity requirements to support the growth of different microorganisms are well documented in the literature. Table 2 of United States Pharmacopeia General Chapter <1112> "Application of Water Activity Determination to Nonsterile Pharmaceutical Products"[6] provides a microbial limit testing strategy based upon product type, water activity, and greatest potential of contaminants. Most microorganisms will not proliferate below a water activity of 0.60. Compressed tablets have a water activity of approximately 0.36, which would not support any microbial proliferation and therefore may be considered for reduced or no microbial limit testing.

During the later stages of development, when the formulation is locked down, a shift in approach is taken in order to accumulate microbial data on the product. These data should be included in regulatory filings as justification for whether routine, reduced/skip lot, or no microbial limit testing is warranted.

Raw materials not included in the USP/NF and drug substance should be tested and the data evaluated to determine the bioburden level, if any, and whether or not the material contains inherent antimicrobial activity.

In the latter stages of development, tablets are typically coated and printed. In those cases, testing of the drug product should be on the coated, printed tablet. A hold study should, however, be performed on the tablet coating solution to determine the initial bioburden load and any change in bioburden level over time. This is especially important if the coating solution will be held for prolonged periods of time before being applied to the tablets. Furthermore, tablet print ink, which can be either aqueous or alcohol based, should be analyzed, as past experience of this author has shown that mold contamination can be an issue.

15.3 ALTERNATIVE RAPID MICROBIOLOGICAL METHODS

Alternative RMMs have advanced over the last decade. In 2000, the Parenteral Drug Association (PDA) published the first guidance document on how to validate and implement alternative RMMs; Technical Report 33 "Evaluation, Validation and Implementation of New Microbiological Testing Methods".[7] The USP and EP also have guidances on alternative methods; United States Pharmacopeia General Chapter <1223> "Validation of Alternative Microbiological Methods"[8] and EP 5.1.6 "Alternative Methods for Control of Microbiological Quality".[9]

Different alternative technologies are available for the rapid detection of microorganisms. EP 5.1.6 breaks out these RMMs into three categories: growth-based methods, direct measurement, and cell component analysis. Growth-based methods detect a signal after a short incubation period in liquid or on solid media. Examples include detection of CO_2 production by colorimetric methods or a change in headspace pressure and detection of adenosine triphosphate (ATP) by bioluminescence. Direct measurement methods can detect cell viability without requiring the growth of the microorganism. One example of a direct measurement method combines fluorescent labeling and laser scanning cytometry to enumerate organisms. The sample containing microorganisms is filtered onto a membrane and treated with a combination of stains to fluorescently label viable organisms without the need for growth. The membrane is scanned by a laser, fluorescent light is detected and a membrane scan map is produced which captures the position of each fluorescent event, which is then verified by visual examination using an epifluorescent microscope. The third type of RMM is cell component analysis or an indirect measurement; expression of certain cell components correlates to microbial presence. One example is amplification of DNA or RNA by the polymerase chain reaction. RMMs are qualitative (presence/absence) or quantitative (enumeration), destructive or nondestructive, and can be applied to filterable or nonfilterable products.

In 2006, the Center for Drug Evaluation and Research of the FDA (CDER) published a paper on the use of alternative microbiological methods.[10] The authors stated that "New microbiology methods can offer advantages of speed and precision for solving microbiological problems associated with materials or environmental influences. Neither corporate economics nor regulatory attitudes should be a barrier to the use of new testing technologies or different measurement parameters."

15.4 RAPID STERILITY TEST METHODS

Sterility testing has evolved since the 1930s when it was first introduced for testing of liquid products (USP XI). The initial sterility test was a 7-day test using one medium at 37 °C targeted for human pathogens. By the early 1940s, an additional incubation temperature of 22–25 °C was added specifically for yeasts and mold with a 15-day incubation period. The sterility test progressed by the mid-1940s to using a Sabouraud-based medium for 10 days instead of 15 days and fluid thioglycollate medium (FTM) for seven days. In the mid-1960s, the incubation conditions for FTM changed to 30–32 °C for seven days. Several changes were incorporated into the test in 1970, including different incubation times for aseptically filled products (14 days) versus terminally sterilized products (seven days), incubation temperature ranges were increased to 30–35 °C for FTM and 20–25 °C for soybean casein digest medium, and the incubation period was used to differentiate the membrane filtration (MF) test (seven days) from the direct inoculation (DI) test (14 days). Harmonization efforts were in progress for several years with the incubation times being harmonized to 14 days in 2004, and by 2009 (USP 32) the remaining portions of the sterility test were harmonized with only a few exceptions.[11]

Although harmonization of the sterility test has been achieved, it is still a lengthy 14-day test. Therefore, the use of a rapid method as an alternative to the traditional sterility method has several advantages. For example, a delay in the recovery of microbial contaminants would hinder the implementation of corrective actions that would prevent any possible cross-contamination to other product batches. Additionally, a shorter incubation time would condense the product release time.

The Center for Biologics Evaluation and Research (CBER) within the FDA evaluated three growth-based rapid sterility methods: two qualitative methods utilizing CO_2 monitoring technologies (BacT/Alert system by BioMerieux and the BACTEC system by Becton Dickinson) and one quantitative method incorporating ATP bioluminescence technology (Milliflex® Rapid Detection System by EMD Millipore). A total of 14 different microbial strains (ATCC and environmental isolates) representing bacteria (Gram-negative, Gram-positive, aerobic, anaerobic, and spore forming), yeast, and fungi were used. The sensitivity of the RMMs was compared to the compendial membrane filtration (MF) and direct inoculation (DI) methods with regard to the observation of growth at various low levels of inoculations. Results showed that the Milliflex® Rapid system was the most sensitive of the methods, the BacT/Alert and BACTEC system methods were more sensitive than the compendial methods, and the compendial MF method was more sensitive than the DI method.[12]

In 2010, Novartis implemented a rapid sterility method consisting of a five-day incubation as compared to the traditional 14-day incubation. Novartis chose the Milliflex® Rapid Detection System developed by EMD Millipore Corporation over other rapid systems because it is growth based, uses MF which is similar to the compendial method, and can detect one colony forming unit (cfu) following incubation.[13]

The Milliflex® Rapid system uses ATP bioluminescence to detect and quantitate microcolonies; ATP is the primary energy carrier in all living microorganisms. The first step is to filter a sample through a Milliflex® filter unit and place the membrane onto a solid media cassette. The media cassette is incubated to allow for the formation of microcolonies and the detection of ATP. The filter is removed from the media cassette and sprayed with an ATP releasing agent that makes the cell wall of the microorganism permeable to ATP. A bioluminescent enzyme reagent is then sprayed, which reacts with the ATP to produce light (photons). The membrane is moved to the detection tower where image processing takes place. The photons are converted into electrons and multiplied in the photomultiplier tube. The location of the photons correlates with the location of the microcolonies. The image forms on a charge coupled device camera, a computer algorithm then processes the data and enumerates the microcolonies in colony forming units (cfus), and 2D and 3D image maps are generated (Fig. 15.1).[14]

Novartis, taking into consideration the compendial guidelines (United States Pharmacopeia General Chapter <1223>, EP 5.1.6), validated the rapid method and was able to demonstrate that it delivered robust, reliable results and demonstrated an equivalent performance to the compendial

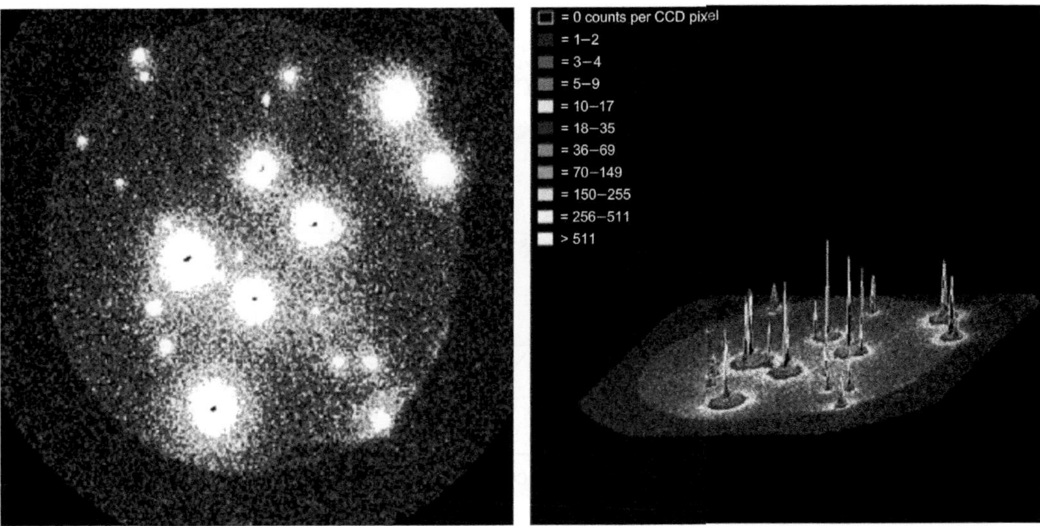

FIGURE 15.1

Milliflex® rapid 2D and 3D images. (For color version of this figure, the reader is referred to the online version of this book.)

Source: Courtesy of EMD Millipore Corporation.

sterility test method in terms of robustness, ruggedness, repeatability, limit of detection, specificity, accuracy, and precision. In 2010, Novartis achieved regulatory approval by the FDA, EMA, and Medicines and Healthcare Products Regulatory Agency (MHRA) to use the alternative method in lieu of the compendial method.[15]

15.5 RAPID BIOBURDEN METHODS

Quantitative rapid methods can be used as alternatives to the traditional bioburden test. One RMM based upon solid phase cytometry is the ScanRDI® by AES Chemunex. This system incorporates fluorescent cell labeling and laser scanning. The ScanRDI® is classified as a direct measurement method which does not require microbial growth for detection and was previously discussed in Section 15.3. Other RMMs that are growth based include the Growth Direct™ system by Rapid Micro Biosystems, and the Milliflex® Quantum by EMD Millipore Corporation.

The Growth Direct System uses proprietary digital imaging technology (Fig. 15.2) that automatically enumerates microcolonies days earlier than the traditional plate counting methods. The sample is filtered through a membrane which is applied to nutrient media and incubated for a shorter time than required by the traditional method. The system captures the autofluorescence that is emitted by the living cells and the microcolonies are detected and enumerated.[16]

The Milliflex® Quantum system is based on two proven technologies: MF and fluorescent staining. MF is the standard method for microbial bioburden testing due to the capacity to remove any inhibitory agents and the ability to process larger volumes. After filtration and a short incubation time

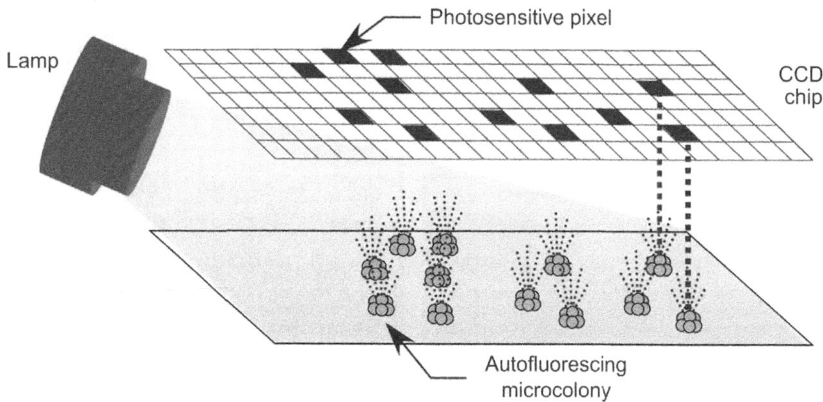

FIGURE 15.2

Growth Direct System digital imaging technology. (For color version of this figure, the reader is referred to the online version of this book.)

Source: Courtesy of Rapid Micro Biosystems, Inc.

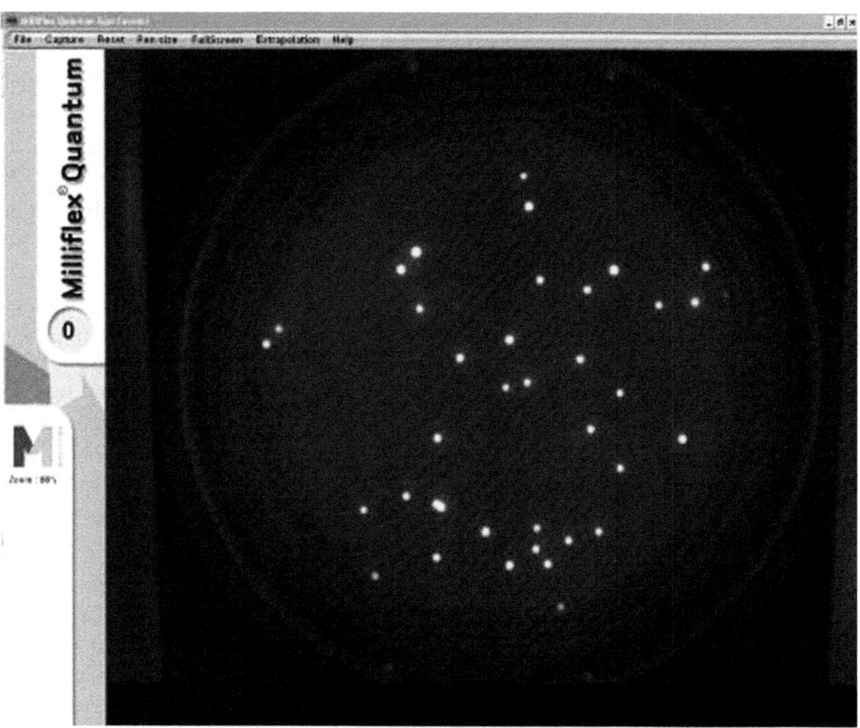

FIGURE 15.3

Milliflex® quantum reader image. (For color version of this figure, the reader is referred to the online version of this book.)

Source: Courtesy of EMD Millipore Corporation.

(approximately one-third shorter than traditional incubation times), the reagent is applied to the membrane and any viable and culturable microorganisms are retained on the filter and stained with a fluorescent marker. The active microbial metabolism of the microorganism causes an enzymatic cleavage of the nonfluorescent substrate and, once cleaved inside the cell, the substrate liberates free fluorochrome into the microorganism cytoplasm. As fluorochrome accumulates inside the cells, the signal is naturally amplified. The cells are then exposed to the excitation wavelength in the Milliflex® Quantum reader so that they can be visually counted (Fig. 15.3).[17]

15.6 VALIDATION OF ALTERNATIVE METHODS

There are several differences between microbiological methods and other analytical methods. United States Pharmacopeia General Chapter <1223> states "Validation studies of alternate microbiological methods should take a large degree of variability into account. When conducting microbiological testing by conventional plate count, for example, one frequently encounters a range of results that is broader (% RSD 15 to 35) than ranges in commonly used chemical assays (%RSD 1 to 3). Many conventional microbiological methods are subject to sampling error, dilution error, plating error, incubation error, and operator error".[8] The USP goes on to state that the characteristics such as accuracy, precision, specificity, detection limit, quantification limit, linearity, range, ruggedness, and robustness are applicable to analytical methods and less appropriate for alternate microbiological method validation. Yet, the general present regulatory expectation is to apply these analytical performance characteristics to alternative RMM validation. Additionally, USP includes these validation parameters in United States Pharmacopeia General Chapter <1223>.

There needs to be a distinction between the validation requirements of the vendor and the validation requirements of the end user. It is more than appropriate for vendors of new alternative technologies to apply these "analytical" performance characteristics during validation. The data generated from the validation testing should be analyzed using statistical tools to show that the method meets the applicable requirements. However, once the technology is validated, the end user should not have to repeat the in-depth validation that was conducted by the vendor. Rather, the end user should focus on whether or not the alternate method will yield results equivalent to, or better than, the results generated by the conventional method when testing their product.

In 2011, the Center for Drug Evaluation and Research (CDER) within the FDA published "A Regulators View of Rapid Microbiology Methods".[18] The author states "While it is important for each validation parameter to be addressed, it may not be necessary for the user to do all of the work themselves. For some validation parameters, it is much easier for the RMM vendor to perform the validation experiments." The author goes on to say that end users would still have to perform their own studies not addressed by the vendor, which include product specific data. Additionally, only the portion of the test utilizing the alternate technology should be validated. An RMM may incorporate portions of the compendial test up to a certain point. For example, a sample may be processed using conventional membrane filtration and the membrane placed on a recovery medium and incubated. However, at that point the presence of viable cells may then be demonstrated by use of some alternative rapid technology. Hence, validation would be required on the recovery portion of the method rather than on the entire test.

When evaluating the range of the method, the vendor needs to ensure that the upper end of the range is challenged. New technologies that enumerate microcolonies verses macrocolonies can count a higher population. Traditional pour plate or MF methods are limited in the numbers of macrocolonies counted, with 300 cfus being the maximum number. New technologies can count much higher cfus in some cases.

15.7 CONCLUSION

Although alternative RMMs have advanced over the last several years, there still needs to be alignment in the validation approach by the end user and what, if any, statistical tools are required to evaluate the data. In the past, microbiological recovery methods applied the criteria of ≥70% to show equivalence. Now, statistics are being applied to the validation of RMMs. Is it reasonable to think that these statistical tools can be applied to microbial data when we know that microbiological methods have increased variability due to the nature of culturing and recovering microorganisms? Another concern is that the typical microbiologist has limited experience in statistics and trying to apply the appropriate statistical tool(s) to microbiological data can be like learning a foreign language. Furthermore, if the validation requirements of the end user are not simplified, then they will never get past the equipment operation qualification and will not be able to implement the RMM.

References

1. *United States Pharmacopeia General Chapter <61>*, United States Pharmacopeial Convention: Rockville, MD, December 2012, 35th revision.
2. *United States Pharmacopeia General Chapter <62>*, United States Pharmacopeial Convention: Rockville, MD, December 2012, 35th revision.
3. *United States Pharmacopeia General Chapter <71>*, United States Pharmacopeial Convention: Rockville, MD, December 2012, 35th revision.
4. *United States Pharmacopeia General Chapter <1111>*, United States Pharmacopeial Convention: Rockville, MD, December 2012, 35th revision.
5. ICH Q6A—Specifications: Test Procedures and Acceptance Criteria for New Drug Substances and New Drug Products: Chemical Substances. 1999, International Conference on Harmonization.
6. *United States Pharmacopeia General Chapter <1112>*, United States Pharmacopeial Convention: Rockville, MD, December 2012, 35th revision.
7. PDA Technical Report No. 33 Evaluation, Validation and Implementation of New Microbiological Testing Methods. *PDA J. Pharm. Sci. Tech.* **2000,** *53* (3). Supplement TR33.
8. *United States Pharmacopeia General Chapter <1223>*, United States Pharmacopeial Convention: Rockville, MD, December 2012, 35th revision.
9. European Pharmacopeia 5.1.6, PhEur 7.5. Council of Europe, 2012.
10. Hussong, D.; Mello, R. Alternative Microbiology Methods and Pharmaceutical Quality Control. *Am. Pharm. Rev.* **2006,** *9* (1), 62–69.
11. Cundell, A. The History of the Development, Applications and Limitations of the USP Sterility Test. *Rapid Steril. Test.* **2011,** *7,* 127–169.
12. Parveen, S.; Kaur, S.; Wilson David, S. A.; Kenny, J. L.; McCormick, W. M.; Gupta, R. K. Evaluation of Growth Based Rapid Microbiological Methods for Sterility Testing of Vaccines and Other Biological Products. *Vaccine* **2011,** *29,* 8012–8013.

13. Gray, J. C.; Staerk, A.; Berchtold, M.; Mercier, M.; Neuhaus, G.; Wirth, A. Introduction of a Rapid Microbiological Method as an Alternative to the Pharmacopoeial Method for the Sterility Test. *Am. Pharm. Rev.* **October 2010**.
14. Millipore Corporation, Milliflex® Rapid System Operator's Manual. 5/2006. Publication No. PF09390 Rev. B.
15. Gordon, O.; Gray, J. C.; Anders, H. J.; Staerk, A.; Schaefli, O.; Neuhaus, G. Overview of Rapid Microbiological Method Evaluated, Validated and Implemented for Microbiological Quality Control. *Eur. Pharm. Rev.* **2011,** *16* (2), 9–13.
16. London, R.; Schwedock, J.; Sage, A.; Valley, H.; Meadows, J.; Waddington, M.; Strauss, D. An Automated System for Rapid Non-Destructive Enumeration of Growing Microbes. *PLoS One* **2010,** *5* (1), e8609.
17. Millipore Corporation, Milliflex® Quantum Rapid Detection System User Guide. 2/2010. Publication No. PF11940 Rev A.
18. Riley, B. A Regulators View of Rapid Microbiology Methods. *Eur. Pharm. Rev.* **2011,** *16* (5), 3–5.

PART

7

Biological Fluids

CHAPTER

Bioanalytical method validation and bioanalysis in regulated settings

16

Krzysztof Selinger*, Eliza N. Fung[†], Peter Bryan[‡]

* *Forest Research Institute, Farmingdale, NY, USA,* [†] *Bioanalytical Sciences, Bristol-Myers Squibb, Princeton, NJ, USA,* [‡] *Mendham, NJ, USA*

CHAPTER OUTLINE

Specification of Drug Substances and Products. http://dx.doi.org/10.1016/B978-0-08-098350-9.00016-3

16.1 INTRODUCTION—BRIEF HISTORY OF HARMONIZATION EFFORTS

The subject of bioanalytical method validation has been extensively debated for several years and a number of papers describing various approaches to validation have been published.[1–20] The need to develop and accept a uniform approach to method validation has been generally recognized in the pharmaceutical industry and regulatory agencies, which resulted in several meetings and guidance documents such as:

- Food and Drug Administration (FDA) 2001,[1] referred to here as Method Validation Guidelines (MVG).
- European Medicine Agency (EMA) 2011,[2] referred to here as Guideline for Bioanalytical Method Validation (GBMV).

Starting in 1990, scientists representing the industry, academia, and regulatory bodies met regularly at conferences devoted to harmonization and refinement of bioanalytical methods and validations. These meetings were taking place initially in Crystal City, VA, and hence the reports from conferences have been called Crystal City I, II, III, etc.[3–5] Numerous workshops and other initiatives have been undertaken by various professional organizations, e.g. American Association of Pharmaceutical Scientists (AAPS) and Society of Quality Assurance (SQA), to work out problems even though so far no consensus has been reached or alternative solutions proposed.[6–13] Other professional associations, e.g. Société Française des Sciences et Techniques Pharmaceutiques (SFSTP), also contributed to the harmonization efforts.[14–16] Finally, numerous independent papers from academia and industry have been published.[17–20] The growing globalization of the economy and the pharmaceutical industry gave more impetus to the harmonization efforts. To this effect, a Global Bioanalytical Consortium (GBC) has been created,[21–25] which includes scientists from different regulatory regions. The goal of the GBC is to merge existing bioanalytical guidances and industrial procedures into a unified document and procedures acceptable to the regulatory authorities in various countries. The resulting document would ensure that the bioanalytical work is done to the highest scientific standards, and that the review process would be uniform and objective in all countries involved.

Additionally, the Global Contract Research Organization Council for Bioanalysis has been formed[26] with broadly similar goals. Their efforts are fully supported by professional organizations such as AAPS, European Bioanalysis Forum, Canadian LC–MS Group/Calibration and Validation Group, and Applied Pharmaceutical Analysis—Boston Society for Advanced Therapeutics.

This chapter describes validation procedures which are based on the two major regulatory documents[1,2] augmented by various meeting reports, White Papers, and other independent publications[27–34]; these will be referred to as consensus papers. The following logic has been applied:

- If MVG and GBMV agree on an issue, this common position is presented in the chapter.
- If there is a difference between the MVG and GBMV, the stricter and more comprehensive position is presented here.
- If one of the documents remains silent on an issue, the position expressed in the other guidance is presented here.
- If both documents remain silent on issues, which in the meantime have been discussed and agreed on in professional forums, the point of view expressed in white papers and workshop reports is presented here.

The reader must be advised that the FDA has announced the publication of a new bioanalytical guidance. It has not been published as of this writing in the Spring of 2013.

The aim of this chapter is to provide in Sections 16.2–16.8 detailed instructions on method validation, which would facilitate acceptance by the regulatory authorities in most developed and developing countries. Method development and validation issues for which an agreement has not yet been reached, or have been under the radar of the bioanalytical community, are discussed in Section 16.7.

Most of the comments are related to chemical (chromatographic) methods; Ligand-Binding Assays (LBAs) are not addressed. The scope is limited to bioavailability, bioequivalence, and pharmacokinetic studies, including also some aspects of toxicology and metabolism.

The term "bioanalytical method validation" can be understood in a narrow or in a broad sense; this chapter will cover the broad sense, which includes:

- Proper validation exercise
- Application of the meetings validated method to routine drug analysis, which includes:
  - **a.** Proper execution of a bioanalytical project
  - **b.** Control of a method during its execution

16.2 GLP, GCP, AND REGULATED BIOANALYSIS

The MVG stipulates that bioanalytical work in support of pharmacology/toxicology and other preclinical studies for regulatory submissions must be done according to good laboratory practice (GLP),[35] while the support for human bioavailability, bioequivalence, pharmacokinetic, and drug interaction studies shall be done according to good clinical practice (GCP).[36] However, the GCP document-21 CFR 320.29 is very short and demands only that a method used to support bioequivalence and bioavailability studies must be accurate, precise, and sensitive. There is another document from the EMA that describes in much greater detail the basic requirements of GCP[37] and requirements on the part of laboratories performing analysis in support of clinical trials. In practice both types of work are done according to the same set of rules with addition of some aspects of cGMP (current Good Manufacturing Practice) and standard industrial practices. For these reasons, the term "regulated bioanalysis" will be used here as it includes the hybrid of GLP, GCP, and cGMP.

16.2.1 Bioanalytical method validation according to the current regulatory and industrial standard

There are several kinds of validation:

- Original validation—the method is validated for the first time, immediately after method development
- Partial validation—when changes are introduced to a validated method, such as new personnel or equipment is used, or other minor changes in the method
- Crossvalidation—when a comparison is made between two laboratories or two procedures

These types of validation will be discussed in detail in Section 16.6. It is worthwhile to point out that the term "qualification" is frequently used. It can refer to a partial validation or a "fit-for-purpose" validation in which only selected parameters, such as accuracy and precision, are evaluated based on the intended use of the method. It has to be understood that the process of method development and validation is a continuum, and that method development and validation has a certain life cycle.

In the current highly cost-conscious pharmaceutical environment the balance of costs and benefits is a critical issue. New drug candidates, frequently from the same chemotype, are screened and tested and it would be wasteful to undertake a full validation for all of them. Hence, some initial pilot toxicology projects can be done using a method that has not been yet fully validated. Usually at the drug discovery stage doses administered to laboratory animals are rather high, enabling simple approaches, e.g. protein precipitation as sample pretreatment and a generic reversed-phase liquid chromatography (LC) method with liquid chromatography–tandem mass spectrometry (LC–MS/MS) detection. This is not to say that such a method is without scientific merit and does not absolve the analyst from understanding the chemistry of the compounds investigated. Testing at least some rudimentary level is needed, such as stability (is the compound stable for one day or one week that is needed to complete this pilot project?) or extraction efficiency (is recovery at least 20–30%?). At this stage a one-day testing procedure consisting of a single or duplicate calibration curve and a set of quality controls (QCs) is sufficient.

A project leaving the discovery stage and entering the development stage requires the regulated-bioanalysis treatment and a fully validated analytical method. Formal toxicology studies are pivotal, these include toxicological studies, chronic toxicology, and toxicokinetics. Although the ADME studies (absorption, distribution, metabolism, and elimination) of the parent drug do not formally require a validated method, such a method usually exists at this rather late stage of drug discovery. Many bioanalytical laboratories perform such studies as per regulated bioanalysis. All human studies submitted to regulatory authorities require a fully validated method.

Every laboratory needs to develop a validation protocol, which describes specific elements of validation in that particular laboratory and its acceptance criteria; this usually takes the form of a Standard Operation Procedure (SOP), a protocol, or a validation plan. The validation exercise has to be auditable and should be summarized in a validation report. A validation report may contain a detailed description of the analytical method, or the method can be described in a separate document (SOP, test method, or a method sheet).

16.3 METHOD DEVELOPMENT AND INITIAL VALIDATION OF THE CHEMICAL METHOD

Validation follows methods development, transfer, or modification, and is followed by a method application, i.e. a bioanalytical project. While validation logically follows the method development, the method application does not necessarily directly follow the validation. Continuity between these two elements is needed which means that if a method is not used on a regular basis it needs to be validated or qualified again before use; such a validation protocol may be abbreviated and limited to a single run. There is no set limit on how much time is allowed to elapse between testing occasions. The project manager of the study will need to use his or her judgment based on the circumstances. In the bioanalytical community, such a run is called a prestudy assay evaluation, a qualification, or even a validation run. Conversely, a method cannot be considered fully validated until it is applied to real clinical or animal samples with good reproducibility. To verify this attribute, an exercise called Incurred Sample Reassay or Reanalysis (ISR) is performed and will be discussed in Section 16.4.5.

A bioanalytical method cannot be developed and validated without considering the ultimate objective of a study. This means that during method development one has to keep in mind basic requirements and numerous details which may have a bearing on the project: range of concentrations needed and targeted lower limit of quantitation (LLOQ), matrix to be used, anticoagulants in the case of blood or plasma (ethylenediaminetetraacetic acid (EDTA)—at what concentration? which salts? heparin—lithium, sodium or potassium? any additives such as citrate, oxalate), volume of blood needed or available per assay (pediatric studies provide small sample volumes, usually <0.5 mL; the same limitation applies to small laboratory animals such as mice), stability, and safety considerations. The stability considerations are of utmost importance. For new chemical entities (NCEs), the stability data are frequently not available and neither is the method of analysis. One has to develop a skeleton of an assay and immediately employ it to evaluate stability in the sample matrix under different conditions (e.g. storage at room temperature, multiple freeze–thaw cycles, or at −20 °C). Once any instability issues are resolved, one has to ensure that appropriate sample collection and storage procedures are used in the clinic or animal rooms.

After a period of trials and tests, a bioanalytical method will be ready for validation. An analyst has to make sure that method development is complete and no additional changes will be introduced during validation; some examples of changes are provided in MVG.[1] Introduction of changes and modifications require starting the validation procedure from the beginning.

The essential parameters that need to be defined to ensure the acceptability of a bioanalytical method as per MVG are precision, accuracy, sensitivity, specificity, response function, recovery, reproducibility, and stability. The GBMV also demands evaluation of carryover, dilution, integrity, and matrix effects.

16.3.1 Precision and accuracy

The definitions of accuracy and precision are presented in Chapter 4. The goal of this chapter is to describe all necessary steps leading to precision and accuracy appropriate for trace analysis in biological matrices. The majority of analytical measurements are relative in nature, which means that a result is obtained by comparison of sample response with a standard response. Hence, the quest for accuracy begins with a reference standard.

The best reference standard for well-established and easily available drugs should be a compendial material (US pharmacopeia (USP), British Pharmacopoeia, European Pharmacopoeia), although USP does not provide a certificate of analysis (CoA). Standards from reputable commercial suppliers with CoAs are also acceptable. NCEs are available only from their originators, who should also provide a CoA. Nevertheless, it is the responsibility of the user to obtain the CoA. Ultimately, if no CoA is available, the user has to prepare a CoA by performing a number of tests confirming the identity and purity of the standard; these tests may be spectral and elemental analysis, high-performance liquid chromatography (HPLC) area summation, water and ash contents (see Chapter 2 for more details). One has to limit testing to a necessary minimum as many compounds are in short supply. It is worthwhile to consider using quantitative nuclear magnetic resonance spectroscopy as one-step analysis for purity. The use of drug formulations (tablets, injections) as a source of analytical standards is strongly discouraged. If a reference standard is difficult to obtain or purchase, and a drug formulation is the only source of the material, then such a secondary standard should be also characterized analytically by other techniques. Although a CoA is not necessary for the internal standards (ISs), it is obvious that the user must make sure that the potential IS is "fit for purpose". At the very least the analyst must ascertain that the IS does not contain any significant amount of the analyte of interest, or produce it by chemical reaction during the analytical procedure.

The next step is to ensure the correctness of calculations. Many substances exist in the form of salts and/or hydrates or solvates at different degrees of purity. Measurements in biological matrices should provide a result expressed in terms of a free base or acid. The calculations should be verified by a second analyst.

Weighing of the reference standard should be carried out on a properly maintained and currently calibrated analytical balance. For a typical weighing of 1–10 mg of material, sensitivity of at least 0.00001 g is needed. In order to avoid bias in the analytical technique, ideally two analysts should prepare a total of at least two weighings and two separate stock solutions. The master stock solutions should preferably be prepared in volumetric flasks of appropriate volume, or by accurately pipetting the required amount of solvent. After dilutions by the primary and secondary analysts, these stock solutions should be compared using an appropriate analytical technique; the technique does not have to be the same as in the final bioanalytical method. The acceptance criteria for stock/spiking solutions should be specified *a priori* in the laboratory's SOP; in general, stock solutions should be within 5% of one another.

The equation to calculate these differences is as follows:

$$(x1 - x2)100\%/\text{mean} \tag{16.1}$$

Should the differences be greater than the acceptable limit, records of preparation should be reviewed. New stock solutions might be prepared either by preparing a new stock solution or dilutions of existing stock solutions reprepared and compared. An outlier must be ultimately identified and eliminated.

Having made sure that the stock solutions accurately reflect the analyte concentration, one of the two stock solutions should be chosen to spike both the calibration standard and QCs. This approach has been endorsed by the Crystal City conference report,[5] although several laboratories in recent years received FDA 483 citations for following this advice.[34] While it is common in the industry to use separate stock solutions for the calibration standards and for the QCs, it is the opinion of the authors that, having verified the correctness of stock solutions prepared as described above, it is no longer necessary to use two sets of stock solutions because the use of separate stock solutions may introduce bias.

Spiking of calibration standards and QCs should involve a primary and secondary analyst. The primary analyst should prepare the calibration curve while the secondary analyst should prepare QC samples, or a combination thereof. Again, the secondary analyst is needed to make sure that a slightly different analytical (i.e. pipetting) technique would lead to the same results. Although the spiking into biological matrices appears trivial, it provides an ample opportunity for the introduction of errors. Here are some examples.

First, one may use either volumetric flasks (volumes restricted to certain values, sometimes difficult to mix efficiently) or deliver volumes using a pipette as biological matrices tend to be precious and should not be wasted; hence the smallest necessary volume should be used. Second, blood, plasma, and serum are viscous and relatively difficult to measure accurately. Third, blood, serum, plasma, and urine tend to foam while mixing; this makes the volume control difficult as the meniscus can be difficult to see. Fourth, stored plasma may contain precipitated proteins and fibrins, while stored urine may contain precipitated salts; both may block pipettes.

Pipettes (electronic or manual) used for dilutions should be maintained and calibrated according to the laboratory SOP. Usually delivering 98–102% of the nominal volume and with <1% coefficient of variation (CV) is acceptable for automated pipetting devices; calibration or performance verification should be performed every 3–6 months or more frequently when required (after maintenance or repair).

There are two schools of thought for the preparation of calibration standards. The first requires spiking of small volumes of standards on each analytical day using freshly prepared or diluted spiking solutions: such standards are not stored and are used on the day of preparation; this approach seems to be favored by the Crystal City I Conference Report I.[3] The second approach permits the preparation of standards in bulk, aliquoting them in separate tubes and storing under the same conditions as the QCs and study samples. In both situations, QCs are spiked in bulk, aliquoted, and stored with the study samples. The justification of the first case is that by always using new standards, the sample stability is monitored, and that calibration standards are distinct from the QCs. The justification for the second case is that this procedure is acceptable if stability data exist to support this approach. It is easier, more productive, and avoids an additional potential for bias to spike once only. Additionally, the difference between calibration standards and QCs is that a calibration curve is forced through the standards, but not the QCs. The authors of this chapter prefer the second approach.

The calibration standards should be prepared in the same matrix as the samples (whenever possible), and separate calibration curves should be established for each analyte. It is recommended to have six to eight nonzero standards for each calibration curve, with single or replicate samples, in each analytical batch. More standards are needed for complex, nonlinear calibration models. One calibration standard should be at the LLOQ; it is good practice to have the second standard at 2 × LLOQ to define the lower end of the curve well, and all the other standards spread over the remaining range of concentrations. It is also good practice to have two levels of standards between the concentration of the high QC sample and the upper limit of quantitation (ULOQ). It is recommended to have three standards per order of magnitude, i.e. a calibration curve over 1–200 ng/mL should have six standards, but a calibration curve over the range 0.1–100 ng/mL may require nine standards. Drug-free matrix (blank) and drug-free matrix with the IS added (standard zero) should be a part of every analytical batch. Calibration standards can be placed either all in the beginning of the run or dispersed throughout the batch. A run should start with system suitability samples. These are prepared by mixing the analytes of interest in the mobile phase or reconstitution solution at certain proportions representing either typical concentrations seen after the extraction, or the concentrations as observed in an LLOQ sample. The system suitability samples serve to verify the retention times, resolution (if critical), sensitivity, and stability of the system. The system suitability samples should meet predetermined acceptance criteria.

One of the issues frequently discussed is whether a calibration curve should be measured once or performed in duplicate in each analytical run. Again, a balance of costs and benefits is necessary. For robust assays with a stable response of the system, a single calibration curve may suffice. If a significant drift in the instrument's response is observed, the second calibration curve placed toward the end of the batch may be needed.

Quality controls should be prepared with at least three concentration levels and analyzed in duplicate with each analytical batch. One set of QCs should be close to the LLOQ (2–3 times higher than the LLOQ). The second set should be at approximately 40–60% of the ULOQ, and the third at 70–90% of the ULOQ. Quality controls should be spread evenly throughout the analytical batch. The selection of the calibration curve range and placement of the QCs must be judicious to represent the concentrations of samples encountered in the study. In other words, clinical or animal sample concentrations should not be clustered around the lower or upper quadrant of the calibration curve only, with one QC that falls within the concentration range of most study samples. Of course, this may not be known at the time of validation, but discovered only after first batches of a study have been analyzed. In this case, addition of the extra QCs of appropriate concentration are suggested.[2]

Assay accuracy is expressed as a percentage of the true value which is calculated according to the formula:

$$\text{Accuracy} = \text{observed value} \times 100\%/\text{true(nominal)value} \tag{16.2}$$

The term "recovery" is sometimes used to describe accuracy; this usage should be discouraged. The true value is assumed to be the nominal value at which the sample has been spiked; the accuracy can be expressed as a percentage of nominal or percent deviation from nominal (i.e. bias). The practice of using the observed values instead of the nominal ones is no longer considered acceptable.[1]

The use of samples from dosed subjects to assess accuracy has been suggested.[3] The rationale is that despite our best efforts, it is virtually impossible to mimic a clinical sample by the simple addition of a standard solution to appropriate medium, as a clinical sample may contain drug metabolites, concomitant medication and its metabolites, as well as endogenous substances, the level of which may be impacted upon by the drug administration. Hence, a pool of clinical samples could be used as an extra QC sample. Such a solution is generally impractical for the reason of availability, yet may be very useful in some situations. For example, drug conjugates (glucuronide and sulfate) are notoriously difficult to obtain. A solution to this problem is to hydrolyze chemically or enzymatically these conjugates and measure the concentration of the free drug. However, a hydrolysis control is needed to make sure that such a process remains reproducible; a pooled subject sample could play this role.

Another example is the possible conversion of unstable metabolites such as N-oxides back to the parent drug during the analytical process.

Precision is a measure of repeatability of a method and can be expressed by the relative standard deviation (RSD); this value is commonly known as the coefficient of variation (CV):

$$\mathrm{RSD} = \mathrm{SD} * 100\%/\mathrm{mean} \tag{16.3}$$

Here the standard deviation (SD) is calculated as per Eqn (2.2).

In today's bioanalytical assays, a CV of $\leq 5\%$ characterizes a very precise method. A CV of 5–10% is probably the most common and represents an industrial norm in terms of precision, while a CV around 15% suggests either a method of extreme difficulty and unusually low LLOQ or some analytical problems; a CV of 20% may be acceptable only around the LLOQ.

The ultimate goal of any method is to assay samples; QCs are the best approximation of clinical samples. Hence, accuracy and precision of a method should be estimated using the percent nominal and CVs calculated for the QCs, and not the back calculated (interpolated) value of calibration standards, although these values should also be reported. A calibration curve is forced through the calibration points, and accuracy and precision based on standards always look somewhat better than those based on QCs. Back calculated standard concentrations are a useful and necessary tool in the evaluation and adherence of the system to the selected mathematical model.

Within-run precision and accuracy are evaluated during the validation by assaying a minimum of five replicate samples independent of standards at concentrations representative for the assay; a separate set of QCs could be used for that purpose.

16.3.2 Limits of quantitation

The LLOQ is frequently confused with the limit of detection (LOD). Both are a measure of the sensitivity. The LLOQ is the lowest concentration of the standard in the calibration curve and is higher than the LOD. The LOD has no practical use in regulated bioanalysis. Various ways of calculating the LLOQ are presented in Chapter 2. The consensus documents recommend a very pragmatic approach to the determination of LLOQ. It is the concentration which provides a CV $\leq 20\%$, and accuracy between 80% and 120%. The way to establish this experimentally is to prepare at least five samples independent of the standards at the concentration of the projected LLOQ, another set of five at concentration $2 \times$ LLOQ, one more at $4 \times$ LLOQ, and so on. These samples should be analyzed with a calibration curve. The concentration, which fits into specification, should be considered the LLOQ, and the lowest calibration standard should be set at this value. The conference also endorsed other approaches to LLOQ, and alternative models of LLOQ are presented in references.[38]

The MVG and GBMV specify that the minimal signal-to-noise ratio should be 5:1, which is a pragmatic recognition of the fact that it is nearly impossible to obtain acceptable precision and accuracy if the signal-to-noise ratio is <5:1. The practical way of calculating the signal-to-noise ratio is presented in Fig. 2.22 of the first version of this book. Many modern analytical computer programs are available to calculate this parameter. Most of the scientists working in the area of biological trace analysis are under constant pressure to improve sensitivity and lower the LLOQ; the question "Can you get lower than this?" is proverbial. As a practical rule of thumb, most bioanalytical assays only need to have sufficient sensitivity to quantify the main analyte of interest at concentrations estimated after five biological half-lives to adequately characterize a compound's pharmacokinetic profile. A chemical or instrumental breakthrough answers today's questions, and more insight into the nature of things invites more questions.

It should be noted that there is not only an LLOQ but also an ULOQ, which is the highest concentration of the standard in the calibration curve. There are several reasons for the existence of the ULOQ: above a certain concentration a calibration curve may no longer be described by a chosen mathematical model (this usually means plateauing), large chromatographic peaks may be truncated if a detector is saturated, chromatographic peaks can be deformed by overloading of the system, or a method simply has not been validated above a certain concentration. How to handle results that are above the ULOQ (above the Upper Limit of Quantitation) is explained in Section 16.4.3.

16.3.3 Specificity

There are two components of specificity. First, a bioanalyst must prove that other components within the matrix do not generate (or contribute significantly to) the measured signal; second, that the signal (chromatographic peak) is indeed generated by the analyte of interest. In chromatographic methods with detectors other than MS, an analyte is identified solely on the basis of its retention time, wavelength in ultraviolet (UV) detectors, combination of excitation and emission wavelengths in fluorescence detectors or applied potentials in electrochemical detectors. These techniques are inherently nonspecific and can only limit the number of compounds which may be seen otherwise in the same time window. Only LC–MS/MS and to some lesser degree also LC–MS provide virtual certainty that the signal observed was generated by the analyte of interest. Extra caution should be exercised when developing methods for compounds with glucuronide and N-oxide metabolites. These metabolites are often difficult to obtain during early phase development and are prone to in-source fragmentation or may not be stable in the biological matrix. Chromatographic separation of these metabolites is required to ensure assay specificity. A good example of a case where chromatographic separation of a potential interfering compound is for analytes containing a glutarimide ring (e.g. thalidomide). The glutarimide ring is prone to hydrolysis and this degradation product often coelutes with the analyte and can be isobaric with the analyte. This problem can easily be solved by the addition of acid to the plasma samples, by only using fresh solvent for the preparation of stock solutions and the addition of dilute acid to the diluted working solutions.

Another type of specificity problem can be caused by the choice of a stable isotope labeled-IS (SIL-IS) for LC–MS or LC–MS/MS assays. Deuterated ISs can be prone to proton exchange with the solvent and need to have a sufficient number of substitutions to ensure there will be no interference in the analyte channel. Typically, a mass increase of four is sufficient to alleviate interference between the analyte and the IS for molecules containing only C, N, and O. The actual interference in selected reaction monitoring (SRM) can be calculated accurately based on chemical structures of the precursor and product ions, labeling positions, and concentration ratio.[39]

Six samples of the drug-free matrix obtained from six individuals should be used to prove lack of significant interference with the intended analyte. These blanks must be obtained from the relevant population, in the simplest case being split equally between the genders. It is becoming a common practice to evaluate hemolyzed and lipidemic plasma either as a part of or in addition to the six lots. Matrices from special populations, such as pediatric, may be evaluated at the discretion of the bioanalyst. The same biological matrix should be used for validation as that in the clinical/animal samples. In the case of blood, plasma, serum, or urine from humans or large animals, the matrix availability does not present a problem. Some matrices, for example control cerebral–spinal fluid (CSF), bone marrow, sputum, bile, or samples from small animals may not be available in sufficient volumes or not at all; a surrogate matrix can be used instead for calibration standards and QC preparation.

A couple of issues require elaboration. There is no such thing as "no peak". If one amplifies electronically the baseline in the area of interest then oscillations of the baseline and minor spikes will become visible and in most cases there is something that could be integrated. The issue is how significant the contribution of the interference is allowed to be. The absolute minimum is that the interference should be <20% of the peak corresponding to the LLOQ. This requirement has to be specified in the appropriate SOP.

On the one hand, some projects involve dosing healthy volunteers whose diet is controlled and who provide samples that are relatively free of interference. On the other hand, in phases II and III of drug development or in oncology programs in which a drug is administered to patients who routinely take concomitant medications and whose general health condition may be poor, concomitant medications can be present in the samples for a variety of reasons and interference check is required. It is also prudent and in fact necessary to test whether or not the common over-the-counter (OTC) medications are present in the samples, even for normal healthy volunteers. A quick glance at the OTC shelves in a local pharmacy would indicate that these could be:

- Common pain killers (aspirin, ibuprofen, acetaminophen)
- Antiacids (cimetidine, ranitidine, famotidine, omeprazole, etc.)
- Antihistamines (loratadine, diphenhydramine, brompheniramine, chlorpheniramine, cetirizine)
- Components of cold medicines (pseudoephedrine, dextromethorphan)
- Caffeine

The situation may be much more complicated as one has to consider not only the parent drugs but also their metabolites, which can be numerous and difficult to obtain.

In the present era, when bioanalysis of small molecules means almost exclusively analysis by LC–MS/MS, the danger of interfering peaks originating from OTC medication is minimal. However, there is a danger of pharmacokinetic interactions between the tested drug and an OTC medication, cimetidine, for example.

In the simplest case, a bioanalytical laboratory could prepare a test mixture for evaluating the interference from OTC medication by mixing the easily obtainable reference standards and at least some of the metabolites at typical concentrations, such as C_{max}.

In order to prove that the substances being quantified are the analytes of interest in assays other than LC–MS/MS, one has to inject a reference solution containing only the intended analytes (drug, and/or metabolite, IS) in pure solutions at the beginning and the end of the run, and compare the obtained retention times with the biological samples. The best solvent is the mobile phase. A solvent in LC stronger than the mobile phase (e.g. methanolic solutions injected at volumes >10–20 μL into a typical reverse phase system) will produce a distorted peak with a shorter retention time. In gas chromatography, injection of simple reference solutions may not provide peaks at all if there are active sites in the system which adsorb analytes. The solution to this is to mix a blank extract with a reference solution or to include a "carrier" substance in large excess in the reference solutions, which would not interfere with the assay, but saturate the active sites. If the sample preparation involves a back extraction from diethyl ether, ethyl acetate, or other relatively water-soluble organic liquids, the solvent in the reference solution should also be saturated with this reagent; otherwise the retention times will differ.

One must realize that today's supersensitive tandem mass spectrometers and even more sensitive detectors of tomorrow will make obtaining of the so-called "clean blanks" more and more difficult as it is very easy to introduce contamination to the samples. For laboratories that support development of a drug for a number of years, it is safe to assume that most of the laboratory benches, laboratory glassware, and LC–MS/MS systems come into contact with the analyte, and perhaps at high concentrations. Of course, the assays that are most prone to contamination are the ones with the LLOQ in the picograms per milliliter range. If there are persistent interfering peaks in the blanks or at lower concentrations, one would be well advised to follow precautions that are typical for the activities in a clean room. The bioanalyst should consider the following practical pieces of advice:

- Devote a separate room to this assay only, if possible
- If not possible, devote a corner of the laboratory to the assay and limit traffic there
- Start every day with a fresh laboratory coat or a disposable laboratory coat
- Clean the bench top with methanol every day or cover it with a paper mat
- Wipe the pipettes with methanol swabs frequently
- Change gloves and wash hands frequently
- Use disposable containers
- If glass volumetric flasks are needed, wash them personally with acid, water, and methanol
- If evaporation is involved in the processing, clean the jets of the evaporator with methanol
- Have devoted office supplies (pens, scissors) to the assay
- Never open sample tubes with stock solutions near the working space; if there is temperature difference a spray can be formed

16.3.4 Recovery

Recovery as defined here is the extraction efficiency. Only very uncomplicated samples in terms of concentration and matrix can be injected directly or after dilution onto an HPLC column; these could be CSF, urine, and saliva. Most biological samples have to be prepared in some way before entering a chromatographic column, on or off-line system. There is no formal requirement regarding how high recovery should be. A bioanalyst always tries to develop a method with recovery as close to 100% as possible. However, a recovery of $\leq$50% is also acceptable if it provides precise and accurate results and it is the best that can be achieved under the circumstances. Low recoveries are frequently associated with poor reproducibility, and are red flags for an analyst or a reviewer to watch for unexpected problems or outliers. A bioanalyst is often forced to work at the sensitivity limit of the system, and can hardly afford poor extraction efficiency, which in turn will decrease the sensitivity of the assay.

The absolute recovery can be calculated by comparing the peak areas (or peak area ratio of analyte/IS) from the extracted QCs (or calibration standards can be used) at three concentration levels in triplicate with those of unextracted solutions. The unextracted solutions should have the same concentration as those of the extracted QCs (or calibration standards). The unextracted solutions can be prepared by diluting neat solutions to concentrations representing 100% recovery with the extracted blank matrix, incorporating all the material losses due to the volume transfer. During the recovery study, all the volume transfers should be done quantitatively unless an IS is used. Another way of establishing recovery is to divide the slope of the extracted calibration curve by the slope of unextracted standard curve representing 100% recovery; this approach applies only to rectilinear calibration curves.

16.3.5 Response function

The theoretical background of the establishment of the appropriate response function between the measured signal and the analyte concentration is described in Section 2.2.2 of the earlier version of this book. In general, the simplest response function should be selected, the fit should be statistically tested, and an appropriate algorithm or graph presented. What it means in practical terms is that during validation and/or study, the response function selected should remain the same, and not be changed from one batch to another.

Another issue related to the response function is the question of whether to use peak height or peak area. From a theoretical standpoint, only the peak area is proportional to the mass of the analyte, and the peak height is related to the mass only at the height of a triangle, which approximates an ideal peak. In today's highly computerized environment, measuring a peak area is very simple, and it should be used unless there are special circumstances, such as the presence of many overlapping peaks or severe tailing, where minor imperfections of the integration may cause significant errors. The alternative use of peak height is worth investigating at the prevalidation stage in order to find an optimal response function and range of calibration curve. The ultimate decision on using peak area or peak height, choice of regression, weighing factors, and linearity assessment belongs to the analyst and should be carefully documented.

16.3.5.1 Acceptance criteria

The validation is accepted or rejected as a whole with expectations that the overall precision and accuracy will be $\leq$15% at concentrations above the LLOQ, and $\leq$20% at the LLOQ.

16.3.6 Reproducibility

The methods used in regulated bioanalysis must be reproducible and repeatable. Briefly, repeatability is precision achieved in the same laboratory by the same operator using the same equipment, while reproducibility is precision in different laboratories by different operators.[40] Validation should be performed using a similar number of samples per batch, as in the study (so-called "run-size evaluation"). In the past, this number used to be considered as being close to 100. Today a discrete analytical run may consist of several 96-well plates as long as each plate contains its own set of six QCs. With typical analysis times of 2–4 min per sample, and some time devoted to the injection process which is highly dependent on the autosampler, one can expect the analysis of 96 samples to take 4–6 h. The rationale for run-size evaluation is to make sure that appropriate precision and accuracy are obtained by an operator (human or robot) challenged with a large number of samples, as well as to see if the system (chromatographic, robotic) performs correctly over the period of time needed to complete an analytical run.

There are no acceptance criteria for individual runs in the validation process. The validation exercise should be limited to a certain number of runs. It is the opinion of the authors that if three acceptable runs are needed to complete validation, no more than five attempts should be allowed. Should the fifth attempt fail to provide three acceptable runs, the method should be sent back for further development, poorly defined parameters identified and optimized, and the validation should start from the beginning. Otherwise, a method would be a kind of game of chance rather than a science-driven process. Regulatory authorities consider this approach as "testing into compliance".

The bioanalyst always needs to maintain a balance of costs and benefits depending on the physicochemical properties of the analyte, concentrations required, and time considerations. The simplest solutions are quick, but do not necessarily provide the most robust methods. The order of extraction techniques from biological matrices according to increasing difficulty and time consumption may be as follows: direct injection, protein precipitation, single liquid–liquid extraction (LLE), simple solid phase extraction (SPE), LLE followed by back extraction, but the order in terms of chromatographic system robustness will be reversed.

Not much is said in MVG and GBMV about the ruggedness or robustness of a method, which is an important parameter and is discussed in Section 16.7.11.

16.3.7 Stability

The most common reasons for instability of drugs in biological matrices are chemical, enzymatic, and photochemical processes.[41] The chemical processes include hydrolysis of esters (diltiazam, aspirin), opening of the lactam ring in β-lactam antibiotics, opening of the lactone ring in campthotecin analogs, oxidation of phenols and naphthols, oxidation, dimerization, and side reactions of captopril, etc. The enzymatic processes include hydrolysis of esters such as procaine, esmolol, irinotecan, and remifentanil by esterases. The light sensitivity affects drugs such as nitrofurantoin, clomiphene, retinoids, and fluoroquinolones.

Stability has to be tested in the matrix of the study, under conditions encountered during the execution of a bioanalytical study. This includes the matrix in the presence of analytes of interest and/or their major metabolites which could potentially convert back to the parent drug, and includes bench top (processing) stability, freeze–thaw stability, on-instrument/autosampler stability, processed samples stability, and long-term storage stability. If a metabolite is included in the assay, then it should be included in the stability evaluation. It is also necessary to establish the stability of the stock and working solutions of the analytes and their ISs under the storage conditions.

Benchtop stability tests simulate situations during the sample collection and analytical processing, where samples typically remain at room temperature for a total of 3–6 h. At the end of that period, stability samples should be analyzed against freshly prepared calibration standards. In practice, one set of QCs is stored at room temperature (or ice bath, if needed) for 3–6 h, another set to 24 h, and then they are extracted along with freshly prepared calibration standards. Samples from HIV-positive patients are deactivated by heating at 56 °C for 3–5 h; hence stability at 56 °C should be also included in the validation exercise, if applicable.

The freeze–thaw stability test mimics the situation where samples undergo multiple freezing and thawing cycles either during sample collection, processing, or repeat analysis. The consensus is to subject the samples to at least three freeze–thaw cycles. These cycles should be at least 12 h apart with one cycle frozen for 24 h, if they are to simulate real-life situations. After the third thawing, the samples should be analyzed against freshly prepared calibration standards. The number of cycles should be adequate to cover the actual situation encountered in the study. The freeze–thaw test should be also performed with a dilution QCs (see section 16.3.9) in addition to the conventional low and high concentration QC samples to ensure that analyte precipitation at very high concentrations is not encountered.

The long-term storage stability test should be performed over a period of time that equals or exceeds the time between the date of sample collection and the date of analysis. Further evaluations may be made at later times; for example, 6 and 12 months of storage at the same storage temperature and in the same containers (geometry, caps) as the study samples. The stored stability samples are compared to freshly prepared calibration standards. What constitutes "fresh" is a matter of heated debate right now.[34] In an ideal situation, new calibration standards and new QCs would be prepared over 1–2 h, extracted, and injected to verify the correctness of the preparation (3–6 h), and then the new set of calibration standards and QCs would be extracted with the stored stability QC samples. From the author's most recent experience, it seems that the FDA insists on the definition of a "fresh standard" as that prepared and used on the same day. Bioanalysts would be well advised to complete the stability evaluation on the same day to avoid potential citation and Form 483, even though the logistics and practical aspects of the process may be challenging. On the other hand, even 24 h may be too long for unstable analytes, and other approaches are needed. In order to speed up the process, the bioanalyst can use the existing stock solutions, if their stability is known, or eliminate the verification of the new calibration standards and QCs.

The MVG[1] advises to use two concentration levels—high and low—in triplicate to evaluate stability, and the regular acceptance criteria of 15% applies.

Even if instability of a drug in the sample matrix is observed, the bioanalyst can take appropriate means to ensure sample integrity. For example, an antioxidant such as ascorbic acid or bisulfite can be added to avoid oxidation, or the pH can be lowered by the addition of citric acid to avoid hydrolysis of esters or lactam ring opening. Esterases can be inactivated by the addition of esterase inhibitors such as fluorides, physostigmine, or dichlorvos.[42] Addition of a derivatizing reagent can yield a stable entity. Lowering of temperatures is a good way to slow down degradation. Samples immediately after collection can be either flash frozen or kept in icy water and a refrigerated centrifuge can be used to harvest plasma. During sample processing, samples may be kept at 0–4 °C and processed quickly. Refrigerated autosamplers are readily available to ensure stability of extracted samples. If samples cannot be stored for any period of time, then samples may need to be analyzed immediately at the clinical site.[43] If the freeze–thaw stability is the problem, the samples can be divided into a number of aliquots at the clinical site, and reassays must be done using only separate aliquots.

An unusual case of instability is the situation in which a metabolite or degradant can convert back to the parent drug by undergoing a chemical reaction during the analytical process.[44] In this particular case, the instability may be detected as an increase in the parent drug concentration with repeat analysis.

16.3.8 Carryover

Carryover should be minimized and evaluated by injecting blank samples after high concentration calibration standards. Carryover should not be >20% of the LLOQ and 5% of the IS. If carryover is significant, then the analytical batch shall be organized in such a way that the carryover would be minimal and not significantly bias the results.[5] It should be stressed that carryover is highly compound and system specific. Some analytes tend to adhere to metal or polymer elements of the system and they may be difficult to eliminate. Sometimes an autosampler of a different design may provide carryover-free injections. Unfortunately, the carryover also depends on the maintenance condition of an autosampler and its history of use. Routine maintenance of the autosampler and components on the flow path such as replacement of worn-out components (e.g. injector syringe), polyether ether ketone tubing, old columns (or guard columns if they are used) will help to keep carryover in check. It should be stressed that carryover is different from contamination. Carryover occurs after injection of a highly concentrated sample, while contamination is usually random in nature, not necessarily following a highly concentrated sample. Bioanalysts should carefully evaluate the situation and tackle the problem accordingly.

16.3.9 Dilution integrity

The process of diluting samples should not impact precision or accuracy. During the validation, the analyst should prepare an "Above the Upper Limit of Quantitation" Quality Control (sometimes called dilution QC) sample, dilute it to the expected concentration (e.g. by a factor of 10) and extract the sample five times. The acceptance criteria should be 15% both for accuracy and precision. Dilution should be done using the same matrix as the sample, although not necessarily from the same individual. If the dilution factor changes or study samples are at concentrations above that of the dilution QC, a new dilution experiment is needed.

16.3.10 Matrix effect

The matrix effect is a phenomenon where the signal of an MS/MS detector is different in the presence of coeluting components of the sample as compared to the neat sample. The signal can be increased or decreased and it is caused by changes in the efficiency of ionization and droplet formation in the MS source.[45] Coeluting interferences known to cause matrix effects include salts, phospholipids, additives (e.g. esterase inhibitors), metabolites, or other endogenous components. Phospholipids are a class of lipids that are commonly found in the blood and plasma. Two common structural classes of phospholipids are glycerophospholipids and sphingomyelins. Extensive work has been reported in the literature on the matrix effect caused by phospholipids.[46–48] Phospholipids can be monitored easily by using these transitions: positive precursor ion scan of *m/z* 184, positive neutral loss scan of 141 Da, and negative precursor ion scan of *m/z* 153. In general, protein precipitation is not effective in eliminating phospholipids since they tend to remain in the organic supernatant. A number of methodologies have been devised to eliminate phospholipids, such as the use of mixed-mode SPE, LLE, and the addition of selective trivalent cations.[49]

While it is desirable to eliminate the phospholipids from the extracted samples, it is more practical to avoid coeluting phospholipids and analytes of interest. This can be easily accomplished by performing a MS scan of the aforementioned transitions to locate the retention times of the phospholipids and adjust the chromatography accordingly (e.g. change the pH of the mobile phase, use a column with a different chemistry, change the gradient). It is also worthwhile to incorporate an organic wash (e.g. 100% Mobile Phase B for 1–2 min) after the elution of the analytes to wash out the phospholipids after each injection. When it becomes impossible to separate them, a more selective extraction method will be needed.

To investigate the matrix effect quantitatively, a bioanalyst should select six individual matrices (no pooling), one lipidemic lot and one hemolyzed lot, extract them as blank samples, and then spike them with the analytes(s) at concentrations of $3 \times$ LLOQ and ULOQ in replicate; the IS should also be included.[2] The matrix factor (MF) will be calculated for each matrix by dividing the peak area in the presence of matrix by the peak area in the absence of matrix.[50] While there is no acceptance criterion for the MF, the analyst should strive to have a consistent MF across all six lots of plasma for small molecules in plasma samples. The MF should be also normalized for the IS by dividing the MF of the analyte by the MF of the IS; the CV of this ratio should be <15% as per GBMV. If the method fails to meet these, more method development work will be needed.

The matrix effect should be studied in greater detail if the formulation administered contains polyethylene glycol, polysorbate, or surfactants. The effect should also be studied in matrices from special populations such as renally/hepatically impaired subjects. It is also recommended by EMA to assess matrix effects from excipients if the drug is administered by the intravenous route.

The matrix effect on the analyte can be compensated with an SIL-IS,[46,47] or minimized by a number of methods, such as a change in chromatography or employing a thorough sample extraction procedure (e.g. SPE and LLE).

In order to avoid revalidation or unexpected events during the sample analysis phase, it is the opinion of the authors that matrix effects should be evaluated carefully during method development and appropriate methodology adopted to eliminate or minimize their impact on the method. During method development, an infusion experiment[51] can be carried out to identify the regions that have severe matrix effects. The results can guide the development of appropriate chromatography. If it is impossible to separate the coeluting interference from the analytes of interest, a more thorough extraction method will be needed.

16.3.11 Template for a method validation

Table 16.1 shows an example of a method validation template that in the opinion of the authors is up to the current regulatory expectations and industrial standards.

16.3.12 Validation report

The validation report should contain at least the following elements:

- Summary of validation performance
- Operational description of the method with literature references, if any
- Description of reference standards (batch, CoA, storage conditions, expiry dates)
- Preparation of standards and QCs (dates, matrix, anticoagulant, storage temperature)
- Acceptance criteria
- Calibration range and response function
- Table of all analytical runs with dates and outcomes (pass–fail)
- Table of calibration standards derived from accepted runs with back calculated concentrations, accuracy, and precision
- Table of QC results derived from accepted runs with accuracy and precision (both within and between run); QCs outside the acceptance criteria must be clearly indicated
- Stability data on stock solutions, working solutions, QCs
- Data indicating selectivity, appropriate LLOQ, carryover, matrix effect, dilution integrity
- Explanation of unexpected results with a description of the action taken
- Deviation from the method or applicable SOPs
- Typical chromatograms and mass spectra
- Results of Incurred Samples Reassay data, if available at that stage

Table 16.1 Method Validation Template

Validation Day	Experiment	Purpose	N
1	System suitability	Verify retention times and sensitivity of the LC–MS/MS system	5
	Blank (pooled matrix)	Quality of processing, carryover	4
	Zero sample	Impact of int. standard	1 or 2
	Calibration standards, at least six levels	Agreement with the calibration model	6 × 1 or 6 × 2
	LLOQ	Precision and accuracy at LLOQ	6
	QC.1, QC.2, and QC.3	Precision and accuracy over the calibration range	3 × 6
	Individual blanks	Specificity	6 × 1
	Lipidemic blank (if plasma or blood)	Specificity in lipidemic matrix	1
	Hemolyzed plasma (if plasma)	Specificity in hemolyzed plasma	1
	QC.1, QC.2, and QC.3, extracted for recovery	Extraction recovery	3 × 3
	QC.1, QC.2, and QC.3, unextracted, for recovery	Extraction recovery	3 × 3
	Above the ULOQ–QC	Integrity of dilution	5
	QC.2	Autosampler stability	3
2	System suitability	Verify retention times and sensitivity of the LC–MS/MS system	5
	Blank (pooled matrix)	Quality of processing, carryover	4
	Zero sample	Impact of internal standard	1 or 2
	Calibration standards, at least six levels	Agreement with the calibration model	6 × 1 or 6 × 2
	LLOQ	Precision and accuracy at the LLOQ	6
	QC.1, QC.2, and QC.3	Precision and accuracy over the calibration range	3 × 6
	QC.1 and QC.3, in the presence of extracted blank	Matrix effect in six individual matrices	2 × 6 × 3
	QC.1 and QC.3, neat	Matrix effect	2 × 3
	QC.2	Autosampler stability	3
3	System suitability	Verify retention times and sensitivity of the LC–MS/MS system	5
	Blank (pooled matrix)	Quality of processing, carryover	4
	Zero sample	Impact of internal standard	1 or 2
	Calibration standards, at least six levels	Agreement with the calibration model	6 × 1 or 6 × 2
	LLOQ	Precision and accuracy at the LLOQ	6
	QC.1, QC.2, and QC.3	Precision and accuracy over the calibration range	3 × 6

Table 16.1 Method Validation Template *(continued)*

Validation Day	Experiment	Purpose	N
	QC. 1 and QC.3	Processing stability at a selected temperature	2 × 3 × 3
	QC.1 and QC.3	Freeze–thaw stability	2 × 3
	QC.1 and QC.3	Interference experiments	2 × 3
	QC.2	Autosampler stability	3
	QC.1 and QC.3	Interference by comedication	2 × 3

N = *number of samples.*

16.4 APPLICATION OF A VALIDATED METHOD TO ROUTINE DRUG ANALYSIS

16.4.1 Organization of the analytical batch

The size of an analytical batch is limited by practical concern such as instrument capacity and stability of the drug or system. In most studies, all clinical samples from the same subject should be analyzed, if possible, in the same run to avoid between-run variability, which tends to be greater than the within-run variability. This is of particular importance in bioequivalence studies.

An analytical batch should be started by injection of a system suitability/reference solution followed by crucial samples (LLOQ, blank, ULOQ) so as to provide an early indication of whether the run is under control, and still allow the analyst an opportunity to take corrective action, if necessary. The corrective actions can include a change of the in-line filter, guard column, analytical column, lamp, or cleaning the source of the MS detector.

As for the placement of calibration standards in the batch, there is no set rule. Some analysts prefer to disperse them evenly throughout the batch, while others would start a batch with the calibration curve. QCs should be prepared at a minimum of three concentration levels and analyzed in duplicate with each analytical batch. One set of QCs should be close to the LLOQ and two to three times higher than the LLOQ. The second set should be at approximately 40–60% of the ULOQ, and the third at 70–90% of the ULOQ. QCs should be spread evenly throughout the analytical batch. There shall be at least six QCs in a batch or 5% of all samples in the batch, whichever is greater. Additional sets of QCs at different concentrations can be prepared to mirror the concentrations of study samples observed in the study.

16.4.2 Acceptance criteria

The commonly used run acceptance criteria are as follows. For the calibration curve, at least six calibration standards constituting at least 75% of the total number of standards must be within ±15% of the nominal concentration; in the case of the LLOQ the difference can be ±20%. This means that if eight calibration standards are extracted, at least six (75%) must be used to set up the calibration curve. If nine standards are extracted, at least seven (78%) must be acceptable for the calibration curve to be valid.

QC samples are the ultimate tool to accept or reject a batch of samples. The rule "4–6–15" is generally accepted, which means that six QCs at three concentration levels in duplicate must be extracted with a batch of study samples (<100), four (4) out of these six (6) must be within ±15% from the nominal, and each QC level must be represented in these acceptable QCs.

Sometimes additional acceptance criteria are included. These, for example, can be:

- Required coefficient of determination (r^2) of the calibration curve of at least 0.99
- Lack of interferences in drug-free samples
- Consistency of the absolute peak area or height of an IS
- Bracketing of samples: if some of the QCs fail, the study samples between those failing shall be rejected and reassayed
- Special QCs, such as hydrolysis QCs, if the assay involves, for example, an enzymatic reaction liberating a drug from its glucuronide or sulfate metabolite
- Use of a dilution QC

16.4.3 Dilutions

There are three reasons for sample dilution in bioanalyses: (a) to bring samples with a concentration above the ULOQ within the calibration range, (b) for parallelism in live blood analysis (LBAs) (not covered in this chapter), and (c) insufficient volume of samples. In all cases, the dilution should be done using the same matrix as the study samples. One has to be careful in performing dilutions of samples with concentrations close to the LLOQ, as the diluted samples may be classified as below the quantitation limit (BQL). For example, if the LLOQ is 1 ng/mL, the dilution factor is 2, and the back calculated concentration (no dilution factor included) is 0.77 ng/mL, the reported concentration should be BQL, and not 1.54 ng/mL.

16.4.4 Reassays

Every company or laboratory should develop and describe in an SOP its policy on repeat analyses. This policy has to be made available before starting a study. First, it must decide who is making decisions with regard to reassays. As for the cases involving some kind of analytical or technical difficulties, the decision should be left in the hands of the bioanalytical personnel. These instances can be:

- Poor chromatography: which may include interfering peaks making the integration impossible; no peaks at all, a chromatographic pattern very different from the expected one, collapse of the stationary phase, etc.
- Lost sample (LS): which may include dropped samples at any step of processing, leaking pipette tips, leaking screw caps, etc.
- Bad processing (BP): which may include any kind of human or robotic error—not adding a reagent, forgetting to add an IS, adding two portions of IS, etc.

Any of the above-mentioned errors should leave an audit trail in the form of a deficient chromatogram, note to the file listing the LSs or describing errors in the processing, or computer printouts in the case of robotic systems. No numerical results are associated with such attempts and they should be repeated as a single sample.

Occasionally, clinical samples exhibit concentrations above the validated range (AQL). Such samples should be diluted with the same matrix and repeated as a single sample.

On the other hand, study samples sometimes provide results which formally and chromatographically look correct, yet defy logic or seriously contradict previous results. The SOP must specify who identifies these potential reassays—the pharmacokineticist or bioanalyst and on what grounds. The goal of a bioanalyst or a pharmacokineticist is neither to squeeze study results into a preconceived model nor to smooth out pharmacokinetic profiles. At the same time, it is a scientific duty to challenge suspected results. One may suspect a pharmacokinetic outlier, if a predose sample from naive subjects contains a measurable drug concentration, if the pharmacokinetic profile exhibits a split or double maximum contrary to known pharmacokinetics, or if concentrations are very different (500–1000%) than expected. Such samples, which could be called "suspected outliers", provide numerical values, and repeats should be done in duplicate. The institution should also develop a comprehensive decision tree dictating a verdict in every foreseeable case to eliminate arbitrary decisions. A very good decision tree has been developed by Lang and Bolton.[52,53] Briefly, a 15% agreement between data is considered a confirmation if the repeats are done in duplicate, or 30% if only one repeat was possible. If results are too far apart, no result is reported.

It should also be noted that infrequent and random outliers do not influence the outcome of a study, if an appropriate number of subjects is selected to ensure appropriate statistical power. Pharmacokinetic reassays are discouraged in bioequivalence studies.

16.4.5 Incurred sample reassay or reanalysis

The issue of bioanalytical method reproducibility has been on the agenda of regulatory authorities since the 1990s, when Health Canada requested reanalysis for bioequivalence and bioavailability studies. This was mentioned also in the Crystal City I conference report.[3] However, in 2003, Health Canada removed this requirement. In the meantime, the FDA gathered evidence and observations based on the analysis of pharmacokinetic repeats and repeats in multianalyte assays. It was quite obvious that there were examples of bioanalytical assays that were not reproducible when applied to individual clinical or animal samples despite being formally acceptable on the basis of calibration standards and Quality Controls. After meetings in 2007 and 2008, the incurred sample reassay or reanalysis (ISR) has been widely adopted by the industry.[13] The goal of the ISR is to provide evidence of the method reproducibility, and detect either poor methods or poor execution of a good method. One has also to keep in mind that a scientist's understanding of the molecule's behavior and its interactions increase as the drug candidate progresses through the drug development process, and crucial information or understanding may not be available at the time of a first study.

Essentially, selected samples from a given study are reanalyzed as soon as possible after their initial analysis so as to detect and correct problems immediately. The ISR runs must be separate from the regular production runs in such a sense that there should not be mixed production-ISR runs. If the sample was initially assayed in dilution, the ISR of this sample should be done with the same dilution ratio. In general, 10% of samples (or a minimum of 20) should be reanalyzed for studies with <1000 samples. For larger studies, other rules can apply aimed at reducing the number of samples selected for ISR. While there is no official guidance on selecting particular samples for a given study, samples should be selected from different subjects (animals), time points, and dose groups. No pooled samples should be used unless the matrix volume is very low.

White papers and EMA[2,13,33] provide enough advice to make correct decisions. The ISR should be performed when a method is applied for the first in human (FIH) studies, in a new target population (disease state, renal and hepatic impairment), all bioequivalence studies, upon major changes to the method, after a method transfer to a new laboratory, or in any study where scientific rationale suggests reassaying. As for drug–drug interaction studies, the opinions are divided, perhaps proving that the coadministered drug and/or its metabolites do not interfere with the analyte of interest is sufficient. In animal studies, the ISR should be performed once per species and matrix. Each laboratory must have an SOP to provide general rules on selecting samples, e.g. samples at C_{max} from different subjects, across different doses, number of samples, and selecting studies for ISR.

The results from initial analysis are compared with the second analysis and calculated according to the equation:

$$\%\text{Diff (variability)} = \frac{\text{Reanalysis concentration} - \text{Original concentration}}{0.5 \times (\text{Reanalysis concentration} + \text{Original concentration})} \times 100\% \quad (16.4)$$

The difference between the two analyses should be <20% for chemical assays (e.g. LC–MS/MS-based assays), and 30% for LBAs for two-thirds of the total number of samples reanalyzed. The ISR results should be included in the bioanalytical study report, and/or in the validation report, if these results are available by the time the report is finalized.

In the event that the results do not meet the acceptance criteria, an unexpected event investigation should be initiated. Based on the findings, resolution plans will be devised and executed. The impact on the quality of the bioanalytical data generated will be assessed based on the findings from the investigation. An investigation report should be compiled to summarize the investigation. Even in successful ISR experiments the bioanalyst is well advised to pay close attention to results outside the acceptance criteria, in particular if these are concentrated in discrete runs.

16.5 BIOANALYTICAL REPORT

A study should end with a report describing the procedure, its performance, and study results, where applicable. The data should be presented as a narrative and in tabulated form, and include:

- Operational description of the assay procedure.
- Information on reference standards (batch, CoA, storage conditions) and ISs.
- Preparation of standards and QCs (dates, matrix, anticoagulant, storage temperature).
- Acceptance criteria.
- Sample tracking—dates of receipt, conditions, storage location, and temperature.
- Table of all analytical runs with dates and outcomes (pass–fail).
- Equations used for back calculating of results.
- Table of accepted runs with the calibration curve parameters and correlation coefficients.

  Parameters should contain a sufficient number of digits to back calculate concentrations accurately. Slope of the calibration curve may change from day to day, yet it remains a valuable diagnostic tool. A consistent value of the slope suggests a solid assay. Dramatic changes may suggest modification to the method, errors, or maintenance done on a detector. Values of the intercept consistently above zero may suggest an interfering peak hidden underneath the peak of interest.
- A table of calibration standards derived from accepted runs with back calculated concentrations, accuracy, and precision.

The table should be complete, with no empty spaces. If a standard has been rejected, its value should be provided anyway in brackets or with an asterisk, and an explanation should be provided. A sample lost or disqualified for whatever reasons should be flagged as such. Interday precision and accuracy should be calculated providing the mean, SD, CV, % of nominal and number of observations. An adequate number of significant digits should be provided, so a reviewer can verify calculations and arrive at the same results.

- A table of QC results derived from accepted runs with accuracy and precision: QCs outside the acceptance criteria must be clearly indicated.
  All the rules specified above apply to this table also. Additionally, all evaluable QC values have to be reported and included in the statistics, whether or not these QCs meet the acceptance criteria. The bioanalyst may provide two sets of statistics: one using all the data, and the other excluding the results outside the acceptance criteria. Precision and accuracy calculated on QC data represent the precision and accuracy of the method.
- Explanation of unexpected results with a description of the action taken.
- A list of requested PK reassays, including the reason, and original and reassay results.
- Reasons for missing samples.
- Deviation from the method or applicable SOPs.
- Documentation for reintegrated data, including the initial and repeat integration results, reason for reintegration, the requestor of reintegration, and the manager authorizing the reintegration.
- Chromatograms from 20% of subjects in the pivotal bioequivalence studies; 5% in other studies.
- ISR results from the study, if applicable.

In addition, the report should contain the list of abbreviations and codes used, reference list, copies of the references, and copies of relevant SOPs.

16.6 VALIDATION, PARTIAL VALIDATION, AND CROSSVALIDATION

There are several kinds of validation:

- Original validation (before study, immediately after method development); described in Section 16.3
- Within-study validation; described in Section 16.4
- Partial validation performed to an already validated method if changes are made to the method Some of the changes may include:
  a. Introducing new analyst(s) to the method
  b. Change of platform (e.g. change in the LC system, mass spectrometer) within the same laboratory
  c. Change in anticoagulant for blood or plasma samples
  d. Change of species within the matrix
  e. Change of volume taken for extraction
  f. Modification to the validated range of concentrations
  g. Modification to extraction procedure, etc.
  A validated method may be altered intentionally or inadvertently. In any case, the change should be described in a note to file, and its potential impact evaluated. Intentional modifications should

be authorized by an analytical director and rationale provided in writing before its implementation, for example, in a form of a method validation amendment.

It is recommended that the analyzing laboratory should have an appropriate SOP to determine when a partial validation, full validation, or revalidation is warranted.

- Crossvalidation: should be performed when two or more methods are used to produce data within the same study or across many studies, or the same method is used to support a study at two different laboratories. Crossvalidation should be performed before committing study samples to analysis under these circumstances. The crossvalidation can be done using spiked QCs or authentic study samples. In the former, the acceptance criteria of ±15% should be used. If study samples are used, the acceptance criteria typical for an ISR study are recommended, i.e. ±20% for at least two-thirds of the samples.

16.7 BIOANALYTICAL METHOD VALIDATION—OTHER PARAMETERS AND ISSUES

Sections 16.1–16.6 presented the current state of the art in the area of method validation and execution of a bioanalytical study. In this section the authors evaluate critically some aspects of validation where either there is no consensus, or an alternative solution may be needed, or issues that at this time have not been adequately addressed by the bioanalytical community.

16.7.1 Chromatography

It may appear strange that chromatography has to be brought up as an important but almost forgotten aspect of bioanalytical method validation. The tandem mass spectrometric detector is such a powerful tool and is used so commonly that the proper chromatography for many can be an afterthought. In many cases the tandem mass spectrometric detector can even correct deficiencies of poorly developed chromatographic methods. Very frequently, a generic gradient system on any C8 or C18 column is applied without much consideration of the analytes to obtain as short a run time as possible. There are several important reasons why chromatography should not be too simplistic:

- Peaks of interest should be separated from the area where most endogenous compounds in the matrix elute. These endogenous materials can be observed as a dip(s) in the Total Ion Current. In these areas of chromatograms there can be a huge competition for ionization, potential drop in the sensitivity, and significant matrix effects. One also cannot forget about late-eluting peaks, which may show up even a long time after the original injection and interfere with subsequent sample injections.[46]
- Conversion of unstable metabolites such as N-oxides or glucuronides, which typically elute before the parent drug, may convert back to the drug at a high temperature in the MS ion source. If there is no chromatographic separation between the parent and metabolites, this will result in artificially elevated concentrations of the parent drug.

It is opinion of the authors that a conscientious bioanalyst should find the right balance between good chromatography and the run time.

16.7.2 Alternative acceptance criteria

The fixed range, commonly used as run acceptance criteria "4–6–15", which means that for a run to be accepted it has to have an acceptable calibration curve and four out of six QCs within 15% of the nominal value with all QC levels being represented, is unfortunately arbitrary, ad hoc,[54,55] and unscientific. The only criterion is accuracy, and an assumption is made that precision will be satisfactory. However, it describes quite well what the industry and regulatory agencies are willing to accept in terms of a balance between the quality and efficiency. The expectation that the overall precision and accuracy obtained in a study based on these acceptance criteria will be always $\leq$15% is arguably over-optimistic and unfounded.

Hartmann et al.[56] calculated that in order to obtain mean values within the limits of ±15% and with a probability of 95%, the bias and RSD (%CV) should be 8% with $n = 5$. The fixed range approach is totally pragmatic, not based on statistical principles, and confuses precision and accuracy. Acceptance criteria should be scientifically valid, able to detect errors and false alarms, easy to use, and provide immediate answers. The MVG recognized that a confidence interval (CI) approach is an alternative for acceptance criteria.

The analyst makes measurements which are related to the concentration, not the actual concentration itself. These measurements provide only a certain probability that the true concentration will be within a certain range.[57] If analytical errors are random they follow the normal Gaussian distribution. Hence, 68% of the results fall within one SD of the mean, 95% within 1.96 (popular 2) of the mean, and 99.7% within 3.09 (popular 3) of the mean.

The 99% CI is equal to

$$99\%\text{CI} = \text{mean} \pm 2.58s \tag{16.5}$$

where s is the SD (see also Chapter 2). Gross errors (e.g. bad chromatography or sample processing) should be eliminated from calculations. The acceptance criterion is simple, all QCs must fall within the CI.

The CI-based acceptance criteria are easy to use and provide an immediate answer, although they do not address accuracy. They can be even more liberal than the fixed range "4–6–15" rule. At low concentrations the RSD of many bioanalytical methods is on the order of 10–15%, and hence the acceptance criteria will be ±25.8–38.7%.

Another concept of run acceptance criteria enjoys at least moderate support of the bioanalytical community. It is called "total error acceptance criteria" and requires the summation of the absolute values of the bias and CV and to apply the rule "4–6–20" or "4–6–25". In simplified form it is used in clinical analysis and has been incorporated into some Laboratory Information Management Systems (LIMS).

Much more sophisticated statistical acceptance criteria for a method validation based on total error was proposed by Hoffman and Kringle.[55] This approach proposes a formal statistical framework for evaluation of a bioanalytical method. The three reports[14–16] prepared by a commission of the SFSTP describing validation procedures and acceptance criteria are also based on the total error concept and rigorous statistics.

16.7.3 Regression selection

The MVG stipulates in at least two places that the relationship between the response and the concentration be established using statistical tests for goodness of fit. Though the bioanalytical community generally treats the text of the MVG with great respect, this requirement in most of the cases is merrily ignored. Most of the laboratories apply without much consideration their common regression and the weighting scheme which in most cases is $1/x^2$ and to a much lesser extent, $1/y^2$. What matters in practice is the quality of the inverse prediction more than the goodness of fit.[16] There are a number of publications describing rational and statistically valid procedures for selecting the proper regression.[58–60]

The MVG leans heavily toward the use of the simplest response function, i.e. linear calibration curve. From the beginning, there was a lot of confusion with regard to linearity of the calibration curve. The bioanalytical community assumed that linearity means a rectilinear curve expressed by the equation

$$Y = ax + b \tag{16.6}$$

where a is the slope, and b is the intercept of the calibration line.

In statistical sense, a function is "linear" if it is a linear combination of its parameters. The quadratic function is also linear in its parameters although its graphical plot is not a straight line.[16]

This emphasis on rectilinearity may cause problems as well. A subjective judgment as to whether or not a set of points represents a linear model may be at variance with statistical tests, and this mistake can be in either direction.[61] A linear calibration curve may be forced on data that are slightly, but clearly nonlinear.

There could be several causes for the curving of calibration lines in chromatographic assays; receptor-binding assays are inherently nonlinear. Certain kinds of detectors provide nonlinear responses, such as the electron capture detector in gas chromatography, or in fact any detector if the range of calibration curve is excessive and covers concentrations of several orders of magnitude. To show the detector linearity one needs to inject increasing amounts of the unextracted analyte solutions and record responses. The analytical process may be also responsible for non-linearity; for example, variable extraction recovery or adsorption. To detect and document non-linearity one may use a number of techniques[62–64]:

- Visual assessment—subjective and requires an expertise in analytical methodology.
- Conventional analysis stemming from least squares regression—several approaches can be used such as components of variance, lack-of-fit testing, quadratic regression.
- Analysis of consecutive differences—simulates the visual assessment of linearity.
- Comparison of observed values against expected results (residuals, see 2.2.2.1).

A very simple test for linearity based on the residuals (% deviation from nominal) is called the "sign test".[61] The signs of residuals should be distributed at random between plus and minus, if no systematic error is involved. If a sequence of signs looks more like –++++-, a curvature of the regression line and a lack of linear fit could be suspected.

Particularly useful as a diagnostic tool is the analysis of consecutive differences, also called "deltas", or rather a variation of it. Peak height (area) ratio or absolute peak height (area) divided by the nominal concentration gives a value which is called a "response factor" or "unit ratio", and is readily available in some LIMS. This value represents the slope of the calibration curve at this point, and should be constant and equal to the overall slope of the rectilinear calibration curve. If a decreasing/increasing trend in the value is visible, the response function cannot be linear. Additionally, if response factors are constant over the whole calibration curve with the exception of the lowest standards, an interference hidden underneath the peak of the analyte should be suspected.

The coefficient of determination (r^2) is used frequently as a measure of the goodness of fit. However, it is a rather poor predictor of the fit. Even poor calibration curves may have coefficients of determination quite high and >0.99; this value is frequently used as an acceptance criterion or at least an indicator for the goodness of fit. Also, as discussed earlier in Chapter 2, one should not rely too heavily on the r^2 value as a measure of linearity because this parameter includes curvature as well as random errors. There is some practical use for the coefficient of determination in one sense that although a high correlation coefficient does not ensure a good calibration curve, but a low one, say <0.99, indicates that the calibration curve is biased with serious errors and probably is unacceptable.

To conclude, an automatic application of linear regression may be as wrong as the use of a complicated model to a simple chromatographic assay.

16.7.4 Blood samples

For convenience, plasma, rather than blood, is the most common matrix in bioanalysis. However, several types of experiments in blood are necessary at or before the stage of validation in order to provide meaningful data.[65]

In most cases, the blood/plasma ratio of analyte concentrations is approximately 1 ± 0.5. It is well known however, that several drugs bind preferentially to red-blood-cell membranes, hemoglobin, or carbonic anhydrase. In these cases the blood-to-plasma ratio is much >1. These, for example, are (blood/plasma ratios indicated in parentheses): pimobendan (3.2–4.5), cyclosporin A (2.0–4.6), tacrolimus (22.6–55.5), methazolamide (241), acetazolamide (2.9), chlorthalidone (30.7–32), chloroquine (3.5), rapamycin (14.3), and ribavirin (~100). The blood/plasma ratio must be known to ensure that plasma is indeed the most appropriate medium for bioanalysis and pharmacokinetic evaluation. A potential consequence of using plasma as a matrix for an analyte with a high blood to plasma ratio is that a small amount of hemolysis can cause an artificially high concentration of the analyte in the sample.

Another issue to consider is the thermodynamics and kinetics of red-blood-cell partitioning. There are several drugs (e.g. cyclosporin A, amitriptyline, and nortriptyline) that partition differentially at different temperatures. The erythrocyte/plasma ratio for amitriptyline at 2–10 °C was 0.3, and 0.5 at 40 °C. For nortriptyline, the ratio was approximately 0.85 at lower temperatures and 1.25 at 40 °C.[66] If the drug partitioning shows such a dependency on temperature, a bioanalyst has to devise a proper plasma-harvesting procedure for the clinical sites. The most common procedures of obtaining the blood, cooling in an ice bath, and centrifuging at 4 °C may not be appropriate, and perhaps centrifuging at 37 °C would provide better results. What is more, applying variable procedures with regards to the centrifuging temperature may introduce immediately a sizable variation of the drug concentration in plasma.

16.7.5 Drug stability in blood

Many drugs are actively metabolized in the blood cells of humans, e.g. chlorpromazine, captopril, haloperidol, heroin, isoproterenol, ribavirin, testosterone, and many more. It is important to devise a plasma-collecting procedure so that the integrity of samples will be maintained and the drug concentration in the plasma will be the same at the time of phlebotomy as at the end of the process when plasma samples are placed in a freezer. The bioanalyst must establish temperature conditions and time limits for blood processing and plasma harvesting. A frequently encountered problem is that a bioanalytical method for plasma exists at this stage, but there are no plans for studies in blood and hence no need to validate the method in whole blood. It is recommended either to apply the validated plasma method to the plasma obtained from whole blood in a controlled stability experiment (so-called "whole-blood stability" evaluation), or to qualify the plasma method for use in whole blood in such an experiment.[34] Methods that involve liquid extraction in most cases will work equally well for plasma and blood. Methods that employ protein precipitation may require more robust precipitation and increasing the ratio of the precipitating agent (methanol, acetonitrile) to blood to about 10. No concentration data are needed, because the peak area ratio of the drug to the IS plotted against the time or temperature should reveal instability.

16.7.6 The "other" matrix effect

In today's bioanalysis dominated by the LC–MS/MS technique, the term "matrix effect" means almost exclusively the effect that endogenous extracted components of the matrix have on the ionization in an MS detector that results in the decreasing or increasing of the instrument's response. However, there are "other types" of matrix effects. Here are some of these as reported in the literature or experienced by the authors:

- Variable protein binding in FVB/N strain of mouse plasma

An LLE method has been developed to quantify reserpine in mouse plasma.[67] The method performed well for the standards and QCs prepared in control plasma, but for the study samples in the FVB/N mouse plasma, the IS was not detected in 30% of samples, and was decreased by a factor of 5–10 in additional 20% of samples. The phenomenon was attributed to specific protein binding in FVB/N plasma, which was eliminated by the addition of sodium EDTA.

- pH of samples

A liquid extraction was performed on plasma samples without buffering them at the physiological pH of plasma (7.4), which was an appropriate pH for that extraction. However, stored plasma releases carbon dioxide, which changes pH. Freeze thawed plasma may reach a pH of 8.5. A methanolic supernatant evaporated and redissolved can have a pH of 9.5.[68] These pH increases can have an adverse effect on the extraction and stability of analytes. Blood also may be affected by the same process but to a lesser extent.

- Ionic strength in the ion-exchange process

Urine samples were injected directly into a column switching system containing an ion-exchange column.[69] Some samples provided suspicious results. It was discovered that these were very concentrated urine samples with much higher levels of salts. The volume of injection was reduced by a factor of 10 and the matrix effect disappeared.

- Protein content

Recovery during protein precipitation is frequently incomplete; solubility of the drug and its protein binding plays a role. Total concentration of protein in human serum varies between 58 and 77 mg/mL depending on age and gender,[70] less in undernourished and sick individuals. In one experiment, recovery of triamterene from serum was measured at 50 ng/mL during precipitation with 10% perchloric acid. The serum was diluted with 0.9% sodium chloride solution in the following ratios (v/v): undiluted, 2:1, 1:1, and 1:2. The recoveries were 64%, 75%, 80%, and 88%. Obviously, the potential for significant errors exists due to differing protein content. An appropriate IS corrected the recovery problems, but the errors might have gone undetected without it.

- Presence of Protease Inhibitor Cocktail (PIC)

On occasion it was observed that in an LC–MS/MS-based assay, the absolute peak areas of the ISs in subject samples were different from those observed in standards and QCs. A hypothesis that the variability of adding a PIC to the samples was responsible for this effect was tested. The targeted concentration of PIC in the plasma samples was 1%. Control samples were spiked into plasma that contained PIC added at 0%, 1%, 2%, 4%, and 6%. The samples were extracted according to the method in triplicate. It appeared that the absolute peak area of the drug was inversely related to the concentration of the PIC in the plasma; the loss was at worst 24%. However, the IS compensated for this effect and the observed changes in the peak area ratio were approximately 5%; well within the precision and the accuracy of the assay.

16.7.7 Hemolyzed plasma

The blood from clinical or toxicology studies on occasions is sometimes partially hemolyzed due to drug action (e.g. ribavirin), disease state, addition of additive used to stabilize the drug, or technical errors while obtaining the blood. Hemoglobin and bilirubin are released from the red blood cells causing the plasma to appear pinkish to deep red. The degree of hemolysis is measured by the concentration of hemoglobin in the plasma. Plasma with concentration <30–50 mg/dL of hemoglobin is considered not hemolyzed, while plasma with hemoglobin >300 mg/dL is considered badly (grossly) hemolyzed. The degree of hemolysis can be easily estimated using the Becton Dickinson scale as presented by Hughes.[71] According to this paper, the impact of hemolysis can be considered a special case of matrix effect, and is caused by either the presence of additional interfering peaks or serious suppression of the MS/MS signal due to the presence of hemoglobin, bilirubin, and other endogenous components of erythrocytes. Better extraction procedures (LLE, or SPE instead of protein precipitation), more selective chromatography, replacing of an analog IS with an SIL-IS, or dilution of hemolyzed plasma eliminated these effects. The authors propose doing validation experiments using simulated hemolyzed plasma that is produced by adding 2% of totally hemolyzed plasma to regular plasma and which corresponds to approximately 550 mg/dL hemoglobin. If the back calculated concentrations of QCs prepared in hemolyzed plasma are outside of the regular ±15% limit such samples cannot be analyzed, and the experiment should be repeated using 1% of blood in plasma.

In the experience of the authors, one of their assays could not provide satisfactory results in plasma containing approximately 1060 mg/dL of hemoglobin. In fact, the precipitation of proteins was not complete with the supernatant being visibly pinkish. The chromatographic column could not handle the protein load and failed in the middle of the run. The experiment was performed successfully at 530 mg/dL of hemoglobin.

16.7.8 Lipidemic plasma

Lipidemic plasma should be considered as part of method validation for several reasons:

- Specificity: Brazilian Agência Nacional de Vigilância Sanitária document demands to run at least one lot of lipidemic plasma and one lot of hemolyzed plasma as a part of the specificity experiment.[72]
- Postprandial plasmas have elevated levels of lipids and proteins; it is very easy to see as the plasma can be very cloudy. The extra lipids can be extracted during the extraction process, and eventually injected into the LC–MS/MS system and cause potential signal suppression, increase in the column backpressure, or even column overload and collapse.
- The extra lipids present in the plasma can be considered as one more reagent or organic solvent in the system. Hence, it may change extraction recovery during the LLE or SPE. An SIL-IS will compensate for it, but an analog IS may not.

16.7.9 Internal standard consistency

The consistency of an IS peak area or height is another parameter of increasing interest in the bioanalytical community and regulatory agencies.[34] The fact is that in today's mostly automated methods there is no good reason for the great variability of the IS, though the response from the tandem mass spectrometer has certain inherent variability and is nowhere close to the stability of simpler UV or fluorescence detectors. It is assumed that the variability is caused by the unidentified matrix effects, and that such a method is perhaps underdeveloped and not sufficiently robust. Two approaches have emerged so far on how to deal with a highly variable IS response:

- Set up a fixed range of 50–150% around the mean of all samples (standards, QC's, zeroes, study samples), and flag and possibly repeat samples outside the range.
- Trend the absolute peak area/height in the known samples (standards, zeroes, QCs) and investigate unknown samples in which the ISs are outside these expected limits.

It must be stressed that the nature of an ideal IS, a C^{13} stable isotope label, is to compensate for all variables in the analytical process. The highly variable IS cannot be an automatic ground for sample or batch rejection, but can trigger an investigation whether or not the IS indeed compensates for these variables. One may use a standard addition method to prove the point.

The most commonly used SIL-ISs are the deuterated analytes, which contain three to six deuterium atoms. The deuterated analytes are a little more hydrophilic as compared to the analyte, their retention times are normally 1–2 s shorter than those of the analytes, and in rare cases such an IS may not behave in the same way as the analyte.[73]

16.7.10 Tubes and containers

In the opinion of the authors the bioanalytical community devotes too little attention to the tubes and containers. Even though proteins and lipids can help to form an emulsion which could aid the solubility, or homogeneity of poorly soluble analytes, or cover the active sites that may bind to a drug, they still do not ensure the avoidance of adsorptive losses.[67] It is even worse in the case of matrices that contain very little or no proteins or lipids, such as urine, CSF, bronchoalveolar lavage, tears, and so on.

In the laboratory of one of the authors, the aqueous solution of an analyte well known for adhering to various surfaces was placed in 11 different types of glass and polymer tubes, at the volume of 1 mL each, at a concentration 100 ng/mL. The analyte was known not to be light sensitive. The solutions were left in the tubes for 1 h and vortexed from time to time. After 1 h, aliquots were injected into a simple HPLC/UV system to ensure the best reproducibility of the response.

The differences were quite drastic. Selecting a wrong type of tube to collect the CSF samples decreased the apparent concentration by as much as 50%. The bioanalyst should propose the proper type of tubes to collect the samples and the proper procedures. In this example, addition of 10% isopropanol to CSF was needed to avoid adsorption to the tubes and maintain the integrity of samples.

16.7.11 Robustness testing

According to the ICH (Q2(R1)), the robustness/ruggedness of an analytical procedure is a measure of its capacity to remain unaffected by small, but deliberate variations in method parameters and provides an indication of its reliability during normal usage. The factors influencing the assay can be quantitative (pH, concentration, temperature, time) or qualitative (batch of HPLC columns).[74] It is impractical to investigate all parameters of a method. The bioanalyst who is the originator of a method certainly knows the crucial factors of the method. At least one or two of them should be investigated. The introduced changes should mimic potential errors in day-to-day laboratory operations. These, for example could be:

- pH of mobile phase

Prepare mobile phase which deliberately is 0.2 pH unit off the target on the positive or negative side.

- Percentage of organic solvent in the mobile phase

Prepare mobile phase which deliberately is slightly off in terms of organic content, such as 62% of methanol instead of 65%.

- Composition of extracting solvent

If for example the extracting solvent is a mixture of hexane/isopropanol at a ratio of 90:10, try extracting with a mixture at a ratio of 95:5.

- If a method calls for the completion of extraction within certain time limits for the reason of stability, set aside a set of samples and complete it within time limits that are 50% longer.

Vander Heydan et al.[74] provided comprehensive and elaborate guidance on robustness testing, including the appropriate statistics. As the bioanalytical community, generally speaking, tends to opt for simple solutions, the pragmatic acceptance criteria of ±15% used for the batch acceptance and stability testing can be applied here. The robustness testing does not need to be extensive or costly. In many cases, the existing extracts can be reused and reinjected under different conditions of the assay.

16.7.12 Bioanalysis in tissues and homogenates

While plasma, serum, and urine are the most common matrices in bioanalysis, it is not uncommon to analyze tissue samples such as liver, brain, heart, and kidney. In general, the workflow "grind-extract-measure" applies to tissue sample analysis. Representative tissue sample is excised from the organ, homogenized, followed by extracting the homogenates and analysis by LC–MS/MS.[75]

It is important to obtain representative tissue samples. Unlike plasma or serum samples, drug concentrations can vary between different parts of an organ, e.g. a drug-coated stent that releases the drug slowly will have a concentration gradient in the surrounding tissue, with higher concentration closer to the stent.[76,77] Therefore, a thorough understanding of which part of the organ provides meaningful drug measurement is needed. For some cases, whole organ (e.g. liver) is required while in other cases, a slice of the whole organ (e.g. brain) is sufficient.

Generally, a buffer such as phosphate buffer or phosphate buffered saline (PBS) is used to homogenize the tissue samples. In some cases, water or plasma or even whole blood is used instead. It is recommended that the bioanalysts carefully evaluate the recovery of the analytes in different reagents. While water and PBS are relatively easy to handle, they may not be able to provide good recovery over the range of the curve due to nonspecific binding.

Another key factor is carryover. An adequate washing procedure should be put in place to avoid carryover of drugs from previous samples during homogenization. The homogenate can be extracted directly with the designated extraction method, or it can be diluted with plasma (e.g. 1:10 or an even higher dilution factor) and then extracted accordingly with the plasma method. The extracted samples will then be analyzed by LC–MS/MS or other techniques.

In general, the matrix effect is more severe with tissue homogenates due to the presence of a large number of endogenous components. It is recommended that the bioanalysts carefully evaluate the matrix effect and adopt an appropriate extraction procedure to ensure ruggedness of the assay. The use of a SIL-IS is highly recommended if available.

In general, there are no "official" acceptance criteria for tissue sample bioanalysis. Zhang et al.[76] reported a validated method on zotarolimus in stented swine arteries. Nonetheless, a "fit-for-purpose" qualification is usually adopted. It can include an accuracy and precision run, recovery and matrix effect, and stability evaluation. It is not necessary to use the same criteria as plasma methods such as $\pm15\%$ from nominal concentration for stability evaluation. The bioanalysts should carefully consider the performance of the assay before deciding on acceptance criteria.

16.8 EMERGING TECHNOLOGIES IN BIOANALYSIS

16.8.1 Dried blood spot

Dried blood spot (DBS) is a microsampling technique that was proposed in 1963 by Guthrie[78] to collect finger-pricked capillary blood from neonates for detecting genetic metabolic disorders. Over the last several years the technique has been applied to drug development to analyze human and animal samples. There are several practical advantages of this approach:

- Reduction in sample volume; very important in pediatric studies and small animals
- Reduction in the number of animals for toxicokinetic studies; no more satellite groups needed
- Reduction in costs of collecting, storing, and shipping
- Better sample stability for certain chemotypes

DBS method development and validation are based on the same principles as regular chemical or chromatographic methods with several technique-specific modifications.[79–81] Before a bioanalyst embarks on employing such a strategy there must be a clear understanding that at this time there has not been any drug submission accepted by regulatory agencies based solely on DBS. At the time of this writing, the FDA considers DBS methods as supporting evidence, and the early adopters must provide the conventional plasma data along with the bridging studies supporting the use of DBS methods.[82]

The special requirements in DBS method development and validation are:

- Selection of the paper for blood collection, both treated with stabilizing agents and untreated; variability of the cards
- Assay robustness related to pipetting variance (10–30 μL)
- Whole-blood stability
- On card stability
- Dilution technique—smaller punch or dilution with control matrix extract
- Application of IS—no IS on the card, IS over DBS, IS under DBS
- Intraspot and interspot homogeneities
- Carryover from punch and from mat
- Hematocrit effect—normal hematocrit is 45%, but the range is 25–75%.

16.8.2 Liquid chromatography–high-resolution mass spectrometry

Recently, there have been numerous discussions at scientific conferences (American Society for Mass Spectrometry, American Association of Pharmaceutical Scientists) on the use of high-resolution mass spectrometers (e.g. time-of-flight (TOF) or orbitrap-based spectrometers) to simultaneously perform qualitative and quantitative analyses of the same sample. This represents a significant paradigm shift. Triple quadrupole-based mass spectrometers with their superior sensitivity are the workhorses for quantitative analysis. High-resolution platforms (e.g. TOF or orbitrap-based mass spectrometers), with their superior mass resolution have been used extensively in qualitative analysis. With the newer generation of high-resolution mass spectrometers that combine high resolution and good sensitivity, and competitive pricing, a single platform can be used to perform both qualitative and quantitative (so-called "Qual/Quan") analysis.[83]

In liquid chromatography–high-resolution mass spectrometry (LC–HRMS), total ion chromatograms (TICs) are acquired over a predefined *m/z* range (e.g. 100–2000 *m/z*) with a preset mass resolution (e.g. 20,000) on the mass spectrometer. Extracted ion chromatograms (EICs) are generated post data acquisition from the TICs with the exact masses of the target analytes and a predefined mass extraction window (MEW). Quantitative information is then obtained from the EICs, similar to that of the triple quadrupole-based method. Unlike triple quadrupole-based methods, in which the mass spectrometers are typically set at unit resolution, with full width at half maximum of 0.7 Da for data acquisition, different mass resolutions are typically available on full-scan mass spectrometers, depending on the type of mass spectrometers used. Higher mass resolution in general provides better selectivity, especially in a complex sample matrix.

There are a couple of advantages to using HRMS. (1) When using HRMS for method development, there is no need to determine the most favorable product ions since data from high-resolution full-scan acquisition should provide sufficient selectivity. This should expedite the method development process, especially when a large number of compounds are monitored, as in the case of discovery settings. (2) In triple quadrupole-based methods, data from selected SRM transitions are acquired. In HRMS-based methods, data from a wide mass range is collected during data acquisition, the data can be "mined" post-acquisition for different analytes of interest such as "unknown" metabolites, phospholipids, etc.

The key parameters for HRMS-based methods are the mass resolution setting (*R*) of the mass spectrometer during data acquisition and the MEW used to extract ion chromatograms during data processing. The interplay between these parameters has been discussed in the literature.[84] Additional work needs to be done to provide further understanding of these parameters and clearly understand their impact on the quality of the data generated. Bioanalysts should carefully investigate these parameters during method development.

Thus far, HRMS has been successfully applied to determine pharmaceutical compounds, pesticides, veterinary drugs, and peptides in both discovery and development settings.[85,86] Based on currently published data, HRMS can provide sufficient sensitivity, selectivity, and ruggedness for routine bioanalysis. At the time of writing this chapter, regulatory agencies have not provided any formal guidance on the use of HRMS in regulated bioanalysis. Another area that needs to be addressed is the amount of data generated from full-scan data acquisition. Thousands of samples are analyzed during the course of development of a drug candidate and the amount of data accumulated at the end will require careful consideration of storage space and retrieval of data for review. Finally, in-depth discussions with regulatory agencies will be needed to gain perspectives and feedback on this new technology platform, in particular with regard to post-acquisition data mining.

16.8.3 Bioanalysis of therapeutic proteins by LC–MS/MS

Another emerging area in the bioanalysis field is quantitative determination of therapeutic proteins (e.g. monoclonal antibodies, domain antibodies) by LC–MS/MS.[87–92] Unlike traditional small molecules with molecular weights <1000 Da, therapeutic proteins in general have molecular weights $>10,000$ Da. This poses different challenges to bioanalysts attempting to use LC–MS/MS for analysis:

- Therapeutic proteins in general have similar physiochemical properties as other endogenous proteins; therefore, traditional sample cleanup techniques for small molecules such as protein precipitation are not suitable to use.
- With their high molecular weights, it is in general not feasible to monitor the intact molecule in their $[M + H]^{+}$ or $[M + 2H]^{2+}$ charge states since it is likely outside of the mass range of the mass spectrometer, especially if triple-quadrupole mass spectrometers are used. On the other hand, proteins in general have multiple charge states (e.g. $[M + H]^{8+}$, $[M + 2H]^{9+}$, etc.), and some of these charge states will fall into the mass range, although the sensitivity may suffer as a result.
- With their relatively large size, traditional narrow-bore HPLC or UHPLC columns used for small molecule analyses may not be ideal candidates for separation of proteins.

The most frequently used strategy for quantitative analysis of proteins is to digest the proteins enzymatically or chemically to form smaller peptides, preferably in the mass range of 1000–3000 Da. These small peptides are used as surrogates for the proteins (and in general referred to as "surrogate peptides") and can be extracted by LLE, SPE, or protein precipitation and analyzed by

LC–MS/MS. Stable isotopically labeled ISs can now be custom synthesized for use in the assay at a reasonable cost.

In terms of sample cleanup from serum samples, there are four major types:

- Immunocapture: an antibody specific to the protein of interest is used to capture the protein, while other proteins in the serum samples are washed out. The protein of interest is then eluted out for enzymatic digestion.
- Differential precipitation by organic solvent: this method explores the different solubility of pegylated proteins and nonpegylated proteins in organic solvents. For example, Wu et al.[91] reported that pegylated proteins are soluble in 0.1% formic acid in 2-propanol while other endogenous proteins are not. The serum samples are thus treated with 0.1% formic acid in 2-propanol to remove endogenous proteins.
- Precipitating out with all other proteins with an organic solvent: in this method, all proteins including the protein of interest and other endogenous proteins are precipitated out with an organic solvent e.g. methanol. The precipitated proteins are then resuspended in a digestion buffer for enzymatic digestion. It is to be noted that this method does not result in clean samples.
- SPE (both on-line and off-line): this is more applicable to peptides and small proteins.

Regarding digestion by endoproteases, a number of endoproteases (e.g. trypsin, chymotrypsin, GluC, AspN, LysC, protease K, and pepsin) have been reported. The most common one thus far is trypsin. It specifically hydrolyzes peptide bonds at the carboxyl side of lysine and arginine residues. Other enzymes hydrolyze peptide bonds at other specific amino acids. The resulting surrogate peptides can then be analyzed by LC–MS/MS. It is the authors' experience that the best surrogate peptides are between 10 and 30 amino acids in length for good retention on reversed-phase LC columns and reasonable sensitivity.

The samples can be further extracted post-digestion by SPE or 2D-HPLC, or simply injected directly to the LC–MS/MS system for analysis. It is recommended that the bioanalysts evaluate these possibilities carefully during method development.

Thus far, a number of LC–MS/MS-based methods for therapeutic protein analysis have been published. It is gaining traction in discovery settings because it mitigates the needs of precious reagents used in LBAs, and thus expedites assay development. Further work needs to be done to make this technology applicable to routine use, especially in the development arena. To name a few, improvement in sample preparation techniques, LC separation, MS sensitivity are some of the areas to be focused on.

Another key area that needs to be addressed is to establish the link between data generated by LBA and LC–MS/MS, which are two fundamentally different but complementary techniques. The data generated by each technique represent unique properties of the protein. For LBA, it relies on the binding of the capturing reagent. For LC–MS/MS, it relies on the generation of a surrogate peptide that is representing the protein of interest. How the two sets of data relate to each other is highly linked to the protein of interest and the capture reagents used in LBA and specific region of the surrogate peptide. A thorough understanding of the link between the two sets of data is needed.

At the time of this writing, regulatory agencies have not provided any formal guidance on the use of LC–MS/MS-based data for filing. However, it is certainly an area that has tremendous growth potential.

16.9 CONCLUSIONS

Bioanalytical method validation and regulated bioanalysis are an integral part of a drug development program. They have evolved over the years in terms of technological platforms and regulations. Different technological platforms have been used to analyze chemical-based drug candidates, from LC–UV and LC–Fluorescence to LC–MS to LC–MS/MS. Guidance documents from regulatory bodies across the globe are revised to reflect the current technologies but are not yet fully harmonized. At the onset of method development, bioanalysts should carefully evaluate the physiochemical properties of the analyte of interest, its metabolites, assay requirements such as LLOQ, matrix, against the currently available technological platforms. Different parameters that can affect the assay performance should be carefully evaluated. Once a desired method is developed, the bioanalysts can then proceed with method validation and bioanalysis in accordance with the different regulatory guidelines and laboratory-specific SOPs. In addition, methods can be amended when new information is available, for example, discovery of a new metabolite that requires monitoring, or as more data are produced, e.g. clinical pharmacokinetic data from an FIH study may drive a lower LLOQ, or other unforeseeable issues. It has to be understood that the process of method development and validation is a continuum. There is a life cycle to a bioanalytical assay as drug development progresses, and it should be science driven.

Acknowledgments

Krzysztof Selinger would like to thank Dr Daksha Desai-Krieger for friendly encouragement and helpful discussions. The authors would like to thank Dr Anne-Francoise Aubry for reviewing the manuscript and providing valuable feedback.

References

1. Food and Drug Administration. *Guidance for Industry: Bioanalytical Method Validation;* US Department of Health and Human Services, FDA Center for Drug Evaluation and Research: Rockville MD, 2001.
2. European Medicines Agency, Guideline on Bioanalytical Method Validation, London, UK, Committee for Medicinal Products for Human Use, 2011.
3. Shah, V. P.; Midha, K. K.; Dighe, S.; McGilveray, I. J.; Skelly, J. P.; Yacobi, A.; Layloff, T.; Viswanathan, C. T.; Cook, C. E.; McDowell, R. D., et al. *Pharm. Res.* **1992,** *9,* 588–592.
4. Shah, V. P.; Midha, K. K.; Findlay, J. W. A.; Hulse, J. D.; McGilveray, I. J.; McKay, G.; Miller, K. J.; Patnaik, R. N.; Powell, M. L., et al. *Pharm. Res.* **2000,** *17,* 1551–1557.
5. Viswanathan, C. T.; Bansal, S.; Booth, B.; DeStefano, A. J.; Rose, M. J.; Sailstad, J.; Shah, V. P.; Skelly, J. P.; Swann, P. G.; Weiner, R. *Pharm. Res.* **2007,** *24,* 1962–1973.
6. Miller, K. J.; Bowsher, R. R.; Celniker, A.; Gibbons, J.; Gupta, S.; Lee, J. W.; Swanson, S. J.; Smith, W. C.; Weiner, R. S. *Pharm. Res.* **2001,** *18,* 1373–1383.
7. DeSilva, B.; Smith, W.; Weiner, R.; Kelley, M.; Smolec, J.; Lee, B.; Khan, M.; Tacey, R.; Hill, H.; Celniker, A. *Pharm. Res.* **2003,** *20,* 1885–1900.
8. Smolec, J.; DeSilva, B.; Smith, W.; Weiner, R.; Kelly, M.; Lee, B.; Khan, M.; Tacey, R.; Hill, H.; Celniker, A. *Pharm. Res.* **2005,** *22,* 1425–1431.
9. Abbott, R. W. *Bioanalysis* **2010,** *2,* 703–708.

10. Abbott, R. W.; Gordon, B.; van Amsterdam, P.; Lausecker, B.; Brudny-Kloeppel, M.; Smeraglia, J.; Romero, F.; Globig, S.; Globig, M.; Knutsson, M., et al. *Bioanalysis* **2011,** *3,* 833–838.
11. Abbott, R. W.; Brudny-Kloeppel, M. *Bioanalysis* **2009,** *1,* 273–276.
12. Tudan, C. Highlights of 4th Regulated Bioanalysis Workshop, SQA Technical Document 2010-2, Charlottesville, VA; Society of Quality Assurance, 2010.
13. Fast, D. M.; Kelley, M.; Viswanathan, C. T.; O'Shaughnessy, J.; King, S. P.; Chaudhary, A.; Weiner, R.; DeStefano, A.; Tang, D. *The AAPS J.* **2009,** *11,* 238–241.
14. Hubert, Ph.; Nguyen-Huu, J.-J.; Boulanger, B.; Chapuzet, E.; Chiap, P.; Cohen, N.; Compagnon, P. A.; Dewé, W.; Feinberg, M.; Lallier, M., et al. *J. Pharm. Biomed. Anal.* **2004,** *36,* 579–586.
15. Hubert, Ph.; Nguyen-Huu, J. J.; Boulanger, B.; Chapuzet, E.; Chiap, P.; Cohen, N.; Compagnon, P. A.; Dewé, W.; Feinberg, M.; Lallier, M., et al. *J. Pharm. Biomed. Anal.* **2007,** *45,* 70–81.
16. Hubert, Ph.; Nguyen-Huu, J.-J.; Boulanger, B.; Chapuzet, E.; Chiap, P.; Cohen, N.; Compagnon, P. A.; Dewé, W.; Feinberg, M.; Lallie, M., et al. *J. Pharm. Biomed. Anal.* **2007,** *45,* 82–96.
17. Braggio, S.; Barnaby, R. J.; Grossi, P.; Cugola, M., et al. *J. Pharm. Biomed. Anal.* **1996,** *14,* 375–388.
18. Dadgar, D.; Burnett, P. E.; Choc, M. G.; Gallicano, K.; Hooper, J. W. *J. Pharm. Biomed. Anal.* **1995,** *13,* 89–97.
19. Wieling, J.; Hendriks, G.; Tamminga, W. J.; Hempenius, J.; Mensink, C. K.; Oosterhuis, B.; Jonkman, J. H. *J. Chromatogr. A* **1996,** *12,* 381–394.
20. James, C. A.; Breda, M.; Frigerio, E. *J. Pharm. Biomed. Anal.* **2004,** *35,* 887–893.
21. Timmerman, P.; Lowes, S.; Fast, D. M., et al. *Bioanalysis* **2010,** *2,* 683.
22. Bansal, S. K.; Arnold, M.; Garofolo, F. *Bioanalysis* **2010,** *2,* 685–687.
23. Lausecker, B.; van Amsterdam, P.; Brudny-Kloeppel, M.; Luedtke, S.; Timmerman, P. *Bioanalysis* **2009,** *1,* 873–875.
24. van Amsterdam, P.; Lausecker, B.; Luedtke, S.; Timmerman, P.; Brudny-Kloeppel, M. *Bioanalysis* **2010,** *2,* 689–691.
25. van Amsterdam, P.; Arnold, M.; Bansal, S.; Fast, D.; Garofolo, F.; Lowes, S.; Timmerman, P.; Woolf, E. *Bioanalysis* **2010,** *2,* 1801–1803.
26. Premkumar, N.; Lowes, S.; Jersey, J.; Garofolo, F.; Dumont, I.; Masse, R.; Stamatiou, B.; Caturla, M. C.; Steffen, R.; Malone, M., et al. *Bioanalysis* **2010,** *2,* 1797–1800.
27. Savoie, N.; Booth, B. P.; Bradley, T.; Garofolo, F.; Hughes, N. C.; Hussain, S.; King, S. P.; Lindsay, M.; Lowes, S.; Ormsby, E., et al. *Bioanalysis* **2009,** *1,* 19–30.
28. Savoie, N.; Garofolo, F.; van Amsterdam, P.; Booth, B. P.; Fast, D. M.; Lindsay, M.; Lowes, S.; Masse, R.; Mawer, L.; Ormsby, E., et al. *Bioanalysis* **2010,** *2,* 53–68.
29. Savoie, N.; Garofolo, F.; van Amsterdam, P.; Bansal, S.; Beaver, C.; Bedford, P.; Booth, B. P.; Evans, C.; Jemal, M.; Lefebvre, M., et al. *Bioanalysis* **2010,** *2,* 1945–1960.
30. Garofolo, F.; Rocci, M. L., Jr.; Dumont, I.; Martinez, S.; Lowes, S.; Woolf, E.; van Amsterdam, P.; Bansal, S.; Gomes Barra, A.; Bauer, R., et al. *Bioanalysis* **2011,** *3,* 2081–2096.
31. Timmerman, P.; Anders Kall, M.; Gordon, B.; Laakso, S.; Freisleben, A.; Hucker, R. *Bioanalysis* **2010,** *2,* 1185–1194.
32. Freisleben, A.; Brudny-Klöppel, M.; Mulder, H.; de Vries, R.; de Zwart, M.; Timmerman, P. *Bioanalysis* **2011,** *3,* 1333–1336.
33. Timmerman, P.; Luedtke, S.; van Amsterdam, P.; Brudny-Kloeppel, M.; Lausecker, B.; Fischmann, S.; Globig, S.; Sennbro, C.; Jansat, J. M.; Mulder, H., et al. *Bioanalysis* **2009,** *1,* 1049–1056.
34. Lowes, S.; Jersey, J.; Shoup, R.; Garofolo, F.; Savoie, N.; Mortz, E.; Needham, S.; Caturla, M. C.; Steffen, R.; Sheldon, C., et al. *Bioanalysis* **2011,** *3,* 1323–1332.
35. Good Laboratory Practice for Nonclinical Laboratory Studies, Code of Federal Regulations, Title 21, Chapter I, Subchapter A, Part 58.

36. Analytical Methods for an in vivo Bioavailability or Bioequivalence Study, Code of Federal Regulations, Title 21, Volume 5, Chapter I, Subchapter D, Section 320.29 (a).
37. EMA/INS/GCP/532137/2010, Reflection Paper on Guidance for Laboratories that Perform the Analysis or Evaluation of Clinical Trial Samples, London, UK, February 2012.
38. a. Kaiser, H. *Anal. Chem.* **1970,** *42,* 24A–38A. b. Kaiser, H. *Anal. Chem.* **1970,** *42,* 26A–58A.
39. Gu, H.; Wang, J.; Aubry, A. F.; Jiang, H.; Zeng, J.; Easter, J.; Wang, J. S.; Dockens, R.; Bifano, M.; Burrell, R., et al. *Anal. Chem.* **2012,** *84,* 4844–4850.
40. International Organization for Standardization, in Accuracy (Trueness and Precision) of Measurement Methods and Results, ISO 5725-1 and 5725-3, 1994.
41. Heizman, P.; Zinapold, K.; Geshke, R. *Methodol. Surv. Biochem. Anal.* **1994,** *23,* 351–357.
42. Fung, E. N.; Zheng, N.; Arnold, M. E.; Zeng, J. *Bioanalysis* **2010,** *4,* 733–743.
43. Scott, D. O.; Bindra, D. S.; Stella, V. J. *Pharm. Res.* **1993,** *10,* 1451–1457.
44. Jersey, J. A.; Duyan, S. A.; Davis, I. M. *Pharm. Res.* **1994,** *11,* S-58.
45. Dams, R.; Huestis, M. A.; Lambert, W. E.; Murphy, C. M. *J. Am. Soc. Mass Spectrom.* **2003,** 1290–1294.
46. Xia, Y.; Jemal, M. *Rapid Commun. Mass Spectrom.* **2009,** *23,* 2125–2138.
47. Liang, Z. *Bioanalysis* **2012,** *4,* 1227–1234.
48. Silvester, S.; Smith, L. *Bioanalysis* **2012,** *4,* 879–895.
49. Wu, S. T.; Schoener, D.; Jemal, M. *Rapid Commun. Mass Spectrom.* **2008,** *22,* 2873–2881.
50. Matuszewski, B. K. *J. Chromatogr. B* **2006,** *830,* 293–300.
51. De Nardi, C.; Bonelli, F. *Rapid Commun. Mass Spectrom.* **2006,** *20,* 2709–2916.
52. Lang, J. R.; Bolton, S. *J. Pharm. Biomed. Anal.* **1991,** *9,* 357–361.
53. Lang, J. R.; Bolton, S. *J. Pharm. Biomed. Anal.* **1991,** *9,* 435–442.
54. Kringle, R. *Pharm. Res.* **1994,** *11,* 556–560.
55. Hoffman, D.; Kringle, R. *Pharm. Res.* **2007,** *24,* 1157–1164.
56. Hartmann, C.; Massart, D. L.; McDowall, R. D. *J. Pharm. Biomed. Anal.* **1994,** *12,* 1337–1343.
57. Thompson, M.; Howarth, R. J. *Analyst* **1980,** *105,* 1188–1195.
58. Kimanani, E. K. *J. Pharm. Biomed. Anal.* **1998,** *16,* 1117–1124.
59. Kimanani, E. K.; Lavigne, J. *J. Pharm. Biomed. Anal.* **1998,** *16,* 1107–1115.
60. Singtoroj, T.; Tarning, J.; Annerberg, A.; Ashton, M.; Bergqvist, Y.; White, N. J.; Lindegardh, N.; Day, N. P. J. *J. Pharm. Biomed. Anal.* **2005,** *11,* 219–227.
61. Thompson, M. *Analyst* **1982,** *107,* 1169–1180.
62. Tholen, D. W. *Arch. Pathol. Lab. Med.* **1992,** *116,* 746–756.
63. Karnes, H. T.; March, C. *J. Pharm. Biomed. Anal.* **1991,** *9,* 911–918.
64. Krouwer, J. S.; Schlain, B. *Clin. Chem.* **1993,** *39,* 1689–1693.
65. Hinderling, P. H. *Pharmacol. Rev.* **1997,** *49,* 279–295.
66. Fisar, Z.; Fuksová, K.; Sikora, J.; Kalisová, L.; Velenovská, M.; Novotna, M. *Neuro. Endocrinol. Lett.* **2006,** *27,* 307–313.
67. Ke, J.; Yancey, M.; Zhang, S.; Lowes, S.; Henion, J. *J. Chromatogr. B Biomed. Sci. Appl.* **2000,** *9,* 369–380.
68. Fura, A.; Harper, T. W.; Zhang, H.; Fung, L.; Shyu, W. C. *J. Pharm. Biomed. Anal.* **2003,** *14,* 513–522.
69. Morris, D. M.; Selinger, K. *J. Pharm. Biomed. Anal.* **1994,** *12,* 255–264.
70. Lentner, C., Ed. Ciba-Geigy Ltd: Basel, Switzerland, 1984.
71. Hughes, N. C.; Bajaj, N.; Fan, J.; Wong, E. Y. *Bioanalysis* **2009,** *1,* 1057–1066.
72. *Guide for Validation of Analytical and Bioanalytical Methods.* Resolution-RE n. 899; ANVISA, May 29, 2003.
73. Wang, S.; Cyronak, M.; Yang, E. *J. Pharm. Biomed. Anal.* **2007,** *17,* 701–707.
74. Vander Heydena, Y.; Nijhuisb, A.; Smeyers-Verbekea, J.; Vandeginsteb, B.G.M.; Massart, D.L. Guidance for Robustness/Ruggedness Tests in Method Validation, http://www.vub.ac.be/fabi/tutorial/guideline.pdf.

75. Smith, K. M.; Yan, X. *Bioanalysis* **2012,** *4,* 741–749.
76. Zhang, J.; Reimer, M. T.; Ji, Q. C.; Chang, M. S.; El-Shourbagy, T. A.; Burke, S.; Schwartz, L. *Anal. Bioanal. Chem.* **2007,** *387,* 2745–2756.
77. Ji, Q. C.; Zhang, J.; Rodila, R.; Watson, P.; El-Shourbagy, T. *Rapid Commun. Mass Spectrom.* **2004,** *18,* 2293–2298.
78. Guthrue, R.; Suzi, A. *Pediatrics* **1963,** *23,* 338–343.
79. Evans, C. Current Technology and Use of Dried Blood Spots, 12th Annual Land O'Lakes Bioanalytical Conference, Merrimac, WI, July 2011.
80. Needham, S. Method Development and Validation for Dried Blood Spots, 12th Annual Land O'Lakes Bioanalytical Conference, Merrimac, WI, July 2011.
81. Brewer, E. Special Analytical Challenges and Solutions for Implementation, 12th Annual Land O'Lakes Bioanalytical Conference, Merrimac, WI, July 2011.
82. Viswanathan, C. Regulatory Perspective on Dried Blood Spots, 12th Annual Land O'Lakes Bioanalytical Conference, Merrimac, WI, July 2011.
83. Ramanathan, I. R.; Jemal, M.; Ramagiri, S.; Xia, Y. Q.; Humpreys, W. G.; Olah, T.; Korfmacher, W. A. *J. Mass Spectrom.* **2011,** *46,* 595–601.
84. Xia, Y. Q.; Lau, J.; Olah, T.; Jemal, M. *Rapid Commun. Mass Spectrom.* **2011,** *15,* 2863–2878.
85. Fung, E. N.; Xia, Y. Q.; Aubry, A. F.; Zeng, J.; Olah, T.; Jemal, M. *J. Chromatogr. B Analyt. Technol. Biomed. Life Sci.* **2011,** *1* (879), 2919–2927.
86. Kaufmann, J. A.; Butcher, P.; Maden, K.; Walker, S.; Widmer, M. *Anal. Chim. Acta* **2011,** *700* (1–2), 86–94.
87. Berna, K. M. J.; Zhen, Y.; Watson, D. E.; Hale, J. E.; Ackermann, B. L. *Anal. Chem.* **2007,** *1* (79), 4199–4205.
88. Lu, Q.; Zheng, X.; McIntosh, T.; Davis, H.; Nemeth, J. F.; Pendley, C.; Wu, S. L.; Hancock, W. S. *Anal. Chem.* **2009,** *1* (81), 8715–8723.
89. Li, H.; Ortiz, R.; Tran, L.; Hall, M.; Spahr, C.; Walker, K.; Laudemann, J.; Miller, S.; Salimi-Moosavi, H.; Lee, J. W. *Anal. Chem.* **2012,** *7* (84), 1267–1273.
90. Xu, Y.; Mehl, J. T.; Bakhtiar, R.; Woolf, E. J. *Anal. Chem.* **2010,** *15* (82), 6877–6886.
91. Wu, OS. T.; Ouyang, Z.; Olah, T.; Jemal, M. *Rapid Commun. Mass Spectrom.* **2011,** *30* (25), 281–290.
92. Rauh, P. M. *J. Chromatogr. B* **2012,** *883–884,* 59–67.

Index

Note: Page numbers with "*f*" denote figures; "*t*" tables; and "*b*" boxes.

A

B

C

N

O

P

Q

R

S

T

U

V

W

X

9780080983509